W9-DFR-312

Concepts in Probability and Stochastic Modeling

JAMES J. HIGGINS
Kansas State University

SALLIE KELLER-McNULTY
Kansas State University

An Alexander Kugushev Book

Duxbury Press
An Imprint of Wadsworth Publishing Company
I(T)P™ An International Thomson Publishing Company

Belmont • Albany • Bonn • Boston • Cincinnati • Detroit • London • Madrid • Melbourne
Mexico City • New York • Paris • San Francisco • Singapore • Tokyo • Toronto • Washington

Assistant Editor: Jennifer Burger
Editorial Assistant: Michelle O'Donnell
Production Services Coordinator: Gary Mcdonald
Production Editor: Greg Hubit Bookworks
Print Buyer: Diana Spence
Copy Editor: Technical Editing Services, Inc.
Illustrator: Susan Benoit
Cover Photograph: C. O'Rear/West Light
Cover Design: Vargas/Williams/Design
Compositor: Dusty Foster, WLF Enterprises
Printer: Arcata Graphics / Fairfield

Printed in the United States of America

For more information, contact:

Wadsworth Publishing Company
10 Davis Drive
Belmont, California 94002

International Thomson Publishing
Berkshire House 168-173
High Holborn
London, WC1V7AA
England

International Thomson Publishing GmbH
Königwinterer Strasse 418
53227 Bonn
Germany

Thomas Nelson Australia
102 Dodds Street
South Melbourne 3205
Victoria, Australia

International Thomson Publishing Asia
221 Henderson Road #05-10
Singapore 0315

Nelson Canada
1120 Birchmount Road
Scarborough, Ontario
Canada M1K 5G4

International Thomson Publishing - Japan
Hirakawacho-cho Kyowa Building, 3F
2-2-1 Hirakawacho-cho
Chiyoda-ku, 102 Tokyo
Japan

1 2 3 4 5 6 7 8 9 10—01 00 99 98 97 96 95

Library of Congress Cataloging-in-Publication Data

Higgins, James J.
Concepts in probability and stochastic modeling / James J. Higgins and Sallie Keller-McNulty.
p. cm.
Includes index.
ISBN 0-534-23136-5
1. Probabilities. 2. Stochastic processes. I. Keller-McNulty, Sallie. II. Title.
QA273.H6433 1995
519.2—dc20 94-3497

This book is printed on acid-free recycled paper.

CONTENTS

Our Purpose

This book was developed as a first course in probability for students in computer science, engineering, management science, and the physical sciences. Systems modeling is a theme that is carried throughout the book. This gives continuity and direction to the material and provides interest to students. Probability concepts and tools that are especially useful for modeling stochastic behavior of systems are given prominence in the book. Students are assumed to have had two semesters of calculus, some familiarity with matrix operations, and a working knowledge of a programming language.

Unique Features

Consistent with the systems modeling theme, the book contains unique features not typically found in books at this level. (1) There is an elementary but substantial treatment of Markov chains, queuing processes, and reliability models. Each of these topics receives a chapter of coverage. (2) Computer simulation is used extensively throughout the text. This motivates probability concepts and enables students to experimentally study topics that would otherwise be beyond their mathematical backgrounds. (3) Elementary parameter estimation and data analysis techniques are included as key components in the model-building process. Topics that do not contribute to the systems modeling theme are given less emphasis. For instance, technical derivations involving the moment-generating functions were given less prominence and could be omitted at the instructor's discretion. Topics on the *t*- and *F*- distributions usually found in mathematical statistics courses were not included.

Our Approach

The book takes a hands-on approach to probability. From Section 1.3 onward, students are invited to study random phenomena through computer simulation. Extensive exercises and examples are designed to give students a feel for randomness and variability. Our approach in incorporating computer simulation into the text is to focus on algorithms that describe the stochastic structure of the processes being studied. Unlike the "black-box" approach of a simulation language, our approach helps the student understand the essential mechanisms that generate the random phenomena being studied. We use simulation to study random walks, winning streaks, Markov chains, queuing processes, scheduling problems, systems reliability, lottery problems, and many other random phenomena.

Programming

To give instructors and students flexibility, the presentation is not tied to any particular programming language. The descriptions of the algorithms are in pseudocode, which can be translated into any common programming language. The programming tasks we ask students to do are generally small to moderate in size. By limiting the size of these tasks, students are able to keep up with the flow of the course and do not get distracted with time-consuming programming projects.

Data Analysis

A unique feature of the text is the incorporation of simple data analysis techniques along with simulation. Students learn that interpretation of results is as important as obtaining the simulated data. Such techniques easily transfer to data obtained in real-world experimentation, thus expanding the scope of applicability of the results. In analyzing data, we encourage students to take full advantage of their computing knowledge. For instance, many students will export the data from a simulation to a spreadsheet program where simple descriptive analyses may be carried out. However, no special knowledge of this type is needed. Some students will choose to write their own programs to find sample means, sample standard deviations, and construct histograms. Some may even use pencil, paper, and calculator to analyze the data.

The Mathematics

The modest mathematical level of this book requires some topics to be developed intuitively rather than mathematically. For instance, instead of using differential equations to introduce the Poisson process or queuing processes, we introduce these topics as limiting cases of Bernoulli counting and queuing processes. Coin tossing with biased coins is all that students need to understand the random mechanisms of these processes and to be able to simulate their behavior. A number of topics that are unapproachable through mathematics at this level are introduced through simulation. For instance, students can study nonhomogeneous counting and queuing processes and semi-Markov processes without the necessity of the heavy mathematical machinery often associated with these processes.

Suggested Course Outlines

The book has more than enough material for a 3-hour, one-semester course. It is expected that the instructor will cover a core of material and then select additional topics to meet the needs of the students. Our students have no prior knowledge of probability. We use the following core of topics.

Chapter 1: Basic Probability

- 1.1 Sample Spaces and Events
- 1.2 Assignment of Probabilities
- 1.3 Simulation of Events on the Computer
- 1.4 Counting Techniques
- 1.5 Conditional Probability
- 1.6 Independent Events

Chapter 2: Discrete Random Variables

- 2.1 Random Variables
- 2.2 Joint Distributions and Independent Random Variables
- 2.3 Expected Values
- 2.4 Variance and Standard Deviation
- 2.5 Sampling and Simulation
- 2.6 Sample Statistics
- 2.7 Expected Values of Jointly Distributed Random Variables and the Law of Large Numbers

Chapter 3: Special Discrete Random Variables

- 3.1 Binomial Random Variable
- 3.2 Geometric and Negative Binomial Random Variables
- 3.5 Poisson Random Variable

Chapter 4: Markov Chains

- 4.1 Introduction: Modeling a Simple Queuing System
- 4.2 The Markov Property
- 4.3 Computing Probabilities for Markov Chains
- 4.4 The Simple Queuing System Revisited
- 4.5 Simulating the Behavior of a Markov Chain
- 4.6 Steady-State Probabilities

Chapter 5: Continuous Random Variables

5.1 Probability Density Functions

5.2 Expected Value and Distribution of a Function of a Random Variable

5.3 Simulating Continuous Random Variables

Chapter 6: Special Continuous Random Variables

6.1 Exponential Random Variable

6.2 Normal Random Variable

6.3 Gamma Random Variable

6.6 Method of Moments Estimation

Chapter 7: Counting and Queuing Processes

7.1 Bernoulli Counting Process

7.2 The Poisson Process

7.3 Exponential Random Variables and the Poisson Process

7.4 The Bernoulli Single-Server Queuing Process

7.5 The M/M/1 Queuing Process

Chapter 8: The Distribution of Sums of Random Variables

8.1 Sums of Random Variables

8.2 Sums of Random Variables and the Central Limit Theorem

8.3 Confidence Intervals for Means

Chapter 9: Selected Systems Models

9.1 Distribution of Extremes

9.2 Scheduling Problems

Remaining topics would be selected to round out the course. *To include topics on reliability*, we would cover Section 6.4 on the Weibull distribution, and Sections 10.1 through 10.3 on reliability. *To emphasize queuing processes*, we would cover Sections 7.6 through 7.9 on *k*-server queuing processes. *To include topics on moment-generating functions*, continuous joint distributions, and change of variables, we would add Sections 3.6, 5.4, and 6.5.

For a somewhat slower paced treatment in the spirit of a finite mathematics course, the core would consist of complete coverage of the topics on discrete random variables and processes from Chapters 1 through 4 with additional topics on continuous random variables and processes from Sections 5.1–5.3, 6.1–6.2, 7.1–7.2, and 8.1–8.3.

For a course that would cover traditional topics but would include simulation throughout, an outline would include full coverage of Chapters 1, 2, 3, 5, 6, and 8 (except 8.4). Additional topics would be selected according to the students' interests. For instance, the course could be rounded out with topics on Markov chains from Sections 4.1–4.5, or with the Poisson process and reliability, Sections 7.1–7.3 and 10.1–10.3.

For students who have had an introduction to probability, particularly in an applied statistics course, various courses emphasizing modeling could be constructed depending on the extent of the students' prior knowledge of probability. The core of these courses would consist of Sections 1.3, 2.5, 2.6, and 5.3 on simulation, Chapter 4 on Markov chains, Chapter 7 on counting and queuing processes, Chapter 9 on selected systems models, and Chapter 10 on reliability. Material on discrete and continuous random variables would be included to match the students' prior knowledge of probability.

1

Basic Probability

The uncertainties or chance occurrences that confront us every day are governed by rules of probability. A major focus of this text is to study how the rules of probabilities can be used to model a *system*. A system is a set of objects that interact as a unit. A typical home computer system consists of a desktop computer, a monitor, a printer and other peripherals, and software. A system for delivering packages consists of the equipment, buildings, personnel, and operational procedures needed to get packages to their destination. A *mathematical model* of a system is an abstract representation of the important features of the system in terms of equations, charts, graphs, rules of behavior, and the like. Mathematical models have been constructed for such diverse phenomena as global weather patterns, the flow of chemicals in the bloodstream, computer networks, traffic patterns in cities, wins and losses in professional sports, and long-distance telephone calls.

Many systems have components whose behavior is in some way random. The number of customers waiting to go through the check-out stations at a supermarket varies randomly throughout the day. The number of light bulbs that must be replaced in an office varies randomly from week to week. The time it takes to complete a construction project varies randomly because of uncertain weather. A mathematical model that describes this randomness is called a *stochastic* model. Probability theory is a branch of mathematics that deals with randomness and laws of chance. An understanding of basic probability is necessary to construct stochastic models.

SECTION

1.1 Sample Spaces and Events

Much of statistics and probability is couched in terms of experiments in which outcomes are determined or affected by chance. For example, an experiment might consist of observing the numbers on a winning lottery ticket or of observing the number of queries to a data base system at various times of the day. In a study to determine the best among several varieties of wheat, an experiment could involve measuring yields on experimental plots of ground. A good place to begin an analysis of such experiments is to account for the possible outcomes of the experiment.

Definition 1.1-1

A *sample space* is the set of all possible outcomes of an experiment. The *sample points* are the elements in a sample space.

EXAMPLE **1.1-1** Assume the system under study is a computer lab with four workstations. If the experiment consists of observing the number of occupied workstations, the sample space, S, would be $S = \{0, 1, 2, 3, 4\}$.

EXAMPLE **1.1-2** Consider tossing a single coin. The possible outcomes are heads or tails. The sample space for this experiment is $S = \{H, T\}$.

EXAMPLE **1.1-3** Suppose an experiment consists of tossing a coin three times. If only the number of heads is observed, the sample space would be $S = \{0, 1, 2, 3\}$. However, if the order of occurrence of the heads and tails is of interest, the sample space would consist of outcomes that are 3-tuples corresponding to the possible sequences of heads and tails. The sample space for this experiment could be expressed as

$$S = \{HHH, HHT, HTH, HTT, THH, THT, TTH, TTT\}.$$

For instance, the outcome HHT denotes heads on the first and second tosses and tails on the third toss.

EXAMPLE **1.1-4** Referring to Example 1.1-1, suppose it is of interest to know specifically which workstations are occupied. A sample space for this experiment could be listed as a set of 4-tuples where the first entry corresponds to the status of the first workstation, the second entry corresponds to the status of the second workstation, and so on. The resulting sample space would be

$$S = \{(0,0,0,0), (0,0,0,1), (0,0,1,0), (0,0,1,1), (0,1,0,0), (0,1,0,1), (0,1,1,0), (0,1,1,1), (1,0,0,0), (1,0,0,1), (1,0,1,0), (1,0,1,1), (1,1,0,0), (1,1,0,1), (1,1,1,0), (1,1,1,1)\},$$

where 0 means the workstation is not occupied and 1 means the workstation is occupied.

EXAMPLE **1.1-5** An experiment consists of drawing two numbers, one after the other, out of a hat that contains the numbers 1, 2, and 3. After a number is drawn, it is not replaced. The sample space is

$$S = \{(1,2), (1,3), (2,1), (2,3), (3,1), (3,2)\}.$$

Suppose the experiment is modified so that the after the first number is drawn and recorded, it is replaced and eligible to be drawn on the second draw. In this case the sample space is

$$S = \{(1,1), (1,2), (1,3), (2,1), (2,2), (2,3), (3,1), (3,2), (3,3)\}.$$

The first experiment in this example used a method of selection *without replacement,* and the second experiment used a method of selection *with replacement.*

EXAMPLE **1.1-6** Consider a problem in which the reliability of an electronic component is under study. An appropriate experiment might be to observe the time to failure of the component. The sample space would be

$$S = \{t: t \geq 0\},$$

where t represents the amount of time that passes before the component malfunctions.

When conducting an experiment, our attention is frequently focused on collections of sample points rather than on individual sample points. Such collections are called *events*.

Definition 1.1-2

An *event* is a subset of the sample space.

EXAMPLE 1.1-7 Suppose a coin is tossed three times as in Example 1.1-3. The sample points corresponding to the occurrence of "exactly one head" are HTT, THT, and TTH. This event is the subset of the sample space denoted by

$$A = \{\text{HTT, THT, TTH}\}.$$

The event "at least one head" is the subset

$$B = \{\text{HHH, HHT, HTH, HTT, THH, THT, TTH}\}.$$

EXAMPLE 1.1-8 An experiment consists of observing four workstations. Consider the event "at most one workstation is occupied." If the sample space denotes only the number of occupied workstations as in Example 1.1-1, then the event "at most one occupied workstation" would be denoted $\{0, 1\}$. If the sample space denotes which workstations are occupied as in Example 1.1-4, then this event would be denoted $\{(0,0,0,0), (0,0,0,1), (0,0,1,0), (0,1,0,0), (1,0,0,0)\}$.

EXAMPLE 1.1-9 Suppose officials need to determine whether a request for the installation of a second toll booth at the entrance to a certain park is justified. To investigate this question, an experiment is conducted to observe the number of vehicles waiting to enter the park at some predetermined time. The sample space is

$$S = \{0, 1, 2, 3, 4, \ldots\}.$$

This is an example of a sample space with a countable number of points. Some events of interest might be the event "at least 2 vehicles waiting to enter the park" or the event "at most 2 vehicles waiting to enter the park." These events can be denoted as $\{2, 3, 4, \ldots\}$ and $\{0, 1, 2\}$, respectively.

The concept of an experiment and corresponding sample space will be at the foundation of all systems we study. Since events of interest for an experiment are defined to be subsets of the sample space, it is natural to perform various set operations on events.

EXAMPLE 1.1-10 Let an experiment consist of drawing one card from a deck of 52. Let A be the event "ace," which consists of four cards, and let B be the event "diamond," which consists of 13 cards. Then $A \cap B$ is the one-card event "ace of diamonds." The event $A \cup B$ consists of 16 cards that are either aces or diamonds. The complement of A, denoted A^c, is the event consisting of the 48 cards that are not aces.

Definition 1.1-3

Two events, A and B, are mutually exclusive if they have no sample points in common; that is, $A \cap B = \varnothing$.

EXAMPLE 1.1-11 An experiment consists of tossing a coin twice. The sample space is $S = \{\text{HH, HT, TH, TT}\}$. Let A be the event "at least one head is observed" and let B be the event "at least one tail." Then $A = \{\text{HH, HT, TH}\}$ and $B = \{\text{HT, TH, TT}\}$. The events A and B are not mutually exclusive because $A \cap B = \{\text{HT, TH}\} \neq \varnothing$. The complement of A is the event $A^c = \{\text{TT}\}$ or "no heads," and the complement of B is the event $B^c = \{\text{HH}\}$ or "no tails." The events A^c and B^c are mutually exclusive since $A^c \cap B^c = \varnothing$, that is, tossing two heads and two tails at the same time is an impossible event.

Exercises 1.1

1.1-1 A box contains a red marble, a blue marble, and a yellow marble. An experiment consists of drawing two marbles at random out of the box, one after the other without replacement.

- **a** List the sample space.
- **b** How many sample points are in the event "exactly one marble is red"?
- **c** Repeat parts (a) and (b) for an experiment that draws the two marbles at random from the box, one after the other with replacement.

1.1-2 An experiment consists of observing the number of hits made by each of two children in a softball game. Each child will be up to bat three times.

- **a** List the sample space.
- **b** List the sample points in the event "both children have the same number of hits."
- **c** List the sample points in the event "the total number of hits for the two children is at least 3."

1.1-3 A trained taste tester samples and ranks three new brands of chocolate Swiss almond ice cream. A rank of 1 is assigned to the best brand, a 2 is assigned to the second best brand, and a 3 is assigned to the worst brand.

a List the sample space.

b In how many sample points will each brand be ranked "best"?

1.1-4 Five computers are connected together on a network. At any time any computer could be up or down. The status of the computers is observed at a particular time.

a List the sample space.

b List the sample points in the event "at least two computers are up."

1.1-5 A box contains four electronic components. Two are defective and two are good. Two components are drawn one after the other at random from the box, without replacement.

a List the sample space by denoting the number of defective components in the sample.

b List the sample space by denoting the status of each component as it is drawn.

c Suppose the components are numbered 1 through 4, with components 1 and 2 defective and components 3 and 4 good. List the sample space by denoting the numbers corresponding to the components drawn.

d In parts (a), (b), and (c), how many sample points are in the event "all the components are good"?

e In parts (a), (b), and (c), how many sample points are in the event "all the components are defective"?

1.1-6 A game consists of rolling a pair of dice and moving a game piece the number of spaces according to the total number of dots on the dice. In order to move the game piece on a player's first turn, the player must roll a 1 or a 6 on at least one die. Give a sample space for this experiment, and list the sample points associated with the event "moving on the first turn."

SECTION

1.2 Assignment of Probabilities

An intuitive understanding of probability often affects our daily actions. When driving on Anderson Avenue in Manhattan, Kansas, one is usually careful not to speed because the probability of a police officer lurking around the corner is high. Whether or not one carries an umbrella will depend on the probability that it will rain. Investors' decisions to buy or sell stock will depend on their assessments of the probability that the market will rise or fall.

Probability is a measure that expresses the likelihood that uncertain events will occur. The scale of measurement for probability is between 0 and 1. Events with probability near 0 are unlikely to occur, whereas those with probability near 1 are almost certain to occur. If A is an event, then the probability that A will occur is denoted by $P(A)$. In the following discussion we consider the problem of assigning probability to events.

Suppose we have an experiment, such as rolling dice or tossing a coin, that can be repeated many times. Let $N(A)$ be the number of times an event A occurs in N repetitions of the experiment. The *relative frequency* of the event A is defined to be the fraction $N(A)/N$. For instance, if a coin is tossed 100 times and heads occurs 45 times, then the relative frequency of heads is 45/100.

Experience has shown that after many repetitions of an experiment, the relative frequency will tend to "settle down" or converge to a fixed value. In the tossing of a fair coin, we would expect the relative frequency to converge to 1/2. Assuming that the limiting relative frequency exists, it is this value that one assigns to the probability of an event. That is,

$$P(A) = \lim_{N \to \infty} \frac{N(A)}{N}.$$

As a practical matter, we may never be able to perform a sufficient number of experiments to discern exactly the value of $P(A)$. For instance, if a coin were biased so that the true probability of heads were .50000001, it is doubtful that we would be able to repeat the experiment enough times to find this out. Thus, other ways are needed to assign probabilities to events.

Probabilities are assigned to events in three common ways. The first way uses a *general mathematical model*. One choice for the mathematical model is to treat the outcomes as equally likely events. The idea is to list the points of the sample space in such a way that all the outcomes have the same chance of occurring. If this can be done, then the probability of an event A from a sample space S is

$$P(A) = \frac{\text{Number of points in } A}{\text{Number of points in } S}.$$

EXAMPLE **1.2-1** Consider the experiment of tossing a fair coin three times. The sample space is

$$S = \{\text{HHH, HHT, HTH, HTT, THH, THT, TTH, TTT}\}.$$

Since the coin is fair, each of these eight outcomes is equally likely. Consider the event "exactly one head." The probability of this event is 3/8 because three out of the eight sample points correspond to that event.

The second method of assigning probabilities to events is *empirically*, that is, by experimentation. To assign the probability empirically, we must have an experiment that can be repeated many times under the same conditions. The empirical probability of an event A is defined as the relative frequency of the event A in N repetitions of the experiment. The only difficulty with this assignment is that two people may perform the same experiment N times and come up with different empirical probabilities for $P(A)$. However, if the number of repetitions is large enough, any two empirically determined values of $P(A)$ will be close to each other.

EXAMPLE 1.2-2 A manufacturer wants to find the probability that its widgets are defectively made. An experiment is conducted in which 1000 widgets are selected at random from the production process. Each widget is tested for defects and five are found to be defective. Thus, the empirical probability of a defective widget is $5/1000 = .005$.

Third, probability can be assigned *subjectively* to an event. In this situation the probability of an event represents a personal degree of belief that the event will occur.

EXAMPLE 1.2-3 Suppose we wish to bet on a World Series involving the Kansas City Royals and the St. Louis Cardinals. If we say that the Royals have probability 2/5 of winning the World Series, we are stating our degree of belief that the Royals can win. If the series could be repeated many times exactly under the same conditions, the 2/5 probability means we believe that the Royals would win an average of two out of five series in the long run. Obviously, not everyone would assign this same subjective probability to this event.

The assignment of probabilities must satisfy the following *probability axioms.*

Probability Axioms

Let S be a sample space and $A, E_1, E_2, \ldots, E_n, \ldots$ be events in S.

Axiom 1 $0 \le P(A) \le 1$, for every A.

Axiom 2 $P(S) = 1$.

Axiom 3 If $E_1, E_2, \ldots, E_n, \ldots$ are mutually exclusive events, then
$$P(E_1 \cup E_2 \cup \cdots \cup E_n \cup \cdots) = P(E_1) + P(E_2) + \cdots + P(E_n) + \cdots.$$

In words, Axiom 1 states that probabilities are numbers between 0 and 1, Axiom 2 says that some outcome in the sample space is certain to occur, and Axiom 3 states that the probability of the whole is the sum of the probability of its parts.

EXAMPLE 1.2-4 Consider the experiment of rolling a single fair die. The sample space can be denoted as $S = \{1, 2, 3, 4, 5, 6\}$, where the sample points are the number of dots on the die. Since the die is fair, all six possible outcomes are equally likely to occur. Hence, a probability of 1/6 is assigned to each outcome. Axiom 3 can be used to find the probability of an event in the following way. Let A be the event that the roll was even. Then $A = \{2\} \cup \{4\} \cup \{6\}$ and $P(A) = P(\{2\}) + P(\{4\}) + P(\{6\}) = 1/6 + 1/6 + 1/6 = 1/2$.

EXAMPLE 1.2-5 Suppose an experiment consists of observing two identical machines and noting whether each is operable (UP) or inoperable (DOWN). The sample space can be listed as

$$S = \{(\text{UP},\text{UP}), (\text{UP},\text{DOWN}), (\text{DOWN},\text{UP}), (\text{DOWN},\text{DOWN})\},$$

where the first entry in each pair denotes the status of the first machine and the second entry denotes the status of the second. Suppose it has been determined empirically that machines of this type have $P(\text{UP}) = 5/6$ and $P(\text{DOWN}) = 1/6$. We now need to determine the associated probabilities for the events in S. If the events were equally likely to occur, a probability of 1/4 would be assigned to each, but this does not seem realistic. To determine the more realistic probabilities, we can *simulate* or imitate the experiment with an equivalent experiment for which probabilities can be easily determined.

Assume that the status of each machine can be determined as if we were tossing a fair die. If the numbers 1 through 5 occur on the die, we say the machine is UP, whereas if 6 occurs we say the machine is DOWN. By tossing a pair of dice we determine the status of both machines. This dice-tossing experiment is a simple model that describes the random nature of the failures of these machines. In particular, this model implies that the status of one machine does not affect the status of the other.

The sample space for the dice-tossing experiment consists of the 36 points

$$S = \{(1,1), (1,2), \ldots, (1,6), (2,1), \ldots, (6,6)\},$$

where the elements of the ordered pairs denote the face values on the two dice. Each sample point has probability 1/36 because all points are equally likely to occur. The probabilities of the sample points in S can be determined by associating them with the sample points in S as shown in Figure 1.2-1.

FIGURE 1.2-1
Dice-tossing Experiment

Faces of the dice					S
(1,1) (1,2) (1,3) (1,4) (1,5)	(2,1) (2,2) (2,3) (2,4) (2,5)	(3,1) (3,2) (3,3) (3,4) (3,5)	(4,1) (4,2) (4,3) (4,4) (4,5)	(5,1) (5,2) (5,3) (5,4) (5,5)	(UP,UP)
(1,6)	(2,6)	(3,6)	(4,6)	(5,6)	(UP,DOWN)
(6,1)	(6,2)	(6,3)	(6,4)	(6,5)	(DOWN,UP)
(6,6)					(DOWN,DOWN)

Since there are 25 pairs out of the 36 that correspond to the outcome (UP, UP), then $P(\text{UP,UP}) = 25/36$. Likewise, $P(\text{UP,DOWN}) = 5/36$, $P(\text{DOWN,UP}) = 5/36$, and $P(\text{DOWN,DOWN}) = 1/36$.

With these probabilities we can find the probability of various other events. Let A be the event "at least one machine is operable," B the event "the first machine is operable," and C the event "the second machine is operable." Then

$$\begin{aligned} P(A) &= P(\{(\text{UP,UP}), (\text{UP,DOWN}), (\text{DOWN,UP})\}) \\ &= \frac{25}{36} + \frac{5}{36} + \frac{5}{36} \\ &= \frac{35}{36} \end{aligned}$$

$$\begin{aligned} P(B) &= P(\{(\text{UP,UP}), (\text{UP,DOWN})\}) \\ &= \frac{25}{36} + \frac{5}{36} \\ &= \frac{30}{36} \end{aligned}$$

$$\begin{aligned} P(C) &= P(\{(\text{UP,UP}), (\text{DOWN,UP})\}) \\ &= \frac{25}{36} + \frac{5}{36} \\ &= \frac{30}{36} \end{aligned}$$

It is not always necessary to list all the sample points in an event to find its probability. Specific probability laws can be used.

Probability Law 1

Let A and B be events from the sample space S. Then,

$$P(A \cup B) = P(A) + P(B) - P(A \cap B).$$

This law can be verified by considering the events depicted in Figure 1.2-2.

FIGURE **1.2-2**
Venn Diagram of $A \cup B$

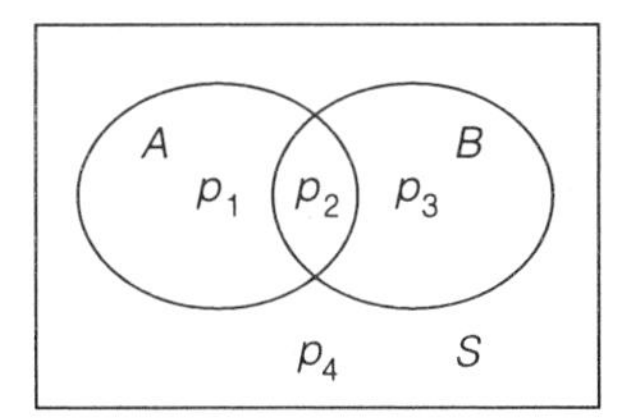

Now,

$$P(A) = p_1 + p_2,$$
$$P(B) = p_2 + p_3,$$

and

$$P(A \cap B) = p_2.$$

Therefore,

$$\begin{aligned} P(A \cup B) &= p_1 + p_2 + p_3 \\ &= (p_1 + p_2) + (p_2 + p_3) - p_2 \\ &= P(A) + P(B) - P(A \cap B). \end{aligned}$$

EXAMPLE **1.2-6** Consider the experiment and events defined in Example 1.2-5. Now,

$$\begin{aligned} P(\text{at least one terminal is up}) &= P(B \cup C) \\ &= P(B) + P(C) - P(B \cap C) \\ &= \frac{5}{6} + \frac{5}{6} - \frac{25}{36} \\ &= \frac{35}{36} \end{aligned}$$

EXAMPLE **1.2-7** The experience of a car dealer shows that 60% of the customers purchase a stereo as an option with their cars, 50% purchase tinted windows, and 35% purchase both. The probability that a customer will purchase at least one of these two options is

$$\begin{aligned} P(\text{"stereo" or "tinted windows"}) &= P(\text{stereo}) + P(\text{tinted windows}) - P(\text{both}) \\ &= .60 + .50 - .35 \\ &= .75. \end{aligned}$$

Probability Law 2

Let A be an event from the sample space S. Then, $P(A^c) = 1 - P(A)$.

This law is verified by recognizing that $S = A \cup A^c$ and that the events A and A^c are mutually exclusive. By Probability Axioms 2 and 3,

$$1 = P(S) = P(A \cup A^c) = P(A) + P(A^c)$$

EXAMPLE **1.2-8** A particular brand of television must go through a series of five inspections during the manufacturing process. Let A be the event "a set passes all five inspections." Experience has shown that $P(A) = .85$. The complement of the event A is "a set fails at least one of the inspections." By Probability Law 2,

$$P(A^c) = 1 - P(A) = 1 - .85 = .15.$$

Exercises 1.2

1.2-1 Let $P(A) = .3$, $P(B) = .8$, and $P(A \cup B) = .9$. Find the probabilities of the events A^c, B^c, and $(A \cap B)^c$.

1.2-2 An experiment consists of tossing a fair coin four times.

a List the sample space so that the sample points are equally likely to occur.

b What is the probability that heads appears on both the first and third tosses?

c What is the probability that heads appears on at least three of the four tosses?

d Find the probability of obtaining exactly three heads in a row.

1.2-3 There are eight microchips numbered 1 to 8 in a box. Numbers 1, 2, and 3 are defective but the others are good. Two microchips are selected randomly, one after the other, without replacement.

a List the points in the sample space.

b Find the probability of selecting two defective chips.

c Find the probability of selecting two good chips.

d Find the probability of selecting one good chip and one defective chip.

e Find the probability of selecting a defective chip on the second draw.

f Would the probabilities in parts (b) – (e) change if the chips were not numbered?

g Would the probabilities in parts (a) – (e) change if the two chips were selected at the same time instead of one after the other?

1.2-4 Refer to Exercise 1.1-1.

a Assign probabilities to the sample points.

b Find the probability that "one marble is red."

1.2-5 For Exercise 1.1-6, find the probability of "moving on the first turn."

1.2-6 One card is selected from a deck of 52. Let A be the event "ace" and let B be the event "diamond."

a Find $P(A)$ and $P(B)$.

b Find $P(A \cap B)$.

c Find $P(A \cup B)$.

d State in words the event $A^c \cap B^c$.

e Find $P(A^c \cap B^c)$.

1.2-7 A particular system will undergo two development stages. After each stage it will be tested. The probability that it fails after the first stage is 1/2. The probability that it fails after the second stage is 1/3, regardless of the outcome of the first stage.

a Assume that a sample point is an ordered pair representing the outcomes of the two tests. List the sample space.

b Using a dice-tossing model as in Example 1.2-5, find the probabilities for all the points in the sample space from part (a).

c Find the probability that the number of failed tests is exactly 1.

1.2-8 Three friends play a game of chance to determine who buys coffee. Each tosses a penny. If one person has an outcome (heads or tails) different from the other two, that person has to buy. If all the pennies match, the game is repeated. What is the probability that the game ends on the first round?

1.2-9 In a television game show, contestants are asked to match baby pictures of three celebrities with their respective adult pictures. Suppose a contestant matches the pictures at random.

a What is the probability that all the pictures match? 1/3 Ask 1/6

b What is the probability that none of the pictures match? 2/3 3/6

1.2-10 Show for events A, B, and C from the sample space S that

$$P(A \cup B \cup C) = P(A) + P(B) + P(C) - P(A \cap B) - P(A \cap C) - P(B \cap C) + P(A \cap B \cap C).$$

SECTION

1.3 Simulation of Events on the Computer

Suppose that a company is trying to decide between one of two maintenance policies for a machine. One policy would be to service the machine only when it fails, while the other would require preventative maintenance at some predetermined interval of time. The company would like to know which method is most cost effective, and if preventative maintenance is the choice, how often maintenance should be done. A mathematical model of this system can help the company make a decision. A model might take the form of an equation that could be used to calculate costs for the two maintenance policies. However, in complex systems such equations are often difficult to derive. In these cases, a model might consist of a computer program whose function is to mimic (or simulate) the behavior of the system. A simulation for this example would calculate the costs incurred as if the actual system were undergoing failures and repairs. With such information, the company could pick the more desirable of the two maintenance policies.

To simulate systems that exhibit random behavior, we must have some mechanism for generating random events. In Example 1.2-5, the machines being UP or DOWN were simulated by rolling a pair of fair dice. In practice, it is common to use computer-generated random numbers to simulate stochastic behavior.

As a starting point for simulating random events, we will assume that we can select numbers at random from the unit interval 0 to 1. We will call such numbers *unit random numbers*. In practice, the numbers are selected from a large, finite subset of equally spaced points contained in the unit interval. For example, the subset might consist of the 100,000 five-decimal place numbers beginning with .00000 and ending with .99999. To obtain more than one unit random number, selections are made with replacement. For instance, if .23445 is selected on the first draw, it will again be a candidate for selection on the next and subsequent draws.

Conceptually, unit random numbers have the following property. If A is a subinterval of the unit interval, then the probability of selecting a unit random number from A should be equal to the length of A. However, when we select from a finite subset of the unit interval, the probability of selecting a unit random number from some intervals is only approximately equal to the interval length. For example, if we select a five-decimal place unit random number from .00000 to .99999, the probability that the number is in the interval $[0, 1/3)$ is $33333/100000 = .33333 < 1/3$. Such approximations are satisfactory for our purposes. Thus, if A is a subinterval of the unit interval, we will assume that the probability of selecting a number from A is equal to the length of A.

Table 1 in Appendix A is a table of random numbers. Placing a decimal point in front of each number converts the entries in the table to five-decimal place unit random numbers. To use this table, begin by selecting a row and column arbitrarily. To obtain more than one random number, move sequentially down the column or across the row. Unit random numbers can be used in various ways to simulate events that have specific probabilities. We will start with a simple example.

EXAMPLE **1.3-1** Suppose the experiment we wish to simulate is a toss of a single fair coin. The sample space is $S = \{H, T\}$. Each outcome in the sample space has probability of .5 of occurring. Simulating this experiment using the random numbers in Table 1 of Appendix A is done as follows. Pick a number from this table, say the number in column 9 and row 1, and convert this number to a number between 0 and 1. The result is .62590. To determine which event, heads or tails, a particular number represents, use the following rule:

Let u be a unit random number.

If $0 \leq u < .5$, then the event is tails;
else the event is heads.

Therefore, the outcome corresponding to .62590 is heads. Note that by generating the events in this way, $P(\mathrm{T}) = .5$ because there are 50,000 possible five-decimal place numbers in the interval $[0, .5)$, each with equal probability of 1/100,000. Likewise, $P(\mathrm{H}) = .5$. If we were to simulate this experiment many times, we would expect the resulting empirical probabilities for heads and tails to be approximately .5.

It is interesting to see what frequencies actually occur in a sequence of 30 simulations of the experiment in Example 1.3-1. Suppose we continue down column 9 of Table 1 in Appendix A for the 30 simulations of the coin toss. The following sequence of events occurs:

$$(\mathrm{H, H, T, H, T, H, T, H, T, T, H, H, T, T, H,}$$
$$\mathrm{T, T, H, T, H, H, H, T, T, H, H, H, H, T, H}).$$

The empirical probabilities for the events tails and heads based on the 30 simulations are $13/30 = .433$ and $17/30 = .567$, respectively. It may be surprising that these empirical probabilities are not closer to .5. In general, the empirical probabilities tend to get closer to the true probabilities as the number of simulations is increased. It is not uncommon for simulated experiments to be repeated several thousand times.

Since it is not very efficient to generate a large number of random events using a random number table, a computer is often used to generate such numbers. Scientific subroutine libraries contain computer programs that can be accessed from most programming languages. Unit random number generators are included in most of these libraries. It is not always necessary to access a scientific subroutine library when you need a unit random number generator. Frequently, the program language compiler will contain a built-in function that generates unit random numbers. For example, a Pascal compiler has a built-in function called RANDOM. When RANDOM is called, a unit random number is returned. These numbers, while not truly random, exhibit many of the characteristics of random numbers. For instance, they tend to occur with equal frequency.

Some people refer to unit random numbers as *pseudorandom numbers* and to the functions that generate the numbers as *pseudorandom number generators*. One should not take computer-generated unit random numbers for granted. Pseudorandom number generators generate sequences of numbers using deterministic algorithms. Depending on the quality of the algorithms, some sequences may appear more random than others. In practice, one should always carefully assess the adequacy of the random number generator they plan to use. The details of creating and testing random number generators are beyond the scope and purpose of this text. The reader is referred to Kennedy and Gentle (1980) and Knuth (1973) to read more about pseudorandom number generators.

Due to the variety of programming languages and computers one could use to simulate an experiment, it is impractical for this text to include computer code for simulations. Instead, algorithms that outline the basic logic used to generate the random phenomenon will be presented. The next example gives two alternative methods for generating the random phenomenon in Example 1.2-5.

EXAMPLE 1.3-2 An algorithm that could be used to simulate the experiment of observing a single machine to determine if it is UP or DOWN, where $P(\text{UP}) = 5/6$ and $P(\text{DOWN}) = 1/6$, is as follows:

1 Generate u, a unit random number.
2 If $u < 1/6$, machine is DOWN;
  else machine is UP.

Note that any interval of length 1/6 could be used in step 2 to denote the outcome DOWN. For example, step 2 could be replaced with the following.

2 If $1/3 \le u < 1/2$, machine is DOWN;
  else machine is UP.

If instead we wished to denote the outcome UP directly, step 2 could be as follows:

2 If $u < 5/6$, machine is UP;
else machine is DOWN.

In Example 1.2-5, we were interested in the status of two machines. The events in the sample space are $S = \{(\text{UP,UP}), (\text{UP,DOWN}), (\text{DOWN,UP}), (\text{DOWN,DOWN})\}$. An easy way to simulate this experiment would be to repeat twice steps 1 and 2, once for each machine. An alternative way to simulate this experiment is to simulate the ordered pairs from S directly. This is done by partitioning the interval $[0, 1)$ in four disjoint subintervals of lengths 25/36, 5/36, 5/36, and 1/36. These subintervals correspond to the outcomes (UP,UP), (UP,DOWN), (DOWN,UP), and (DOWN,DOWN), respectively. This is illustrated in Figure 1.3-1.

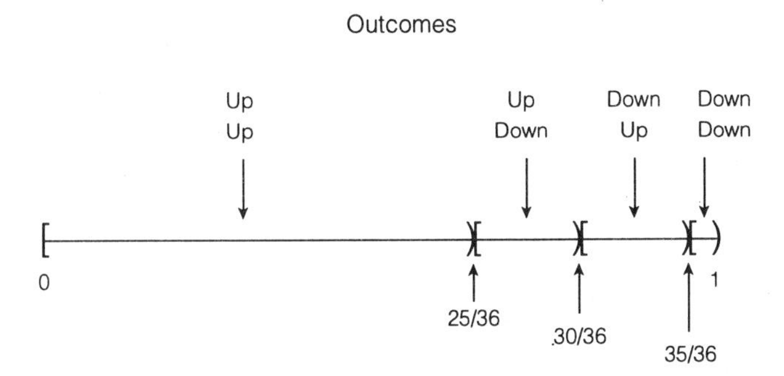

FIGURE **1.3-1**
Division of the Unit Interval for Example 1.3-2.

The corresponding algorithm is as follows:

---

1 Generate u, a unit random number.
2 If $u < 25/36$, outcome is (UP,UP);
else if $u < 30/36$, outcome is (UP,DOWN);
else if $u < 35/36$, outcome is (DOWN,UP);
else outcome is (DOWN,DOWN).

EXAMPLE 1.3-3 Simple games of chance can lead to surprising results. Suppose that a coin is tossed 5000 times. Each time heads occurs, we win a dollar, otherwise we lose a dollar. Let $S(n)$ be our accumulated winnings after n tosses. For instance, if the sequence HHHTT occurs in the first five tosses, then $S(5) = \$1.00$, whereas if the sequence HTTTT occurs, $S(5) = -\$3.00$. Let us consider how many times during the 5000 tosses $S(n)$ will go from a positive balance to a negative balance, or vice versa.

People often guess that a change of sign will occur several thousand times, using the idea that the number of heads and the number of tails should somehow even out.

To see what can happen, 5000 tosses of a coin were simulated on the computer. For each simulated toss of the coin, $S(n)$ was computed by adding one to $S(n-1)$ if a head occurs and subtracting one from $S(n-1)$ if a tail occurs. A change in sign will occur on the nth toss in one of two ways: $S(n-2) = 1$, $S(n-1) = 0$, and $(S(n) = -1)$; or $S(n-2) = -1$, $S(n-1) = 0$, and $S(n) = 1$. This is equivalent to $S(n-2) + S(n-1) + S(n) = 0$. The following algorithm describes the steps involved in the simulation.

Let n be the number of tosses, $S(n)$ the number of heads minus the number of tails in n tosses of the coin, and C the number of times $S(n)$ changes signs.

Initialize $n = 0$, $S(-1) = 0$, $S(0) = 0$, and $C = 0$.

Repeat the following steps 5000 times.

1 Generate u, a uniform random number. Increment $n = n+1$.

2 If $u < 1/2$ (i.e., tails occur), set $S(n) = S(n-1) - 1$;
else set $S(n) = S(n-1) + 1$.

3 If $S(n) + S(n-1) + S(n-2) = 0$, then increment $C = C+1$.

The results of 5000 tosses are shown in Figure 1.3-2. The horizontal axis is n, the number of the toss, and the vertical axis is $S(n)$. In this simulation there were nearly 50% heads and 50% tails, as would be expected, but there were only 13 sign changes in the 5000 tosses.

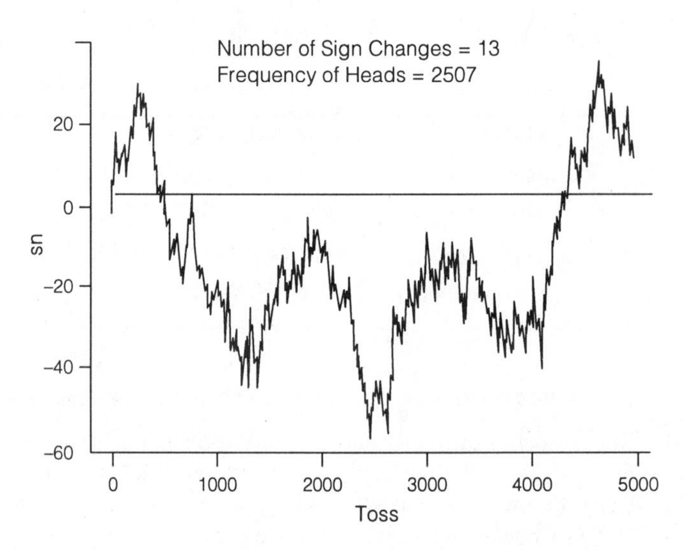

FIGURE **1.3-2**
Sign Changes in 5000 Tosses of Coin

Exercises 1.3

1.3-1 Use a table of random numbers to simulate 20 tosses of a pair of dice.

a What fraction of the 20 tosses resulted in a 7?

b Compare the empirical probability from part (a) to the true probability of observing a 7.

1.3-2 Write a computer program to simulate 200 tosses of a pair of dice.

a From the 200 simulations, what is the empirical probability of observing a 7?

b Compare the empirical probability from part (a) to the true probability of observing a 7.

c Repeat parts (a) and (b) for 1000 tosses.

1.3-3 A box contains the numbers $\{1, 2, 3, 4\}$. A number is drawn from the box at random and replaced. This procedure is repeated until all four numbers have been selected at least once.

a Simulate this experiment several times using a table of random numbers. How many draws did you make each time?

b Using a computer simulation, compute an empirical probability for the event "it takes more than eight draws to obtain all four numbers." Use at least 100 repetitions of the experiment.

1.3-4 The probability that an airline flight is on time is .80.

a Simulate 10 such flights using a table of random numbers.

b Using a computer simulation, compute an empirical probability for the event "at most 5 out of 10 flights are on time."

1.3-5 Simulate the coin toss game in Example 1.3-3 on the computer. Calculate the maximum amount of money and the minimum amount of money accumulated during the course of the game. Calculate the frequency of heads, the frequency of tails, and the number of times $S(n)$ changes sign.

1.3-6 Winning streaks and losing streaks in sports are fascinating (or perhaps frustrating) to the fan. Suppose that your favorite major league baseball team, which plays 162 games a year, has probability .5 of winning each game it plays.

a Without making any calculations, what do you think would be the length of a typical winning streak or losing streak for your team?

b Use a table of random numbers to simulate a single season of play. What was the longest winning streak?

c Write a computer program to simulate the play of a 162-game season. Calculate the longest winning streak and losing streak for 100 seasons of play. How do the results of the simulation compare with your answer to part (a)?

d How many of the 100 seasons had winning streaks longer than seven games?

1.3-7 If the probability of winning in exercise 1.3-6 is changed from .5 to a larger number p, we would expect to see the length of the winning streaks get longer. Simulate the length of the winning streaks for various values of p, say $p = .6, .7, .8$, and $.9$. For what values of p does it become obvious that the probability of winning is no longer .5?

1.3-8 Suppose you begin a game of chance with $\$100$. With each toss of a fair coin, you will win

\$100 if heads appears and lose \$100 if tails appears. The game will continue until you either have \$1000 or are broke.

a Guess how long it would take to play a typical game. Express your answer in terms of the number of tosses of the coin.

b Write a computer program to simulate this game. Record how long it takes to play. Repeat this procedure 20 times, and compare the results of the simulation with your answer to part (a).

c What do you think are your chances of going broke? Use the results of your simulation to help answer this question.

SECTION 1.4 Counting Techniques

In this section we will consider the problem of counting the number of outcomes that make up various events without explicitly listing the entire sample space. For an illustration consider a game of five-card poker. If the game is fair, all possible five-card hands have an equal chance of occurring. Therefore, the probability of drawing, say, four of a kind would be equal to the number of possible four-of-a-kind hands divided by the total number of possible five-card poker hands. If we could make a list of all possible five-card poker hands, computing the probability of four of a kind would just be a matter of counting the number of hands in which four of a kind occurs. However, there are 2,598,960 possible five-card poker hands, so listing the sample space and counting the number corresponding to four of a kind would be very tedious. Thus, techniques for counting the possibilities without making an explicit list of the possible outcomes can be useful.

Counting the ways something can be done is useful in many situations. For example, suppose a company is buying a computer system consisting of a workstation, a color monitor, and a printer. If there are several choices for each of these components, it might be of interest to know how many possible systems the company has to choose from. In another setting, suppose an operating system is being designed for a computing facility that will service 30,000 users. Given a certain number of bits will be used for passwords, it would be useful to have a simple counting technique to determine the number of unique passwords that can be formed without listing them.

The counting techniques that we will consider are based on the Multiplication Rule.

Multiplication Rule

If procedure A can be done in m ways and, for each way of doing procedure A, procedure B can be done in n ways, then A followed by B can be done in mn ways.

The Multiplication Rule extends directly to A followed by B followed by C, and so on.

EXAMPLE 1.4-1 A traveler may drive any one of three routes from Kansas City to Chicago and any one of four routes from Chicago to New York as depicted in Figure 1.4-1. If we count the number of routes from Kansas City to New York through Chicago, we see there are a total of $(3)(4) = 12$ routes to choose from

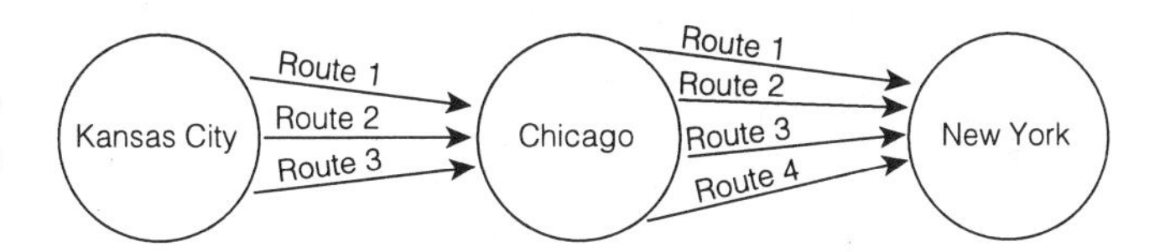

FIGURE 1.4-1 Routes from Kansas City to New York through Chicago

EXAMPLE 1.4-2 A company is planning to buy a computer system consisting of a workstation, a monitor, and a printer. Managers are deciding between two different workstations, four different monitors, and three different printers. Assuming all the equipment is compatible, the number of possible systems they have to choose from is $(2)(4)(3) = 24$.

EXAMPLE 1.4-3 An eight-digit (8-bit) computer word is made up of a sequence of eight 0's and 1's. To count the number of possible words of this type, imagine making a list in which we write down the first digit, and having done this, we write down the second digit, and so forth, until a word has been completed. Thinking of listing the possible words in this way, we see that the Multiplication Rule is applicable. The number of eight-bit words is $(2)(2)(2)(2)(2)(2)(2)(2) = 2^8 = 256$.

Suppose all 8-bit words are equally likely to occur. The probability of any such word would be $1/256$. There are nine different arrangements of the 0's and 1's corresponding to the compound event "at least seven of the digits are 1." Thus, the probability of randomly selecting a word with seven or more 1's is $9/256$.

EXAMPLE 1.4-4 A bank has three tellers whom we observe every hour for 8 hours. Each time we make an observation, we note which tellers are busy and which are not. Each observation can be denoted by the 4-tuple (time, status of teller 1, status of teller 2, status of teller 3), where the status is either busy or not busy. The number of possible observations is $(8)(2)(2)(2) = 64$.

EXAMPLE **1.4-5** Suppose we choose a committee of three people from five where one person is to be named president, another vice-president, and another secretary. How many such committees are there? Regard committees consisting of the same people but in different offices as different committees. The Multiplication Rule applies directly to this problem. There are five choices for president, and having made this choice there are four remaining choices for vice-president, and finally there are three remaining choices for secretary. Therefore the number of possible committees is $(5)(4)(3) = 60$.

We call such an arrangement of individuals as described in Example 1.4-5 an *ordered arrangement*. This terminology is used to denote the fact that the order in which the individuals are selected is important in counting the possible outcomes.

Definition 1.4-1

An ordered arrangement of r items selected from n distinct items is called a permutation of n distinct items taken r at a time.

By applying the Multiplication Rule as in the committee selection in Example 1.4-5, it can be seen that the number of permutations of n distinct items taken r at a time is

$$n(n-1)\cdots(n-r+1) = \frac{n!}{(n-r)!}.$$

EXAMPLE **1.4-6** For a particular phone exchange the first three digits of all phone numbers are the same. We wish to count the number of possible phone numbers in this exchange in which the last four digits are distinct. To solve this problem, imagine selecting four distinct numbers from the set $\{0,1,2,...,9\}$, one after the other, and writing them in a row. Clearly this will produce a four-digit number in which all numbers are different. The order in which a number is selected is important because each ordering will produce a different phone number. There are $(10)(9)(8)(7) = 10!/(10-4)! = 5040$ permutations of 10 distinct numbers taken 4 at a time, producing 5040 phone numbers in which the last four digits are distinct.

EXAMPLE **1.4-7** An operating system for a computing facility that will serve 30,000 users is being designed. It has been decided that the passwords will consist of sequences of distinct letters. Suppose it is required that all the passwords be at least three letters long. If every user picked a password exactly three letters long, there would not be enough passwords to go around because the number of permutations of three distinct letters from 26 is $(26)(25)(24) = 26!/(26-3)! = 15{,}600$. The designers can either hope that approximately one half of the users choose passwords longer than three letters or they must change their rule. If they require passwords to have at least four distinct letters, then the users will have at least $(26)(25)(24)(23) = 26!/(26-4)! = 358{,}800$ unique passwords to choose from.

EXAMPLE **1.4-8** A sequential file consists of 30 distinct items arranged in a row. The number of arrangements of the file is $(30)(29)(28)\cdots(2)(1) = 30!$. In general, the number of permutations of n distinct items taken n at a time is $n!$.

Consider selecting a committee of r people from n where there is no distinction in duties among the members. Such committees are considered *unordered arrangements* of the individuals.

Definition 1.4-2

An unordered arrangement of r items selected from n distinct items is called a *combination.*

EXAMPLE **1.4-9** Consider selecting a committee of three people from five with no distinction of duties. The number of such committees can be counted by relating the number of unordered arrangements to the number of ordered arrangements. For each unordered arrangement of three people, there are $(3)(2)(1) = 3! = 6$ possible ways to put these people in order (first, second, third). Thus, in this case

$$\text{Number of ordered arrangements} = 6\,(\text{number of unordered arrangements}).$$

We have already computed the number of ordered arrangements to be $(5)(4)(3) = 60$. Therefore, the number of unordered arrangements is $60/6 = 10$. If the individuals are numbered 1 to 5, the specific committees are

$$\{1\,2\,3\},\ \{1\,2\,4\},\ \{1\,2\,5\},\ \{1\,3\,4\},\ \{1\,3\,5\},$$
$$\{1\,4\,5\},\ \{2\,3\,4\},\ \{2\,3\,5\},\ \{2\,4\,5\},\ \{3\,4\,5\}.$$

The argument used to count the number of unordered arrangements or combinations for the preceding committee example can be applied in general. The number of ordered arrangements of n distinct items taken r at a time is $n!/(n-r)!$, and for each unordered arrangement there are $(r)(r-1)\cdots(1) = r!$ ways of placing the items in order. Therefore,

$$\begin{aligned}\text{Number of ordered arrangements} &= \frac{n!}{(n-r)!}\\ &= r!\ (\text{number of unordered arrangements}).\end{aligned}$$

It follows that the number of combinations of n distinct items taken r at a time is

$$\frac{n!}{r!(n-r)!}$$

Notationally, the number of n distinct items chosen r at a time is

$$\binom{n}{r} = \frac{n!}{r!(n-r)!}$$

EXAMPLE 1.4-10 From a class of 15 people, 6 will be chosen for an experimental course in engineering. Since the order of selection is not important, the number of possibilities is

$$\binom{15}{6} = \frac{15!}{6!\,(15-6)!} = 5005.$$

If six students wish to be chosen together, the probability is $1/5005$ that this occurs, assuming all 5005 combinations are equally likely to occur.

EXAMPLE 1.4-11 How many 8-bit computer words are there in which there are exactly three 1's and five 0's? A word can be formed by selecting three of the eight positions for the 1's with the remaining positions being assigned 0. Alternatively, five of the eight positions could be selected for the 0's with the remaining positions being assigned 1. The number of words with this configuration is $\binom{8}{3} = \binom{8}{5} = 56$

EXAMPLE 1.4-12 A mail-order company has 20 cameras on hand. Unknown to the company, 5 of the 20 are defective. If we purchase three, what are our chances of getting at least one defective? To work this problem, we will use the rule for finding the probabilities of complementary events. The probability of getting at least one defective camera in three is equal to 1 minus the probability of getting three good cameras. The number of ways we can select 3 good cameras from the 15 good cameras is $\binom{15}{3} = 455$. The total number of ways we can select 3 cameras from the 20 is $\binom{20}{3} = 1140$. Assuming the mail-order house randomly selects 3 out of the 20 cameras to ship, the probability of getting at least one defective is $1 - 455/1140 = .6$.

Suppose we want the probability of getting exactly one defective camera. To have exactly 1 defective, we must select 1 defective camera from the 5 defective cameras and 2 good cameras from the 15 good cameras. Selecting one from 5 can be done in $\binom{5}{1} = 5$ ways. Selecting 2 from 15 can be done in $\binom{15}{2} = 105$ ways. Using the Multiplication Rule the number of ways to select one defective and two good cameras is $(5)(105) = 525$, and the probability of this event is $525/1140 = .46$.

EXAMPLE 1.4-13 In a class of 30 students, what is the probability that no one in class has the same birthday? This problem can be easily solved with the counting techniques we have just developed. Let us label the students 1 to 30. The sample points of the experiment are 30-tuples where the ith entry represents the birthday of the ith person. Therefore, the total number of sample points (if we assume no one was born during leap year) is 365^{30}. Now we need to determine how many of the sample points correspond to the event of no matching birthdays. This is simply a permutation of 30 unique birthdays selected from 365 or $(365)(364)\cdots(336) = 365!/335!$. Assuming all the sample points are equally likely to occur, we have

$$P(\text{no matching birthday}) = \frac{365!/335!}{365^{30}} = .2937.$$

It follows that the probability of at least one matching birthday in a class of 30 is $1 - .2937 = .7063$. It is surprising to many people that this probability is so high. In fact, in a class of 23 students there is about a 50–50 chance of a matching birthday.

Exercises 1.4

1.4-1 In remodeling a kitchen, the home owners have decided to replace all the major appliances. They have narrowed down their choices to four microwaves, three refrigerators, two ranges, and six dishwashers. How many possible selections of these appliances are there?

1.4-2 The basic unit for storing information in a digital computer is a bit that can be designated as either a 1 or a 0. In converting images to a form that can be transmitted electronically, an element called a pixel is used. Each pixel is quantized into gray levels and coded using strings of 0's and 1's. For example, a pixel with four gray levels can be coded using 2 bits as 00, 01, 10, or 11.

a How many gray levels can be quantized using a 4-bit code?

b How many bits are necessary to code 32 gray levels?

1.4-3 How many 16-bit computer words have at least fifteen 1's? How many have exactly twelve 1's and four 0's?

1.4-4 Customers at a fast food restaurant have the option of placing the following ingredients on their hamburgers: lettuce, tomato, onion, ketchup, mustard, mayonnaise, pickles, cheese, bacon, and mushrooms. How many different ways can customers top their hamburgers?

1.4-5 A box of candy has 30 pieces. Twenty-six are made of chocolate and four are made of vanilla. Five pieces are selected at random.

a What is the probability of getting four chocolates and one vanilla?

b What is the probability of getting all chocolates?

c What is the probability of selecting at least one chocolate?

1.4-6 A fraternity consists of 8 freshmen, 10 sophomores, 6 juniors, and 13 seniors. A committee of six is chosen.

a How many committees can be chosen consisting of one freshman, one sophomore, two juniors, and two seniors?

b How many committees are there consisting of at least four upperclassmen (juniors and seniors)?

1.4-7 How many four-digit lock combinations can be made for the following?

a The first digit is a 1 or a 2, and all the digits are unique.

b The first digit is a 1 or a 2, and digits may be repeated.

c The first digit is any number but 0, and the digits may be repeated.

1.4-8 Suppose 5 cards are dealt at random from a well-shuffled 52-card deck.

a What is the probability of getting four aces?

b What is the probability of getting three aces and two kings?

c What is the probability of getting four of a kind?

d What is the probability of getting a full house (three of one kind and two of another kind)?

1.4-9 Using simulation, compute the empirical probability of being dealt 4 aces out of 5 cards from a well-shuffled deck of 52 cards. Do a simulation to compute the empirical probability of being dealt three aces and two kings. Compare your empirical probabilities to parts (a) and (b) of Exercise 1.4-8.

1.4-10 Compute the probability of a matching birthday among 23 randomly selected people. Do this also for 25, 35, and 50 people.

1.4-11 Simulate the birthdays of 30 randomly selected people by selecting numbers at random from 1 to 365 to represent the birthdays. Repeat this procedure 100 times.

a Does the frequency of times you obtained a matching birthday agree with what you would expect from Example 1.4-13?

b Based on your simulation, what is the empirical probability of more than two people having the same birthday?

c Repeat this exercise for 50 people.

1.4-12 Find the probability that 10 numbers selected at random from the integers 1, 2, ..., 100, with replacement, are all different. How many numbers would need to be selected at random for the probability of at least one match to exceed 1/2?

1.4-13 A true–false exam has 15 questions of which 5 are true and 10 are false. A student randomly selects seven questions and answers those false. The remainder of the questions the student answers true. The best possible score the student could obtain on this exam is 12 correct answers. What is the probability of this happening?

1.4-14 How many different arrangements are there of the letters AAABBCC? Note that this is a permutation of letters that are not all distinct. (*Hint*: First work the problem as if the letters were subscripted. That is, $A_1A_2A_3B_1B_2C_1C_2$. For each unsubscripted arrangement, compute the number of subscripted arrangements.)

1.4-15 Suppose we have n items of which r_1 are of one kind, r_2 are of a second kind, . . ., r_k are of a kth kind. Show that the number of arrangements of these items in a row is $n!/(r_1!r_2!\cdots r_k!)$. Use this result to determine the number of arrangements of the letters in MISSISSIPPI.

1.4-16 Lotteries with large prizes are popular moneymakers for state governments. Some may find it highly unlikely that one person could win a big prize more than once, but it has happened. Explain why this may not be such a rare event after all.

SECTION

1.5 Conditional Probability

The probability of obtaining an ace in a single draw from a shuffled deck of 52 cards is 4/52 = 1/13. However, if the card is drawn from the 16 cards consisting of aces, kings, queens, and jacks, then the probability of an ace is 4/16 = 1/4. In the latter case, the probability is computed under the condition that the card is selected from a subset of the original sample space. For this reason, it is called a *conditional probability,* and the subset from which the card is selected is called the *conditioning set.* In the language of probability, we are *given* that the outcome is either an ace, a king, a queen, or a jack, and the conditional probability that it is an ace is 4/16 = 1/4.

To see the distinction between conditional probability and unconditional probability, it is helpful to think in terms of equally likely outcomes. If the unconditional probability of B is .3, then B would make up 30% of the points in the sample space. If the conditional probability of B given A is .3, then the number of points in B that are also contained in A would make up 30% of the points in A. This suggests a mathematical relationship between conditional and unconditional probability that leads to a formal definition.

Definition 1.5-1

Let A and B be events in a sample space S, and let $P(B|A)$ denote the conditional probability of B given A. Then

$$P(B|A) = \frac{P(A \cap B)}{P(A)},$$

provided $P(A) > 0$.

EXAMPLE 1.5-1 To demonstrate that Definition 1.5-1 is reasonable, consider Figure 1.5-1 depicting a sample space with events A and B. The nine points in the sample space are equally likely to occur.

FIGURE 1.5-1 Illustration of Conditional Probability

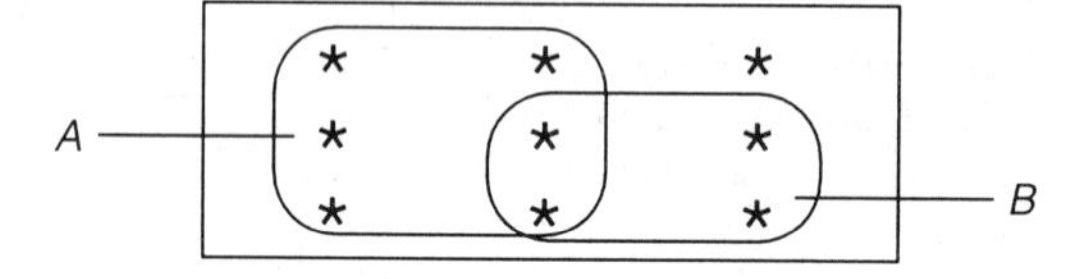

Applying Definition 1.5-1, we have $P(A) = 6/9$ and $P(A \cap B) = 2/9$, so

$$P(B|A) = \frac{P(A \cap B)}{P(A)} = \frac{2/9}{6/9} = \frac{1}{3}.$$

This corresponds to our intuitive notion of conditional probability. That is, there are six equally likely points in A and two of those are also in B, so

$$P(B|A) = \frac{2}{6} = \frac{1}{3}.$$

EXAMPLE 1.5-2 Suppose a pair of dice is tossed. Given that the sum of the face values is 6, we will find the probability that the face value on each die is 3. The conditioning set for which the sum of the face values equals 6 is $\{(1,5), (2,4), (3,3), (4,2), (5,1)\}$. Therefore,

$$P(\text{both dice are 3} \mid \text{sum of face values is 6}) = \frac{1}{5}$$

EXAMPLE 1.5-3 In automotive repair, experience has shown that a rough-running engine can be attributed to bad ignition wires 35% of the time, bad spark plugs 80% of the time, and both problems 20% of the time. The probabilities of these events are diagrammed in Figure 1.5-2.

FIGURE **1.5-2**
Automotive Repair Problems

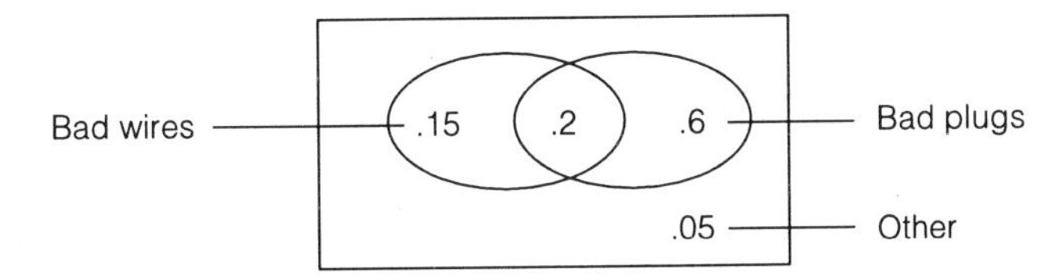

These events could also be expressed in the form of a *contingency table* as illustrated in Figure 1.5-3.

FIGURE **1.5-3**
Contingency Table for Automotive Repair Problems

		Ignition Wires		
		Bad	Good	
Spark Plugs	Bad	.2	.6	.8
	Good	.15	.05	.2
		.35	.65	1.0

Suppose a mechanic begins to diagnose the problem by checking the spark plugs first and finds them to be bad. With this information, what is the probability that the wires are also bad?

$$P(\text{bad wires} \mid \text{bad plugs}) = \frac{P(\text{bad wires} \cap \text{bad plugs})}{P(\text{bad plugs})}$$

$$= \frac{.2}{.8}$$

$$= .25.$$

If the spark plugs are found to be good, what is the probability that the wires are bad?

$$P(\text{bad wires} \mid \text{good plugs}) = \frac{P(\text{bad wires} \cap \text{good plugs})}{P(\text{good plugs})}$$

$$= \frac{.15}{.20}$$

$$= .75.$$

It follows that if the plugs are found to be good, there is three times the probability that the wires are bad than if the plugs are found to be bad.

In many situations the probability of A and the conditional probability of B given A can be readily computed from intuitive considerations. These quantities can be used to compute $P(A \cap B)$.

Multiplication Rule for Conditional Probability

$$P(A \cap B) = P(A)\,P(B|A).$$

Examples that typically require use of this Multiplication Rule are those in which the experiment is done as a sequence of two steps, where A is an event associated with the first step and B is an event associated with the second step.

EXAMPLE **1.5-4** A box contains six good items and two defective items. If two items are selected at random from the box one after the other without replacement, find the probability that both are good. Let A be the event that a good item is selected first, and let B be the event that a good item is selected second. The event of interest is $A \cap B$. Since there are six good items out of eight for the first draw, $P(A) = 6/8$. Given that a good item is selected first, there would be five good items and two defective items remaining in the box. Thus, the conditional probability that a good item is selected second given that a good item is selected first is $P(B|A) = 5/7$. Therefore,

$$P(2\text{ good items}) = P(A \cap B) = P(A)\,P(B|A) = \left(\frac{6}{8}\right)\left(\frac{5}{7}\right) = .5357.$$

The Multiplication Rule extends directly to more than two events. For instance with three events A, B, and C,

$$P(A \cap B \cap C) = P(A)\,P(B|A)\,P(C|A \cap B).$$

EXAMPLE **1.5-5** Suppose that three items are drawn at random one after the other without replacement from a lot consisting of six good items and two defective items. Using the extended Multiplication Rule, we find that the probability of getting three good items is $(6/8)(5/7)(4/6) = .3571$. The probability of getting exactly two good items and one defective item can be computed by keeping track of the order of selection. Figure 1.5-4 shows the three ways in which two good items and one defective item can be chosen.

FIGURE 1.5-4
Selection Probabilities for Two Good Items and One Defective Item

First Selection	Second Selection	Third Selection	Probability
Good	Good	Defective	(6/8)(5/7)(2/6)
Good	Defective	Good	(6/8)(2/7)(5/6)
Defective	Good	Good	(2/8)(6/7)(5/6)

Notice that these probabilities are the same. The probability of the desired event is the sum of these three probabilities or .5357.

Combining the Multiplication Rule with Probability Axiom 3 yields the Law of Total Probability.

Law of Total Probability

If $E_1, E_2, \ldots, E_k$ are a collection of mutually exclusive events such that

$S = \bigcup_{i=1}^{k} E_i$, then for any event A in S,

$$P(A) = \sum_{i=1}^{k} P(E_i)P(A|E_i).$$

The next example demonstrates the use of the Law of Total Probability. The proof of this law is left as an exercise.

EXAMPLE 1.5-6 A manufacturer of springs has three production machines. Machines M_1, M_2, and M_3 produce 25%, 40%, and 35% of the springs, respectively. One percent of the springs produced by M_1 are defective, 5% produced by M_2 are defective, and 2% produced by M_3 are defective. A summary of the manufacturer's production is given in Figure 1.5-5.

FIGURE **1.5-5**
Daily Production of Springs

Machine	Percent of Daily Production	Percent Defective Springs
M_1	25%	1%
M_2	40%	5%
M_3	35%	2%

At the end of the day, the springs from all three machines are mixed in a bin. Suppose an inspector selects a spring at random from this bin. We will compute the probability that the spring is defective. To solve this problem, we need to recognize what information is contained in Figure 1.5-5. For simplicity, let M_i denote the event "the selected spring was produced by M_i" and the event D denote the event "the selected spring is defective." In terms of probability, we have

$$P(M_1) = .25 \quad \text{and} \quad P(D|M_1) = .01,$$
$$P(M_2) = .40 \quad \text{and} \quad P(D|M_2) = .05,$$
$$P(M_3) = .35 \quad \text{and} \quad P(D|M_3) = .02.$$

Therefore,

$$\begin{aligned} P(D) &= P(D \cap M_1) + P(D \cap M_2) + P(D \cap M_3) \\ &= P(M_1)\,P(D|M_1) + P(M_2)\,P(D|M_2) + P(M_3)\,P(D|M_3) \\ &= (.25)\,(.01) + (.40)\,(.05) + (.35)\,(.02) \\ &= .0295. \end{aligned}$$

Now consider the situation where the inspector notices that the spring is defective. The probability that the spring was produced by M_1 is

$$\begin{aligned} P(M_1 | D) &= \frac{P(D \cap M_1)}{P(D)} \\ &= \frac{(.25)\,(.01)}{.0295} \\ &= .0847. \end{aligned}$$

Similarly, $P(M_2 | D) = .6780$ and $P(M_3 | D) = .2373$. These computations are special cases of what is known as Bayes' formula (see Exercise 1.5-13).

EXAMPLE **1.5-7** The birthday problem from Example 1.4-13 can be solved by using the multiplication rule for conditional probability. Recall that we were interested in the probability of no matching birthdays among a class of 30 students. The first person has 365 out of 365 birthdays to choose from. Given the first person's birthday, the second person has 364 out of 365 birthdays to choose from. Given the first and second persons' birthdays, the third person has 363 out of 365 birthdays to choose from, and so on. This gives an expression for the probability as

$$P(\text{no match in 30 people}) = \left(\frac{365}{365}\right)\left(\frac{364}{365}\right)\left(\frac{363}{365}\right)\cdots\left(\frac{336}{365}\right) = .2937.$$

It follows that the probability of at least one match is $1 - .2937 = .7063$.

Exercises 1.5

1.5-1 A pair of fair dice is tossed once. Find the probability of a 3 and a 4, given that the sum of the face values is 7.

1.5-2 Two fair coins are tossed. Find the probability that both are heads given that at least one is heads.

1.5-3 Compute the conditional probability of Exercise 1.5-2 empirically through simulation as follows. Simulate two coin tosses repeatedly. Discard the outcomes that have both tails and keep the rest. That is, keep the outcomes that have at least one head. The empirical conditional probability is the fraction of these outcomes that have two heads.

1.5-4 There are five Democrats, five Republicans, and four independents from which a committee of three is to be selected at random. What is the probability that there is one of each political persuasion? Work the problem using the Multiplication Rule for conditional probability.

1.5-5 Students at a certain college were cross-classified according to two characteristics: (1) amount of time spent using the computer; (2) major area of study. The fractions of individuals in each category are given below. Suppose a student is selected at random.

		Time in Hours per Week			
		Less than 1	1–4	More than 4	Total
Area	Technical	.05	.40	.15	.60
	Nontechnical	.20	.15	.05	.40
	Total	.25	.55	.20	1.00

a Given that a student is in a technical area, what is the probability that he or she uses the computer more than 4 hours a week?

b Given that a student uses the computer more than 4 hours a week, what is the probability that he or she is in a technical area?

c Given that a student uses the computer 4 hours or less a week, what is the probability that he or she is in a nontechnical area?

1.5-6 Box I has two red marbles and two blue marbles in it, and box II has three red marbles and two blue marbles. A marble is selected at random from box I and placed in box II. A marble is then selected at random from box II. Find the probability that this marble is red.

1.5-7 Compute the probability of a red marble in Exercise 1.5-6 empirically through simulation. Base your solution on 500 repetitions of the experiment.

1.5-8 A survey has revealed that 75% of all college students study. It is also known that 85% of all students who study will graduate, while only 35% of those students who do not study will graduate.

a If a student is randomly selected, what is the probability that he or she will graduate?

b A randomly selected student is observed to graduate. What is the probability that the student studied?

1.5-9 Box 1 contains 10 cookies of which 6 are broken and 4 are good. Box 2 contains 12 cookies of which 7 are broken and 5 are good.

a A box is selected at random and a cookie is selected at random from the box. What is the probability that the cookie is broken?

b A box is selected at random and two cookies are selected at random from the box. What is the probability that the cookies are both broken?

1.5-10 A survey of a community taken before a vote on a school bond issue shows the following results. Sixty percent of the population favor the bond issue, 30% have children in school, and 25% both favor the bond issue and have children in school.

a Fill in the following contingency table with the percentages of individuals from the population who fall into the categories.

		Children in School		
		Yes	No	
Favor Bond Issue	Yes			
	No			

b Suppose a person is selected at random from the population. Given this person favors the bond issue, what is the probability that this person has children in school?

1.5-11 We have six boxes numbered 1, 2, 3, 4, 5, and 6. The ith box contains one dime and i pennies, $i = 1, 2, \ldots, 6$. That is, box 1 contains one dime and one penny, box 2 contains one dime and two pennies, and so on. A die is tossed, and a coin is randomly selected from the box whose number coincides with the outcome of the die.

a What is the probability that a dime is selected?

b Given that a dime is selected, what is the probability it came from box 1?

1.5-12 Prove the Law of Total Probability.

1-5-13 Let $E_1, E_2, \ldots, E_k$ be a collection of mutually exclusive events such that $S = \bigcup_{i=1}^{k} E_i$.

Bayes' formula states

$$P(E_j|A) = \frac{P(E_j)P(A|E_j)}{\sum_{i=1}^{k} P(E_i)P(A|E_i)}$$

Prove Bayes' formula.

SECTION

1.6 Independent Events

An important idea in the theory of probability is that of *independent events*. We will consider both the mathematical definition of independent events as well as the intuitive meaning behind the idea of independence.

Definition 1.6-1

The events A and B are independent if

$P(A \cap B) = P(A)P(B)$.

The intuitive meaning behind independence can be inferred from the following theorem.

THEOREM 1.6-1 If A and B are independent events and if $P(A) > 0$ and $P(B) > 0$, then the following conditions are equivalent.

i $P(B|A) = P(B)$.

ii $P(A|B) = P(A)$.

iii $P(A \cap B) = P(A)P(B)$.

If any of the conditions in Theorem 1.6-1 hold, then all of them will hold. For example, consider the equivalence of conditions (i) and (iii). From the definition of conditional probability, $P(B|A) = P(A \cap B)/P(A)$. If (iii) is true, then

$P(B|A) = P(A)P(B)/P(A) = P(B)$. Conversely, if (i) is true, then

$$P(A \cap B) = P(A)P(B|A) = P(A)P(B).$$

Intuitively, $P(B|A) = P(B)$ says that the occurrence of A does not affect the probability of B. That is, A and B behave independently of one another. The notion of mutually exclusive events should not be confused with the idea of independent events. If two events with nonzero probability are mutually exclusive, then $P(A \cap B) = 0$ but $P(A)P(B) > 0$; thus they cannot be independent events.

EXAMPLE 1.6-1 An experiment consists of tossing a fair coin twice. Let A be the event that a head appears on the first toss and B the event that a head appears on the second toss. Whether we observe A or not, the probability of B is 1/2. The tosses of a coin are independent of one another.

EXAMPLE 1.6-2 A single card is drawn at random from a deck of 52 cards. Let A be the event "an ace is drawn," and let B be the event "a diamond drawn." We have $P(A) = 1/13$, $P(B) = 1/4$, and $P(A \cap B) = P(\text{ace of diamonds}) = 1/52$. Since $P(A \cap B) = P(A)P(B)$, these events are independent.

EXAMPLE 1.6-3 Consider a coin-tossing experiment involving two coins where one is fair and the other has two heads. The experiment is conducted as follows. The fair coin is tossed first. If a head appears, the fair coin is tossed again, but if a tail appears the two-headed coin is tossed instead. We will show that the events "heads on the first toss" and "heads on the second toss" are not independent. The outcomes with their respective probabilities are given in Figure 1.6-1.

FIGURE 1.6-1 Outcomes of the Coin Toss Experiment

First Toss	Second Toss	Probability
H	H	$(1/2)(1/2) = 1/4$
H	T	$(1/2)(1/2) = 1/4$
T	H	$(1/2)(1) = 1/2$
T	T	$(1/2)(0) = 0$

The probability that a head appears on the second toss is $1/4 + 1/2 = 3/4$. However, the probability that a head appears on the second toss given that a head appears on the first toss is not equal to $3/4$. Rather,

$$\begin{aligned} P(\text{heads second} \mid \text{heads first}) &= \frac{P(\text{heads first} \cap \text{heads second})}{P(\text{heads first})} \\ &= \frac{1/4}{1/2} \\ &= \frac{1}{2} \\ &\neq P(\text{heads on second}) \\ &= \frac{3}{4} \end{aligned}$$

This verifies that the two events are dependent. Intuitively this is reasonable since the second toss depends on the first.

EXAMPLE 1.6-4 Suppose two components, A and B, are connected in series as diagrammed in Figure 1.6-2.

FIGURE 1.6-2
Components Connected in Series

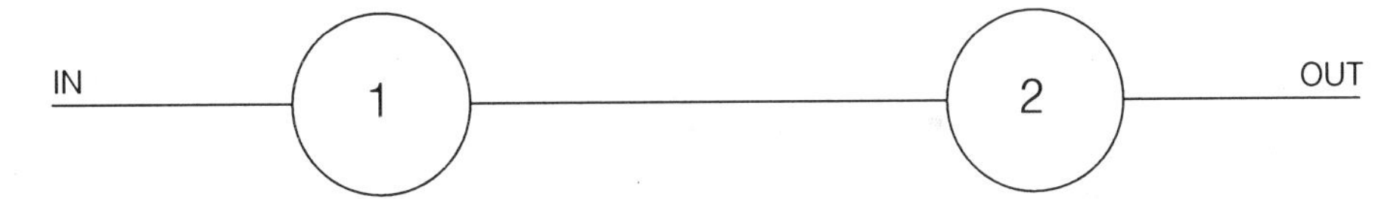

For the system to work, both components must work. Suppose that the components function independently of one another. Further suppose $P(A \text{ works}) = .9$ and $P(B \text{ works}) = .8$. The probability that the system works is computed as

$$\begin{aligned} P(\text{system works}) &= P(A \text{ works} \cap B \text{ works}) \\ &= P(A \text{ works})\, P(B \text{ works}) \\ &= (.9)(.8) \\ &= .72. \end{aligned}$$

EXAMPLE 1.6-5 Suppose two components, A and B, are connected in parallel as shown in Figure 1.6-3.

FIGURE 1.6-3
Components Connected in Parallel

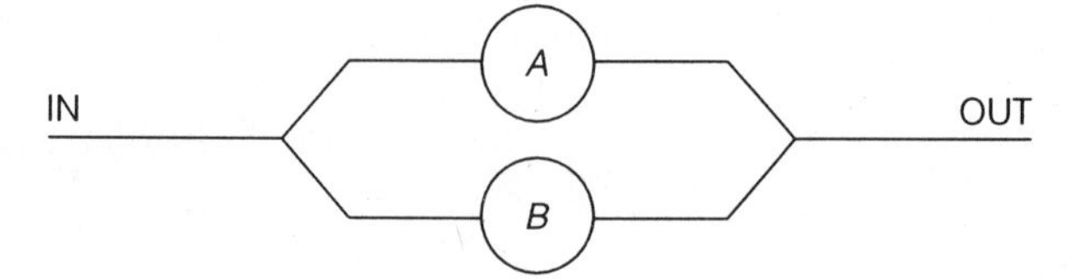

In this case, at least one component must work for the system to work. Again suppose that the components function independently, with $P(A \text{ works}) = .9$ and $P(B \text{ works}) = .8$. Now,

$$\begin{aligned} P(\text{system works}) &= P(A \text{ works} \cup B \text{ works}) \\ &= P(A \text{ works}) + P(B \text{ works}) - P(A \text{ works} \cap B \text{ works}) \\ &= .9 + .8 - .72 \\ &= .98. \end{aligned}$$

The concept of independence can be extended to more that two events.

EXAMPLE 1.6-6 Let the events $E_1, E_2, E_3, \ldots$ all have positive probability. These events are mutually independent if for any finite set of distinct indices $i_1, i_2, \ldots, i_k$

$$P\left(E_{i_1} \cap E_{i_2} \cap \ldots \cap E_{i_k}\right) = P\left(E_{i_1}\right)P\left(E_{i_2}\right)\ldots P\left(E_{i_k}\right).$$

Intuitively, events are mutually independent if the act of observing any set of them has no effect on the probabilities of occurrence of any other events.

EXAMPLE 1.6-7 A system consists of five components configured as shown in Figure 1.6-4.

FIGURE 1.6-4
Five-component System

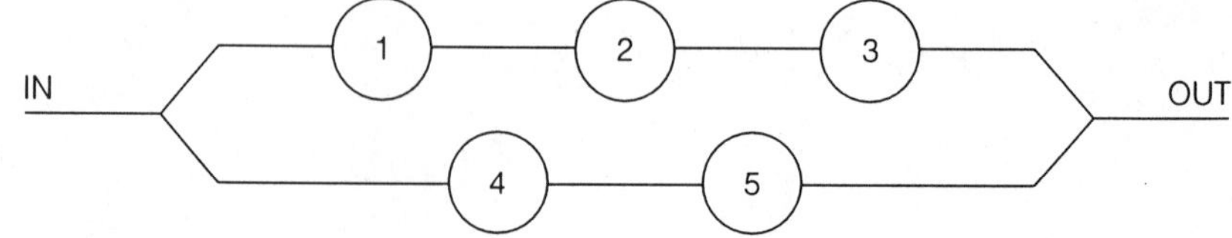

The system works if components 1, 2, and 3 all work or if components 4 and 5 both work. Assume the components are mutually independent and each component has a probability of .8 of working. Let W_i denote the event "component i works." In set notation, the event "the system works" is $(W_1 \cap W_2 \cap W_3) \cup (W_4 \cap W_5)$. From the mutual independence of the components,

$$P(W_1 \cap W_2 \cap W_3) = (.8)(.8)(.8) = .512,$$

$$P(W_4 \cap W_5) = (.8)(.8) = .64,$$

$$P(W_1 \cap W_2 \cap W_3 \cap W_4 \cap W_5) = (.8)(.8)(.8)(.8)(.8) = .32768.$$

Thus,

$$\begin{aligned} P((W_1 \cap W_2 \cap W_3) \cup (W_4 \cap W_5)) &= P(W_1 \cap W_2 \cap W_3) + P(W_4 \cap W_5) \\ &\quad - P(W_1 \cap W_2 \cap W_3 \cap W_4 \cap W_5) \\ &= .512 + .64 - .32768 \\ &= .82432. \end{aligned}$$

Pairwise independence of three events, A, B, and C, does not imply mutual independence. Consider the following example.

EXAMPLE 1.6-8 An experiment consists of rolling a pair of fair dice. Denote the sample space as the 36 ordered pairs (i, j), where i denotes the outcome of the first die and j denotes the outcome of the second die. Let A be the event "an even number occurs on the first die," B the event "an even number occurs on the second die," and C the event "the sum of the face values is 7." Then $P(A) = 1/2$, $P(B) = 1/2$, and $P(C) = 1/6$. These events are pairwise independent since

$$\begin{aligned} P(A \cap B) &= P(\text{even on both dice}) \\ &= \frac{1}{4} \\ &= P(A)\,P(B), \end{aligned}$$

$$\begin{aligned} P(A \cap C) &= P((2, 5), (4, 3), (6, 1)) \\ &= \frac{1}{12} \\ &= P(A)\,P(C), \end{aligned}$$

and

$$P(B \cap C) = P((5,2),\ (3,4),\ (1,6)) = \frac{1}{12} = P(B)\,P(C).$$

However,

$$P(A \cap B \cap C) = P(\text{even on both dice and the sum is 7}) = 0 \neq P(A)\,P(B)\,P(C).$$

Hence, these three events are not mutually independent.

Exercises 1.6

1.6-1 Which of the following pairs of events A and B do you think would be independent?

a A is the event that it rains today, and B is the event that it rains tomorrow.

b There are two computers in an office. A is the event that computer 1 fails today, and B is the event that computer 2 fails tomorrow.

c A is the event that the cost of living rises at least 5%, and B is the event that teachers' salaries increase at least 4% next year.

d Two fair coins are tossed. A is the event that there are two heads, and B is the event that there are two tails.

1.6-2 In a particular board game a player rolls a pair of fair dice on each turn. In order to move a game piece the player must roll a 5 on at least one of the die or a sum of 5 on the two dice. Let A be the event "a 5 on at least one die," and let B be the event "the sum of the face values is 5." Find $P(A)$, $P(B)$, and $P(A \cap B)$. Are A and B independent events?

1.6-3 Let $P(A|B) = .5$, $P(B) = .25$, and $P(A \cup B) = .75$.

a Find $P(A \cap B)$, $P(A)$, and $P(B^c|A^c)$.

b Are A and B independent events?

1.6-4 A coin is tossed repeatedly. Find the probability that heads appears for the first time on the fourth toss. How many tosses must be made so that the chance of at least one head appearing among the tosses is greater than .9?

1.6-5 A computer program consists of three modules all of which must work correctly in order for a job to be completed. When a job is under way, module 1 has probability .1 of failing, module 2 has probability .2 of failing, and module 3 has probability .3 of failing. Assuming that the modules fail independently of one another, find the probability that a job is completed.

1.6-6 An electronic device in a space satellite consists of two components. Component 1 has probability .05 of failing, and component 2 has probability .01 of failing. In order for the device to work, at least one of the components must not have failed. Assuming that the components function independently, what is the probability that the device works?

1.6-7 Components in a system are located in four positions as shown in the figure. In order for the system to operate, either the components in positions 1 and 2 must be operable or those in positions 3 and 4 must be operable. Suppose there are four components labeled A, B, C, and D. Assume the components function independently with $P(A \text{ operable}) = .9$, $P(B \text{ operable}) = .5$, $P(C \text{ operable}) = .8$, and $P(D \text{ operable}) = .7$. Put the components in the positions that give the maximum probability for the system to be operable.

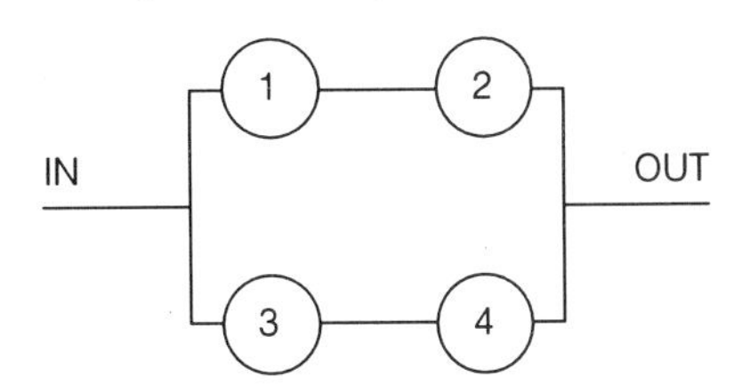

1.6-8 Write a computer program to simulate whether the four components in Exercise 1.6-7 are operable. Use this information to simulate whether the system as configured in your solution above is operable. Repeat the simulation 1000 times and compute the empirical probability that the system is operable. Compare this to the true probability that the system is operable.

1.6-9 Consider a system of three components configured in parallel as shown. In order for the system to function, at least one of the components must work. Assume the components function independently, and let p_i be the probability that the component i works, $i = 1, 2, 3$. Find the probability that the system functions. (*Hint*: Find the probability that the system fails and take the complement.)

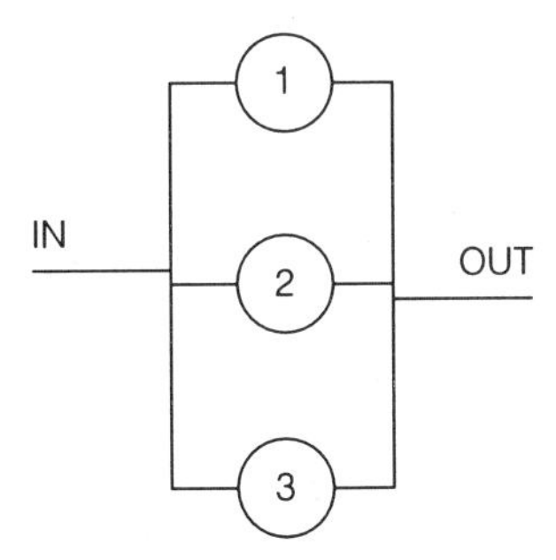

1.6-10 Consider a system like Exercise 1.6-9 with n components configured in parallel as shown. Assume the components function independently, and let p_i be the probability that the component i works, $i = 1, 2, \ldots, n$. Find the probability that the system functions.

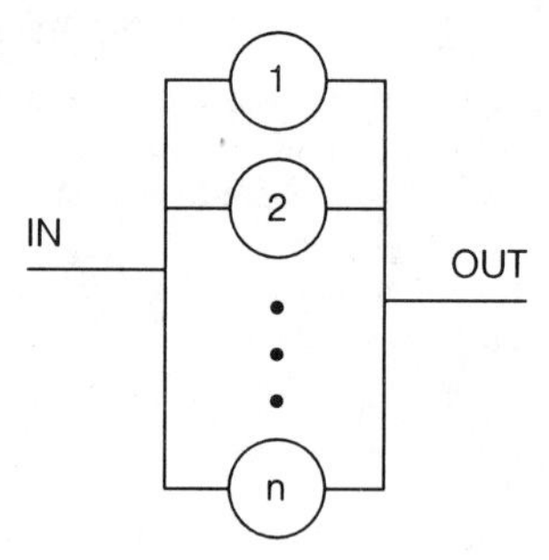

1.6-11 In a manufacturing plant, items undergo three independent inspections. The probability that a bad item is detected at each inspection is .75. What is the probability that a bad item passes all three inspections?

2

Discrete Random Variables

SECTION

2.1 Random Variables

A variable whose value is determined in some way by chance is called a *random variable*. For instance, the height and weight of a randomly selected individual are random variables. Other examples of random variables include the number of people in line at a drive-up bank teller, the time to failure of an electrical component, and the number of wins recorded by a baseball team.

A random variable can be thought of as a function that maps the points of the sample space into the set of real numbers as illustrated in the example below.

EXAMPLE 2.1-1 Let an experiment consist of three tosses of a fair coin. The sample space can be listed as

$$S = \{\text{HHH, HHT, HTH, HTT, THH, THT, TTH, TTT}\}.$$

Let X be the random variable that denotes the number of heads occurring in the three tosses. The possible values for X are 0, 1, 2, or 3. The random variable X maps the points in S to the real numbers as shown in Figure 2.1-1.

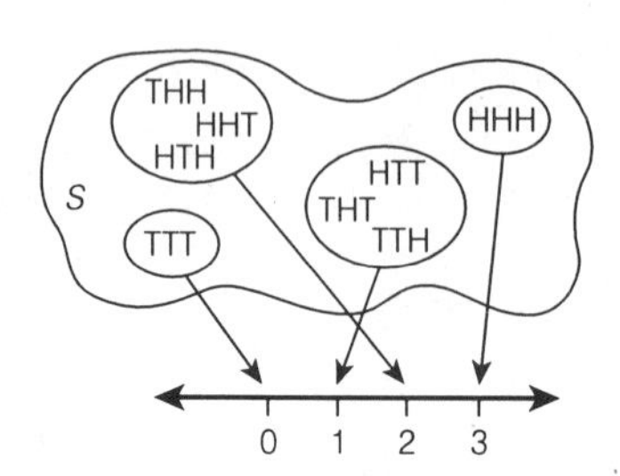

FIGURE 2.1-1
Mapping of the Sample Space to the Real Numbers

We could also define other random variables for this sample space. For instance, let the random variable Y be 0 if no heads occur and 1 if at least one head occurs. This random variable maps the sample point TTT to 0 and the other seven sample points to 1.

We will consider two types of random variables, *discrete* and *continuous*. A *discrete random variable* is one in which the possible values are countable. Two examples of discrete random variables are the number of heads in 10 tosses of a coin and the number of defective automobiles in a fleet of 100. A discrete random variable has its probability concentrated on a finite or countably infinite set of real numbers. For example, the random variable representing the number of defective automobiles in a fleet of 100 has probability concentrated on the points $0, 1, 2, \ldots, 100$. A *continuous random variable* is one in which the probability is spread continuously over intervals of real numbers. For instance, the time to failure of an electrical component and the atmospheric temperature measured at noon each day are examples of continuous random variables. This chapter deals with discrete random variables. Continuous random variables are discussed in Chapter 5.

It is often of interest to know the *probability distribution* of a random variable—that is, how probability is distributed across the possible values of a random variable. For a discrete random variable, this probability distribution is described by the *probability mass function.*

Definition 2.1-1

The probability mass function of a discrete random variable X is the function that assigns a probability to each of the possible values of X. The probability mass function of X is denoted as $p(x)$ and is defined by

$$p(x) = P(X = x).$$

EXAMPLE 2.1-2 Let the random variable X denote the number of heads in three tosses of a fair coin. The probability mass function, $p(x)$, is given below.

x	$p(x)$
0	$p(0) = P(X = 0) = P(\text{TTT}) = 1/8$
1	$p(1) = P(X = 1) = P(\text{HTT}) + P(\text{THT}) + P(\text{TTH}) = 3/8$
2	$p(2) = P(X = 2) = P(\text{HHT}) + P(\text{HTH}) + P(\text{THH}) = 3/8$
3	$p(3) = P(X = 3) = P(\text{HHH}) = 1/8$

The probability mass function is frequently given in tabular or graphical form as shown in Figure 2.1-2. The bar chart in Figure 2.1-2 is drawn such that the bars are centered at the values of X with positive probability, and the height of the bars represent $P(X = x)$. This barchart is frequently called a *histogram*.

FIGURE 2.1-2 Probability Mass Function Displays

x	0	1	2	3
$p(x)$	1/8	3/8	3/8	1/8

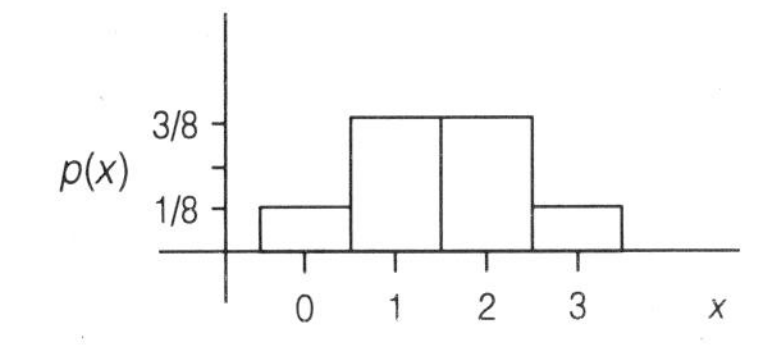

EXAMPLE 2.1-3 Let the random variable X denote the number of automobile accidents per week at a busy intersection in a large city. In a 100-week period, there were no accidents in 50 of the weeks, one accident in 30 of the weeks, two accidents in 10 of the weeks, and three accidents in 10 of the weeks. Based on this information, an empirically determined probability mass function for the number of accidents per week can be defined as

x	0	1	2	3
$p(x)$	$\frac{50}{100} = .5$	$\frac{30}{100} = .3$	$\frac{10}{100} = .1$	$\frac{10}{100} = .1$

In many cases, the probability mass function of one random variable can be derived immediately from the probability mass function of another random variable.

EXAMPLE 2.1-4 Let X denote the number of vacant seats on a commuter airline flight, and let the probability mass function be

x	0	1	2	3
$p(x)$	.50	.30	.15	.05

Suppose the random variable Y denotes the amount of revenue lost due to vacant seats. If each vacant seat costs the airline \$70 in lost revenue, then Y can be expressed in terms of X as

$$Y = 70X.$$

The probability mass function of Y is obtained immediately from the probability mass function of X. For instance, the probability that $X = 2$ is the same as the probability that $Y = 70(2) = 140$. Thus, the probability mass function for Y is

y	0	70	140	210
$p(y)$	.50	.30	.15	.05

EXAMPLE 2.1-5 Let the random variable X denote the number of heads minus the number of tails in three tosses of a fair coin. The probability mass function of X is

x	3	1	-1	-3
Sample points	HHH	HHT, HTH, THH	HTT, THT, TTH	TTT
$p(x)$	1/8	3/8	3/8	1/8

We will find the probability mass function of the random variable $Y = X^2$. There are two possible values for Y: $Y = 1$ when X is 1 or -1 and $Y = 9$ when X is 3 or -3. Thus, the probability mass function of Y is

y	1	9
$p(y)$	6/8	2/8

The probability mass function must satisfy two properties.

Properties of Probability Mass Functions

Let X be a discrete random variable with probability mass function $p(x)$. Then

i $0 \le p(x) \le 1$

ii $\sum_{x} p(x) = 1.$

The first property states that individual values of the probability mass function must be between 0 and 1. The second property states that the sum of the probability mass function over all the possible values of the random variable must equal 1. These properties follow immediately from the probability axioms.

EXAMPLE 2.1-6 At a government installation is an assembly line that manufactures gyroscopes to be used in the guidance systems of small missiles. During production each gyroscope is tested as it comes off the assembly line. From observations of this production over a long time, it has been determined that approximately one-tenth of the gyroscopes produced is defective. Assume for this problem that the defects occur independently of one another. Let the random variable X denote the number of gyroscopes tested until a defective one is found. The possible values of X are 1, 2, ... The random variable X is a discrete random variable with a countable number of values. The probability mass function of X can be determined as the following:

x	Outcome of the Assembly Line Tests	$p(x)$
1	First gyro defective	.1
2	(First good) ∩ (second defective)	(.9)(.1)
3	(First good) ∩ (second good) ∩ (third defective)	(.9)(.9)(.1)
⋮		$= (.9)^2 (.1)$

In general,

$$p(x) = (.9)^{x-1}(.1), \quad x = 1, 2, 3, \ldots .$$

The two mass function properties hold for this random variable. All the probabilities clearly are greater than or equal to zero with the largest probability being .1. The second property is easily verified by recalling that the sum of a geometric series is

$$\sum_{i=0}^{\infty} r^i = \frac{1}{1-r}, \quad |r| < 1.$$

For this mass function we have

$$\sum_{x=1}^{\infty} (.9)^{x-1}(.1) = (.1)\sum_{i=0}^{\infty}(.9)^i$$

$$= \frac{.1}{1-.9}$$

$$= 1.$$

Another convenient function for expressing the various probabilities assigned to the values of a random variable X is the *cumulative distribution function* of X.

Definition 2.1-2

For each real number x, the function $F(x) = P(X \le x)$ is called the cumulative distribution function of X.

EXAMPLE 2.1-7 Let the random variable X denote the number of heads in three tosses of a coin as in Example 2.1-2. The values of $F(x)$ for $x = 0, 1, 2,$ and 3 are

$$F(0) = P(X \le 0) = P(0 \text{ or fewer heads}) = \frac{1}{8}$$

$$F(1) = P(X \le 1) = P(1 \text{ or fewer heads}) = \frac{1}{8} + \frac{3}{8} = \frac{1}{2}$$

$$F(2) = P(X \le 2) = P(2 \text{ or fewer heads}) = \frac{1}{8} + \frac{3}{8} + \frac{3}{8} = \frac{7}{8}$$

$$F(3) = P(X \le 3) = P(3 \text{ or fewer heads}) = \frac{1}{8} + \frac{3}{8} + \frac{3}{8} + \frac{1}{8} = 1.$$

It is also possible to compute $F(x)$ for a value of x not equal to 0, 1, 2, or 3. For example, although $p(1.5) = 0$,

$$F(1.5) = P(X \le 1.5) = P(1.5 \text{ or fewer heads})$$

$$= p(0) + p(1)$$

$$= \frac{1}{8} + \frac{3}{8}$$

$$= \frac{1}{2}.$$

The graph of the cumulative distribution function in the discrete case is a step function. For this example, the graph is given in Figure 2.1-3. Note that the size of the step at x is equal to the probability mass function $p(x)$. For instance, the size of the step at $x = 1$ is $1/2 - 1/8 = 3/8 = p(1)$.

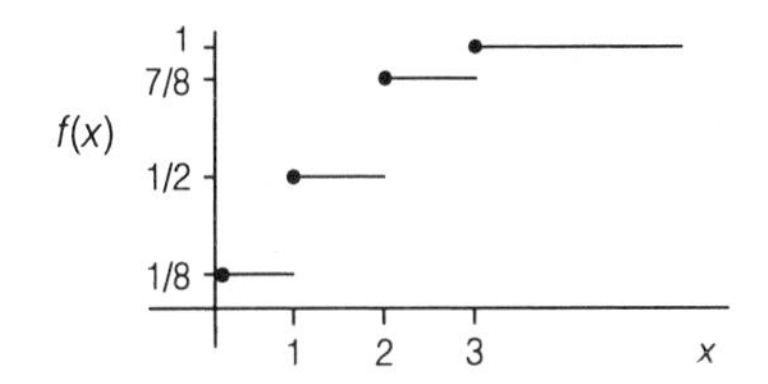

FIGURE **2.1-3** Cumulative Distribution Function for the Number of Heads in Three Tosses of a Fair Coin

EXAMPLE **2.1-8** Let the random variable X denote the number of people that are absent during any given week in an office with a total of five people. Suppose the cumulative distribution function of X is

x	0	1	2	3	4	5
$F(x)$	.30	.75	.90	.96	.99	1.00

The cumulative distribution function can be used to compute various probabilities of interest. For example, the probability that exactly one person is absent can be found as

$$\begin{aligned} P(X=1) &= P(X \le 1) - P(X \le 0) \\ &= F(1) - F(0) \\ &= .45. \end{aligned}$$

The probability that more than two are absent is

$$\begin{aligned} P(X>2) &= 1 - P(X \le 2) \\ &= 1 - F(2) \\ &= .10. \end{aligned}$$

The probability mass function of X can be derived directly from the cumulative distribution function of X as

$$\begin{aligned} p(0) &= F(0) = .30; \\ p(1) &= F(1) - F(0) = .75 - .30 = .45; \\ p(2) &= F(2) - F(1) = .90 - .75 = .15; \\ p(3) &= F(3) - F(2) = .96 - .90 = .06; \\ p(4) &= F(4) - F(3) = .99 - .96 = .03; \\ p(5) &= F(5) - F(4) = 1 - .99 = .01. \end{aligned}$$

We see that it makes no difference whether probabilities are given in terms of the cumulative distribution function or the probability mass function. One can be obtained from the other. The choice of which one to use is usually a matter of convenience.

Exercises 2.1

2.1-1 A fair die is rolled twice. Find the probability mass function of the following random variables.

a The sum of the face values.

b The maximum of the face values.

2.1-2 A box has seven items, four good and three defective. Three items are selected at random without replacement. Let X be the number of good items. Find the probability mass function and the cumulative distribution function of X.

2.1-3 In Exercise 2.1-2, suppose that we have a loss of \$20 for each defective item selected. Let the random variable Y denote the loss. Find the probability mass function of Y.

2.1-4 A multiple choice test has three possible answers for each question, one of which is correct. Let X denote the number of correct answers in an exam consisting of four questions. Find the probability mass function of X, assuming the student guesses on each question.

2.1-5 Let X denote the number of busy servers at the checkout counters in Bargain Barn at 5 P.M. The cumulative distribution function of X is

x	0	1	2	3	4
$F(x)$	.20	.50	.80	.90	1.00

a Find the probability mass function of X.

b Find the probability that three or more servers are busy.

2.1-6 Let Z denote a random variable with probability mass function

z	−2	−1	0	1	2	4
$p(z)$	.1	.3	k	.05	.25	.21

a What is k?

b Find the cumulative distribution function of Z.

c Find $P(-1 \le Z \le 1)$, $P(Z > 0)$, $P(Z \le 2)$, and $P(-2 \le Z < 2)$.

2.1-7 Find the cumulative distribution function of the random variable in Example 2.1-6.

2.1-8 It is known that there is a defective chip on a computer board that contains eight chips. A technician tests the chips one at a time until the defective chip is found. Assume the chip to be tested is selected at random without replacement. Let the random variable X denote the number of chips tested. Find the probability mass function of X.

2.1-9 A fair coin is tossed until heads occurs. Let the random variable Y denote the number of tosses. Find the probability mass function of Y. Verify that the two properties for discrete random variables hold for Y.

SECTION

2.2 Joint Distributions and Independent Random Variables

In Section 2.1 we dealt with individual random variables. Now we consider two or more random variables jointly.

Definition 2.2-1

The *joint probability mass function* of the discrete random variables X and Y is a function $p(x, y)$ defined by

$$p(x, y) = P(X = x, Y = y),$$

where $(X = x, Y = y)$ denotes the intersection of the events $(X = x)$ and $(Y = y)$.

EXAMPLE **2.2-1** A consumer testing agency classifies automobile defects as minor and major. Let X denote the number of minor defects and Y the number of major defects in a randomly selected automobile. Let the joint probability mass function of X and Y be defined as in the table.

Minor Defects x

Major Defects y	$p(x, y)$	0	1	2	3
	0	.1	.2	.2	.1
	1	.05	.05	.1	.1
	2	0	.01	.04	.05

The entries in the table represent the joint probabilities of X and Y. For example, $p(2, 1) = .1$ is the probability a randomly selected automobile has two minor and one major defects.

From the joint probability distribution the distributions of X and Y can be found. For example, the probability of having one minor defect is found by summing all entries in the table for which $X = 1$. Thus,

$$\begin{aligned} P(X=1) &= P((X=1, Y=0) \cup (X=1, Y=1) \cup (X=1, Y=2)) \\ &= p(1,0) + p(1,1) + p(1,2) \\ &= .26. \end{aligned}$$

Similarly, the probability of having one major defect is found by summing all the entries in the table for which $Y = 1$. We find

$$\begin{aligned} P(Y=1) &= P((X=0, Y=1) \cup (X=1, Y=1) \\ &\qquad \cup (X=2, Y=1) \cup (X=3, Y=1)) \\ &= .05 + .05 + .1 + .1 \\ &= .3. \end{aligned}$$

Conditional probabilities as defined in Section 1.5 are also of interest when we are dealing with more than one random variable at a time. For example, the probability of three minor defects given one major defect is

$$\begin{aligned} P(X=3 \mid Y=1) &= \frac{P(X=3, Y=1)}{P(Y=1)} \\ &= \frac{.1}{.3} \\ &= \frac{1}{3}. \end{aligned}$$

As noted in the example, probabilities such as $P(X = 1)$ and $P(Y = 1)$ were found by summing entries in the table over a row or a column, hence obtaining the marginal totals. For this reason, in the context of joint probability distributions, the probability mass functions of the individual random variables are called *marginal probability mass functions*.

Definition 2.2-2

Let X and Y be discrete random variables with joint probability distribution $P(X = x, Y = y)$. The marginal probability mass functions of X and Y are

$$p_X(x) = P(X = x) = \sum_y P(X = x, Y = y)$$

and

$$p_Y(y) = P(Y = y) = \sum_x P(X = x, Y = y).$$

In Example 2.2-1, we computed the $P(X = 3 \mid Y = 1)$ directly from the definition of conditional probability. Similarly, for Example 2.2-1 it is possible to obtain the conditional probabilities for each value of X given $Y = 1$. These probabilities define the *conditional probability mass function* of X given $Y = 1$.

Definition 2.2-3

The conditional probability mass function of X given $Y = y$ is defined for all y such that $p_Y(y) > 0$ as

$$P(X = x \mid Y = y) = \frac{p(x, y)}{p_Y(y)}.$$

The conditional probability mass function of Y given $X = x$ is defined for all x such that $p_X(x) > 0$ as

$$P(Y = y \mid X = x) = \frac{p(x, y)}{p_X(x)}.$$

EXAMPLE **2.2-2** Computer software can be thought of as an ordered string of operators and operands. During execution of the software, a random number of the operators and operands will be accessed. Let X denote the number of distinct operators and Y the number of distinct operands accessed during an execution of the software. Consider a piece of computer software that has four operators and three operands. The joint probability mass function and the marginal mass functions for X and Y are displayed in the table.

Operators x

	$p(x, y)$	1	2	3	4	$p_Y(y)$
Operands y	1	1/32	1/32	1/16	1/32	5/32
	2	1/16	3/16	3/8	1/16	11/16
	3	1/32	1/32	1/16	1/32	5/32
	$p_X(x)$	1/8	1/4	1/2	1/8	1

With this information, specific probabilities about X and Y can be found. Consider the following statements.

$$
\begin{aligned}
P(X+Y \le 3) &= P(\text{software uses at most three distinct operators and operands}) \\
&= p(1,1) + p(1,2) + p(2,1) \\
&= \frac{1}{32} + \frac{1}{16} + \frac{1}{32} \\
&= \frac{1}{8},
\end{aligned}
$$

$$
\begin{aligned}
P(X \le 3) &= P(\text{software uses at most three distinct operators}) \\
&= p_X(1) + p_X(2) + p_X(3) \\
&= \frac{1}{8} + \frac{1}{4} + \frac{1}{2} \\
&= \frac{7}{8},
\end{aligned}
$$

$$
\begin{aligned}
P(X \le 3, Y = 2) &= P(\text{software uses at most three distinct operators} \\
&\qquad\qquad \text{and exactly two distinct operands}) \\
&= p(1,2) + p(2,2) + p(3,2) \\
&= \frac{1}{16} + \frac{3}{16} + \frac{3}{8} \\
&= \frac{5}{8}.
\end{aligned}
$$

Suppose we know that the software will use exactly two distinct operands, and we would like to know the probability that it will use at most three distinct operators during execution. This probability is

$$
\begin{aligned}
P(X \le 3 \mid Y = 2) &= \frac{P(X \le 3, Y = 2)}{p_Y(2)} \\
&= \frac{5/8}{11/16} \\
&= \frac{10}{11}.
\end{aligned}
$$

In evaluating a piece of software, the total number of distinct operators and operands accessed may be of interest. For this example, let $W = X + Y$. The probability mass function for W can be determined from the joint probability mass function of X and Y. The possible values for W are $2, 3, \ldots, 7$. The corresponding mass function is

w	2	3	4	5	6	7
$p(w)$	1/32	3/32	9/32	7/16	1/8	1/32

.

The definition of a joint probability mass function can be extended to any number of random variables. For instance, the joint probability mass function for the random variables X, Y, and Z is

$$p(x, y, z) = P(X = x, Y = y, Z = z).$$

The marginal probabilities of X, Y, and Z can be determined from the joint distribution as

$$p_X(x) = P(X = x) = \sum_y \sum_z P(X = x, Y = y, Z = z);$$

$$p_Y(y) = P(Y = y) = \sum_x \sum_z P(X = x, Y = y, Z = z);$$

$$p_Z(z) = P(Z = z) = \sum_x \sum_y P(X = x, Y = y, Z = z).$$

The concept of independence applies to jointly distributed random variables. Intuitively, two random variables are independent if the probabilities associated with one random variable do not depend on the values of the second random variable. The mathematical definition of independent random variables is simply an application of the definition of independent events given in Section 1.6.

Definition 2.2-4

The discrete random variables X and Y are *independent* if, for all values of x and y,

$$p(x, y) = p_X(x)p_Y(y).$$

Otherwise the random variables X and Y are dependent.

In Example 2.2-1, the random variables are dependent because $p(1,1) = .05 \neq p_X(1)p_Y(1) = (.26)(.3)$. Intuitively this is reasonable since experience indicates that the numbers of major and minor automobile defects are related. Often, independence is an assumption that can be made about two random variables based on intuitive considerations. For instance, the random variables denoting the face values that occur when tossing a pair of dice are naturally assumed to be independent.

EXAMPLE 2.2-3 A manufacturing plant has two assembly lines. Let X and Y denote the number of defects produced in an hour by lines 1 and 2, respectively. Since the assembly lines function independently, it will be assumed the random variables X and Y are independent. The probability mass functions for X and Y are

x	0	1	2
$p(x)$	.5	.3	.2

and

y	0	1
$p(y)$	.9	.1

With these mass functions and the assumption of independence, joint probabilities for X and Y can be easily computed. For example, the probability of no defects in 1 hour's production is

$$\begin{aligned} p(0,0) &= p_X(0)p_Y(0) \\ &= (.5)(.9) \\ &= .45. \end{aligned}$$

The probability that the total number of defects is 2 or more in 1 hour is

$$\begin{aligned} P(X+Y \geq 2) &= p(1,1) + p(2,0) + p(2,1) \\ &= (.3)(.1) + (.2)(.9) + (.2)(.1) \\ &= .23. \end{aligned}$$

Just as was true in the case of independent events, the definition of independent random variables can be extended to any number of random variables.

Definition 2.2-5

The discrete random variables $X_1, X_2, \ldots, X_n$ are *mutually independent* if

$$p(x_1, x_2, \ldots, x_n) = p_{X_1}(x_1)p_{X_2}(x_2)\cdots p_{X_n}(x_n).$$

Exercises 2.2

2.2-1 The random variables X and Y have the joint probability mass function given by the table.

		x				
	p(*x*,*y*)	0	1	2	3	4
y	0	.05	.10	.10	.04	.01
	1	.05	.20	.10	.03	.02
	2	.05	.10	.05	.05	.05

a Find $P(X > 2, Y > 1)$ and $P(Y \leq 1)$.
b Find the marginal probability mass function of Y.
c Find $P(Y = 0 \mid X = 0)$.
d Verify that X and Y are dependent.

2.2-2 Let X denote the number of speeding tickets given by Officer Smith on Monday, and let Y denote the number given on Tuesday. The probability mass functions for X and Y are

x	0	1	2	3	4
$p(x)$	.1	.2	.3	.2	.2

and

y	0	1	2	3	4	5
$p(y)$	.1	.1	.2	.3	.2	.1

a Assuming X and Y are independent, find the joint probability mass function of X and Y.
b Find the probability mass function of $X + Y$, the total number of speeding tickets for Monday and Tuesday.

2.2-3 A pair of dice is tossed once. Let X denote the sum of the face values, and let Y denote the absolute value of the difference between the face values.
a Find the joint probability mass function of X and Y.
b Find the conditional probability $P(Y = 1 \mid X = 7)$.

2.2-4 Box 1 has three chips numbered 1, 2, 3. Box 2 has two chips numbered 1 and 2. Suppose a chip is randomly selected from box 1 and placed in box 2. Then a chip is randomly selected from box 2. Let X denote the number of the chip selected from box 1 and Y the number of the chip selected from box 2.
a Find the joint probability mass function of X and Y.
b Find the marginal probability mass function of Y.
c Find $P(X = 1 \mid Y = 1)$.

2.2-5 The probability mass functions of the number of cars sold on a typical day by two used car dealers are as follows:

	Number Sold		
	0	1	2
Slippery Sam	.5	.3	.2
Big Ed's	.8	.1	.1

Assume that the sales at the two dealerships are independent of each other.

a Find the probability that Slippery Sam sells more cars than Big Ed's on a typical day.

b Find the probability that the two dealers sell the same number of cars on a typical day.

2.2-6 Let X denote the total number of cars sold by the two dealers in Exercise 2.2-5.

a Find the probability mass function of X.

b Find the cumulative distribution function of X.

c Find $P(X > 2)$.

SECTION

2.3 Expected Values

In this section, we consider the concept of an *expected value* of a random variable.

Definition 2.3-1

Let the random variable X have probability mass function $p(x)$. The *expected value of* X, denoted as $E(X)$, is defined by

$$E(X) = \sum_x xp(x).$$

An expected value can be thought of as an average value of a random variable. That is, since $p(x)$ can be interpreted as the fraction of time the value x occurs in the long run, $E(X)$ can be interpreted as the long-run average of the values of X. For example, if the random variable is the number of sunny days in January, then the expected value would be the number of sunny days we would see on average in January. The expected value is also called the mean of the random variable and is frequently denoted by μ.

Conceptually, $E(X)$, or μ, is only defined if this quantity is finite. Mathematically speaking, $E(X)$ will only be finite if the sum $\sum_x xp(x)$ converges absolutely. Exercise 2.3-8 explores this technical aspect of the definition of expectation further. Whenever expected values are used in the remainder of this text, we assume that they exist (i.e., are finite).

EXAMPLE **2.3-1** Let X be the number of heads in two tosses of a coin. The probability mass function of X is

x	0	1	2
$p(x)$	1/4	1/2	1/4

The expected value is

$$\begin{aligned} E(X) &= \sum_{x=0}^{2} xp(x) \\ &= 0(1/4) + 1(1/2) + 2(1/4) \\ &= 1. \end{aligned}$$

This implies that there will be 1 head, on average, in 2 tosses of a coin.

EXAMPLE **2.3-2** Suppose that the number of paintings sold in a day by a starving artist has the following probability distribution:

x	0	1	2	3	4	5
$p(x)$	.1	.2	.3	.2	.1	.1

Based on this probability distribution, the average number of paintings sold in a day is

$$\begin{aligned} E(X) &= \sum_{x=0}^{5} xp(x) \\ &= 0(.1) + 1(.2) + 2(.3) + 3(.2) + 4(.1) + 5(.1) \\ &= 2.3 \end{aligned}$$

If the artist were to sell paintings for many days, the number sold would average out to 2.3 per day. Notice that the expected value does not have to be one of the possible values of the random variable.

If probability is thought of as mass located at points along a number line, the expected value is the center of mass, or the fulcrum at which the line balances. In this sense, expected value is a measure of the center of the probability distribution. In the previous example, this fulcrum would need to be placed at 2.3 to balance the line. This is shown in Figure 2.3-1.

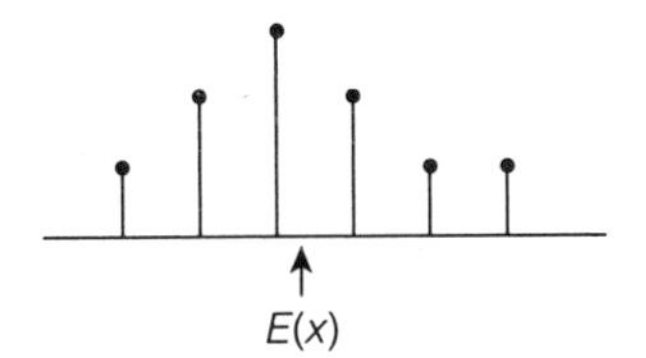

FIGURE **2.3-1** Expected Value Shown as the Center of Mass for Example 2.3-2

If the random variable is a number selected at random from a set of numbers, then the expected value of this random variable is equal to the average of the numbers in the set. This is demonstrated in the next example.

EXAMPLE 2.3-3 The total amount of money for the grand prize on a particular television game show is determined by the spin of a wheel. Suppose the wheel has 10 segments and in each segment appears the amount of money that will be awarded if the wheel stops in that segment. The 10 amounts are \$10, \$10, \$10, \$25, \$25, \$50, \$50, \$100, \$500, and \$1000. Assuming the wheel is fair, each of the 10 segments has probability 1/10. Let the random variable Y be the amount selected by the spin of the wheel. The expected value of Y is

$$\begin{aligned} E(Y) &= 10\left(\frac{3}{10}\right) + 25\left(\frac{2}{10}\right) + 50\left(\frac{2}{10}\right) + 100\left(\frac{1}{10}\right) + 500\left(\frac{1}{10}\right) + 1000\left(\frac{1}{10}\right) \\ &= \$178. \end{aligned}$$

Notice that the average of the \$10 amounts on the wheel is

$$\frac{10 + 10 + 10 + 25 + 25 + 50 + 50 + 100 + 500 + 1000}{10} = \$178,$$

which is the same as the expected value.

It will sometimes be useful to find the expected value of a function of a random variable. Typical functions of interest are $Y = X^2$ and $Y = a + bX$. The value of $E(Y)$ can be computed in two ways. We can find the probability mass function of Y and apply Definition 2.3-1, or we can use the probability mass function of X and apply the following theorem.

THEOREM 2.3-1 If Y is a function of random variable X, where, say, $Y = Q(X)$, then

$$E(Y) = E(Q(X)) = \sum_x Q(x)p(x).$$

EXAMPLE 2.3-4 Suppose the random variable X has probability distribution

x	−2	−1	0	1	2	4
$p(x)$	.05	.20	.30	.30	.10	.05

Consider the random variable $Y = X^2$. The expected value of Y will be computed first by using the probability mass function of Y. The probability mass function for Y is found to be

y	0	1	4	16
$p(y)$	.30	.50	.15	.05

Applying Definition 2.3-1 to the random variable Y, we find

$$E(Y) = \sum_y yp(y)$$
$$= 0(.30) + 1(.50) + 4(.15) + 16(.05)$$
$$= 1.9.$$

Alternatively, by Theorem 2.3-1, the expected value of Y is

$$E(Y) = E(X^2) = \sum x^2 p(x)$$
$$= 4(.05) + 1(.20) + 0(.30) + 1(.30) + 4(.10) + 16(.05)$$
$$= 1.9$$

Notice that $E(X) = .4$ in this example and $(E(X))^2 = .16 \neq E(X^2) = E(Y)$.

Of special interest are linear functions $Y = a + bX$. Expected values in this case are particularly easy to compute.

THEOREM 2.3-2 If $Y = a + bX$, then $E(Y) = a + bE(X)$.

Theorem 2.3-1 can be used to verify this result. That is,

$$E(Y) = \sum_x (a + bx)p(x)$$
$$= a\sum_x p(x) + b\sum_x xp(x)$$
$$= a + bE(X).$$

EXAMPLE 2.3-5 Let Y denote the earnings for the starving artist from Example 2.3-2. If the artist makes \$20 on each painting sold and the daily cost of renting a stall at the market to sell the paintings is \$30, then Y can be expressed in terms of X according to the equation $Y = 20X - 30$. Thus,

$$E(Y) = 20E(X) - 30 = 20(2.3) - 30 = \$16.$$

Thus, one reason our artist is starving is that the artist averages just \$16 a day on painting sales.

Theorem 2.3-2 can be generalized as follows.

THEOREM 2.3-3 If $Y = g_1(X) + g_2(X) + \cdots + g_n(X)$ then

$$E(Y) = E(g_1(X)) + E(g_2(X)) + \cdots + E(g_n(X)).$$

EXAMPLE 2.3-6 Let the random variable X have the following probability mass function:

x	0	1	2	3
$p(x)$	.30	.40	.20	.10

Theorem 2.3-3 can be used to find the expected value of $Y = X^2 + 3X + 1$. Let $g_1(X) = X^2$ and $g_2(X) = 3X + 1$. Then

$$\begin{aligned} E(Y) &= E(X^2) + E(3X + 1) \\ &= E(X^2) + 3E(X) + 1 \\ &= 2.1 + 3(1.1) + 1 \\ &= 6.4. \end{aligned}$$

E(x) = 0.40 + 1.20 + 0.30 = 1.1

E(x²) = 0.40 + 0.80 + 0.90 = 2.1

Exercises 2.3

2.3-1 The number of moves made in a board game is determined by the toss of a die. Let X be the number of moves. Find the expected value of X.

2.3-2 A random variable X has the following probability mass function:

x	0	1	2	3	4
$p(x)$	.24	.41	.26	.08	.01

a Find the expected value of the random variable X.

b Find the expected value of $2X - 7$.

c Find the expected value of X^2.

2.3-3 Sellers bring items to an auction house where buyers may bid for them. Experience has shown that the mean selling price of an article is $\mu = \$100$. The auction house charges \$10 for storage of each item and takes an additional 15% commission on each sale. Let X denote the amount returned to the seller. Find the expected value of X.

2.3-4 A box contains five red marbles and two blue marbles. Marbles are drawn one after the other without replacement. Let X be the number of draws it takes to obtain the first blue marble. Find $E(X)$.

2.3-5 A random variable Y has the following probability mass function:

y	1	2	3	4
$p(y)$	.7	.2	.06	.04

Find $E(Y)$ and $E(1/Y)$. Does $E(1/Y) = 1/E(Y)$?

2.3-6 A service technician visits a factory each week to check for defective fluorescent lights. The technician charges \$20 for each visit, and each defective light costs \$5 to replace. The number of defective lights Y has probability mass function

y	0	1	2	3
$p(y)$	.60	.25	.10	.05

Find the expected cost of the technician's visit.

2.3-7 Let the random variable X have the following probability mass function:

x	10	20	30	40
$p(x)$	.10	.20	.30	.40

Find the expected value of $Y = X^3 + 3X^2 - 4X - 2$.

2.3-8 From calculus it is known that the series $\Sigma 1/n^2$ converges. Let c denote the sum of this series. It follows that the function $p(n) = 1/cn^2$, $n = 1, 2, \ldots$, defines a probability mass function (why?). Show that the expected value of a random variable with this probability mass function does not exist (i.e., is infinite).

2.3-9 Prove Theorem 2.3-1. (*Hint*: Let y be a value of the random variable $Y = Q(X)$, and let $x_1, x_2, \ldots, x_k$ be the values of X such that $y = Q(x_i)$. Note that

$$yP(Y = y) = Q(x_1)P(X = x_1) + Q(x_2)P(X = x_2) + \ldots + Q(x_k)P(X = x_k).$$

Sum over all values of y.)

SECTION

2.4 Variance and Standard Deviation

The expected value of a random variable is only one characteristic of interest. It is not uncommon for two probability mass functions to have the same expected value but to differ in the way in which their probability is dispersed. Consider the following example.

EXAMPLE **2.4-1** The probability mass functions for the random variables X and Y both have an expected value of 2.

x	0	1	2	3	4
$p(x)$	.2	.2	.2	.2	.2

y	0	1	2	3	4
$p(y)$	.1	.2	.4	.2	.1

The histograms of the mass functions for X and Y in Figure 2.4-1 demonstrate that, although X and Y have the same mean, the probability mass of X is more dispersed than the probability mass of Y.

FIGURE **2.4-1**
Histograms for Example 2.4-1

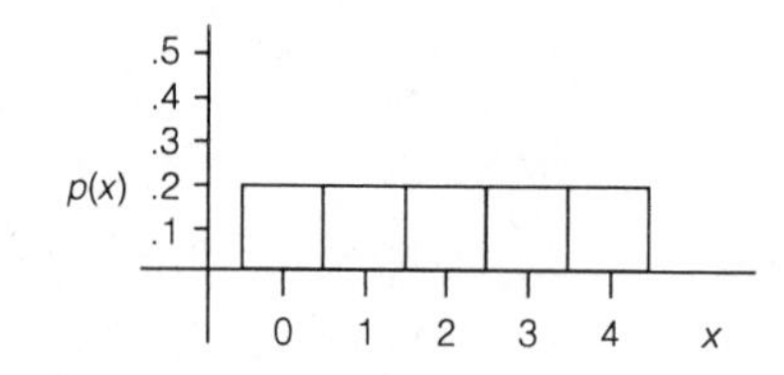

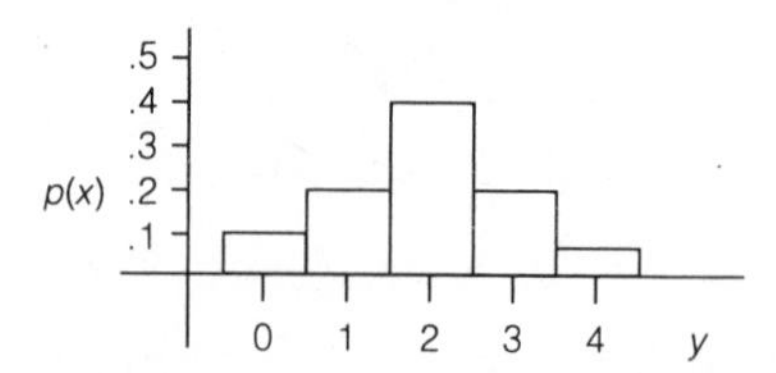

Two common measures of dispersion for a random variable X are variance and standard deviation. These measures are defined in terms of expected values of functions of the random variable.

Definition 2.4-1

For a random variable X, the variance of X, $\text{VAR}(X)$, and the standard deviation of X, $\text{STD}(X)$, are defined as

$$\begin{aligned}\text{VAR}(X) &= E[(X - E(X))^2],\\ \text{STD}(X) &= \sqrt{\text{VAR}(X)}.\end{aligned}$$

The standard deviation of a random variable is in the same units of measurement as the random variable itself, whereas the variance is in the square of the original units. The variance and standard deviation of a random variable are frequently denoted by σ^2 and σ, respectively.

EXAMPLE 2.4-2 The variances and standard deviations for the random variables in Example 2.4-1 are

$$\begin{aligned}\text{VAR}(X) &= \sum_x (x - E(X))^2 p(x)\\ &= (0-2)^2(.2) + (1-2)^2(.2) + (2-2)^2(.2)\\ &\qquad + (3-2)^2(.2) + (4-2)^2(.2)\\ &= 2;\\ \text{STD}(X) &= \sqrt{2} = 1.414;\\ \text{VAR}(Y) &= \sum_y (y - E(Y))^2 p(y)\\ &= (0-2)^2(.1) + (1-2)^2(.2) + (2-2)^2(.4)\\ &\qquad + (3-2)^2(.2) + (4-2)^2(.1)\\ &= 1.2;\\ \text{STD}(Y) &= \sqrt{1.2} = 1.095.\end{aligned}$$

As expected, these measures of dispersion are larger for X than for Y.

EXAMPLE 2.4-3 The variance and standard deviation of the random variable X in Example 2.3-2, the total number of paintings sold in a day by a starving artist, are

$$\begin{aligned} \mathrm{VAR}(X) &= \sum_{x=0}^{5} (x-2.3)^2 p(x) \\ &= (0-2.3)^2(.1) + (1-2.3)^2(.2) + (2-2.3)^2(.3) \\ &\quad + (3-2.3)^2(.2) + (4-2.3)^2(.1) + (5-2.3)^2(.1) \\ &= 2.01; \\ \mathrm{STD}(X) &= \sqrt{2.01} = 1.42 \text{ paintings.} \end{aligned}$$

The following theorem will help with the interpretation of the standard deviation. This theorem is known as *Chebychev's inequality*.

THEOREM 2.4-1 Chebychev's Inequality: Let X be a random variable with $E(X) = \mu$ and $\mathrm{STD}(X) = \sigma$. For any $k > 0$,

$$P(|X-\mu| \le k\sigma) \ge 1 - \frac{1}{k^2}.$$

Note that $|X-\mu| \le k\sigma$ if and only if $\mu - k\sigma \le X \le \mu + k\sigma$. Thus, Chebychev's inequality states that the probability a random variable falls within $\pm k$ standard deviations of its expected value is at least $1 - 1/k^2$. For example, at least $1 - 1/2^2 = .75$ or 75% of the time a random variable will fall within ± 2 standard deviations of its mean, while $1 - 1/3^2 = .89$ or at least 89% of the time will fall within ± 3 standard deviations of its mean. It also follows that the probability that a random variable falls outside $\pm k$ standard deviations of its expected value is at most $1/k^2$. For instance, at most $1/3^2 = .11$ or 11% of the time a random variable will fall outside ± 3 standard deviations of its mean.

For many random variables much more probability lies in these intervals than is indicated by the lower bound of $1 - 1/k^2$. In fact, it is not uncommon to find as much as 95% to 100% of the probability within two or three standard deviations of the mean. We will frequently refer to a two-standard-deviation interval about the mean as a *likely range* of values for the random variable.

Chebychev's inequality will now be proven. The proof will actually show

$$1 - P(|X-\mu| \le k\sigma) = P(|X-\mu| > k\sigma) \le \frac{1}{k^2}.$$

The proof uses the fact that

$$P(|X-\mu| > k\sigma) = P((X-\mu)^2 > k^2\sigma^2).$$

Proof of Chebychev's Inequality: Recall that

$$\sigma^2 = E((X-\mu)^2) = \sum_x (x-\mu)^2 p(x).$$

This proof will manipulate the expression for σ^2 to arrive at the desired inequality.

$$\sigma^2 = \sum_x (x-\mu)^2 p(x)$$

$$= \sum_{\{x:\,(x-\mu)^2 > k^2\sigma^2\}} (x-\mu)^2 p(x) + \sum_{\{x:\,(x-\mu)^2 \le k^2\sigma^2\}} (x-\mu)^2 p(x)$$

$$\ge \sum_{\{x:\,(x-\mu)^2 > k^2\sigma^2\}} (x-\mu)^2 p(x),$$

because we are dropping off the second sum

$$\ge \sum_{\{x:\,(x-\mu)^2 > k^2\sigma^2\}} k^2\sigma^2 p(x),$$

because we are replacing $(x-\mu)^2$ with a lower bound.

$$= k^2\sigma^2 \sum_{\{x:\,(x-\mu)^2 > k^2\sigma^2\}} p(x)$$

$$= k^2\sigma^2 P((X-\mu)^2 > k^2\sigma^2)$$

$$= k^2\sigma^2 P(|X-\mu| > k\sigma)$$

and

$$P(|X-\mu| > k\sigma) \le \frac{1}{k^2}.$$

EXAMPLE 2.4-4 For the random variable X in Example 2.4-1, the two-standard-deviation interval about $\mu = E(X)$ is

$$\mu \pm 2\sigma = 2 \pm 2(1.414),$$

which is the interval $-.828$ to 4.828. Chebychev's inequality states that this interval contains at least 75% of the probability of X. For this random variable, it actually contains 100% of the probability mass of X.

EXAMPLE **2.4-5** Let the random variable W represent the total number of tokens (operators and operands) accessed during an execution of a piece of software. The probability mass function for W given in the table was derived empirically by executing this software 1000 times. We will use this mass function to create a two-standard-deviation interval for W about $\mu = E(W)$.

w	2	3	4	5	6	7
$p(w)$	.031	.094	.281	.438	.125	.031

For this random variable, $\mu = 4.625$ and $\sigma = 1.022$. The two standard deviation interval about the mean is

$$\mu \pm 2\sigma = 4.625 \pm 2(1.022),$$

which is the interval 2.481 to 6.669. This interval contains 93.8% of the probability, which is consistent with Chebychev's inequality of $\geq 75\%$.

We now consider the variance of a linear function $Y = a + bX$.

THEOREM **2.4-2** If $Y = a + bX$, where a and b are constants, then

$$\text{VAR}(Y) = b^2\,\text{VAR}(X)$$

and

$$\text{STD}(Y) = |b|\,\text{STD}(X).$$

Theorem 2.4-2 can be verified by the following sequence of steps:

$$\begin{aligned}\text{VAR}(Y) &= E[(Y-E(Y))^2]\\ &= E[(a+bX-(a+bE(X)))^2]\\ &= E[b^2(X-E(X))^2]\\ &= b^2E[(X-E(X))^2]\\ &= b^2\text{VAR}(X).\end{aligned}$$

Notice that adding a constant to a random variable simply slides the probability mass function to the left or right on the number line, and the dispersion does not change. However, multiplying a random variable by a constant changes the dispersion. If the constant is larger than 1, the variance increases, whereas if it is less than 1 the variance decreases.

EXAMPLE 2.4-6 Referring to Examples 2.3-2 and 2.3-5, our starving artist's profit is $Y = 20X - 30$, where X denotes the number of paintings sold. From Example 2.4-3 we found $\text{VAR}(X) = 2.01$. Using Theorem 2.4-2, we obtain

$$\text{VAR}(Y) = 20^2\text{VAR}(X) = 20^2(2.01) = 804,$$

$$\text{STD}(Y) = \sqrt{804} = \$28.35.$$

The linearity property of expected value can be used to derive an alternative formula for the variance of a random variable. This formula frequently simplifies the computation of the variance.

THEOREM 2.4-3 If X is a random variable and $\mu = E(X)$, then

$$\text{VAR}(X) = E(X^2) - \mu^2.$$

Theorem 2.4-3 is easily verified with the following algebra:

$$\begin{aligned}\text{VAR}(X) &= E[(X-\mu)^2]\\ &= E(X^2-2\mu X+\mu^2)\\ &= E(X^2)-2\mu E(X)+\mu^2\\ &= E(X^2)-2\mu^2+\mu^2\\ &= E(X^2)-\mu^2.\end{aligned}$$

EXAMPLE **2.4-7** Consider the following probability mass function for the random variable X:

x	-3	-1	1	3	5	7
$p(x)$	.1	.2	.3	.2	.1	.1

The mean is $\mu = 1.6$. To compute the variance of X using Theorem 2.4-3, we need to find $E(X^2)$.

$$\begin{aligned} E(X^2) &= (-3)^2\,(.1) + (-1)^2\,(.2) + 1^2\,(.3) + 3^2\,(.2) + 5^2\,(.1) + 7^2\,(.1) \\ &= 10.6. \end{aligned}$$

Thus,

$$\begin{aligned} \mathrm{VAR}(X) &= E(X^2) - \mu^2 \\ &= 10.6 - (1.6)^2 \\ &= 8.04 \end{aligned}$$

and

$$\mathrm{STD}(X) = 2.84.$$

Exercises 2.4

2.4-1 Let the random variable X denote the outcome of a fair die toss.

a Find $E(X)$ and $\mathrm{STD}(X)$.

b Construct a two-standard-deviation interval about the mean. Verify that this interval contains at least 75% of the probability as required by Chebychev's inequality.

2.4-2 Suppose Y has the following probability mass function.

y	-2	-1	0	1	2	3
$p(y)$	.20	.34	.16	.15	.10	.05

a. Find $\mathrm{VAR}(Y)$ using Definition 2.4-1.

b. Find $\mathrm{VAR}(Y)$ using Theorem 2.4-3.

c Find $\mathrm{VAR}(2-3Y)$.

2.4-3 A consultant charges a company a retainer of \$100 a week and \$350 a day for services. The number of days per week the consultant works for this company is the random variable X with the probability mass function

x	0	1	2	3	4	5
$p(x)$	.25	.25	.20	.10	.10	.10

a Find $E(X)$ and $\text{STD}(X)$.

b Find the expected weekly cost of this consultant to the company. Find the standard deviation of this weekly cost.

2.4-4 Let X denote the number of heads in three tosses of a fair coin. Find $E(X)$ and $\text{STD}(X)$.

2.4-5 A doughnut shop has on hand an average of 120 doughnuts with a standard deviation of 12 doughnuts at any time during the day. Suppose a customer arrives and wants to purchase 8 dozen doughnuts. Is it likely that there will be enough doughnuts for this customer? Explain your answer using Chebychev's inequality.

2.4-6 Let X be a random variable with a mean of 1.25 and a variance of .12. What is the likely range of values for the random variable $Y = 2.25 + 1.2X$?

SECTION

2.5 Sampling and Simulation

Science must base its conclusions on data. It is not enough for the scientist to speculate or believe that something is true. Scientific theories and inferences stand or fall on the basis of things that can be measured. Not only pure science but diverse fields such as engineering, environmental monitoring, manufacturing, and economics rely on data. In all these areas, data are collected and analyzed as a basis for decisions or actions.

The process of collecting data is called *sampling*. The concepts of probability give us a way to describe and study this process. Simple games of chance like drawing numbers in a lottery or tossing dice are analogous to how sampling is done in many practical situations. Since we know how probability applies to games of chance, we can use this knowledge to see how probability applies to sampling.

Many sampling procedures can be considered as lottery drawings. Suppose a pollster is interested in the ages of individuals watching a certain television show and by random selection contacts individuals for the study. We can imagine the ages of all individuals in the viewing population being put into a basket. The ages of the individuals actually obtained in the study could be treated as if they were drawn lottery fashion from that basket. A production engineer who is interested in knowing whether the diameters of certain items are within tolerances might randomly select items for inspection. We can imagine that the diameters of all such items are placed in a basket, and the ones that are inspected are selected by a lottery drawing.

Other data collection procedures may be considered as if we were tossing imaginary dice. For instance, consider observing the number of customers who visit a restaurant during a typical business day. We can imagine that the set of possibilities is inscribed on a set of many-sided dice, and the numbers we obtain each day are generated by a roll of these dice. In general, the process of scientific experimentation may be thought of in this way. For instance, in his famous oil-drop experiment, physicist R. A. Millikan set out to determine the charge on an electron. However, because of imprecise techniques, he experienced random variability in his measurements. We can imagine that nature presented Millikan with a many-sided die representing the possible measurements that he could have made and that the data he actually obtained were the result of his rolling of that die.

Suppose we let the random variables $X_1, X_2, \ldots, X_n$ denote n sampling outcomes — that is, measurements or data values. In the dice-tossing treatment of sampling, these random variables are mutually independent. Because the dice are the same each time we toss, the random variables have the same probability distribution. We say that such random variables are *independent and identically distributed.*

The situation is somewhat more complicated in the case of the lottery treatment of sampling. If we do not replace the items in the basket as we sample them one after the other, the selections are not independent. One outcome will affect the next. However, if we replace each item before selecting again, then the selections are independent. In fact, a lottery drawing in which items are replaced is no different than dice tossing from a probability point of view. For instance, if a basket contained the numbers 1 through 6 in equal proportions, then selecting numbers with replacement would turn up numbers in the same way as a toss of a six-sided fair die. If the number of items in the basket from which we are sampling is very large in relation to the number of items being sampled, then there is virtually no difference between selecting these items with replacement or without replacement. That is, removing a few items does not appreciably alter the original composition of the set of items from which we are sampling. Thus, in the lottery model for data collection, if we either replace the items each time or draw relatively few items from a very large set of items, the outcomes of the n draws can be regarded as being independent and identically distributed random variables.

We are now ready to define a *random sample.*

Definition 2.5-1

Random variables $X_1, X_2, \ldots, X_n$ are said to be a *random sample* if they are independent and identically distributed.

Suppose that we are interested in learning something about the probability distribution of a random variable X. From a random sample of observations whose distributions are the same as X, we can learn about the probability distribution of X through the empirical probabilities of the random sample. This is illustrated in the following example.

EXAMPLE **2.5-1** An administrator is interested in the number of days a week college students make use of the university recreational center. A random sample of 60 students was selected, and the number of days per week each student used the facility was recorded. The observed random sample for $X_1, X_2, \ldots, X_{60}$ is

$$\{0,1,1,2,2,4,5,1,2,3,2,7,4,5,5,2,1,0,0,1,0,2,1,1,0,3,4,4,6,1,$$
$$1,4,3,2,3,1,0,0,2,4,5,3,6,1,2,0,1,3,3,1,0,1,3,2,5,3,1,4,6,0\}$$

Two displays that summarize the information contained in the random sample are given in Figure 2.5-1. The first is a table of empirical probabilities for the usage of the recreational center. The second display is a histogram of the empirical probabilities. The histogram is a simple but effective device to visualize the information in the random sample. For instance, we see that the majority of the students use the recreational center two or less days a week, whereas 15% use the center five or more days a week.

FIGURE **2.5-1** Empirical Probabilities and Histogram for Usage of Recreational Center

Center usage (days per week)	0	1	2	3	4	5	6	7
Empirical probabilities	.17	.25	.17	.15	.12	.08	.05	.02

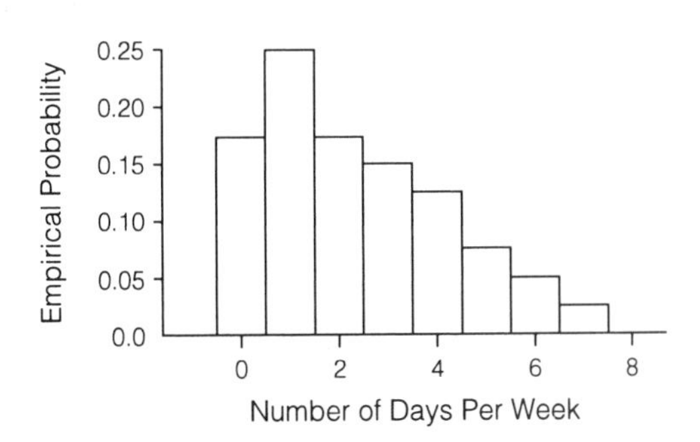

When probabilities are determined from a random sample as in Example 2.5-1, the resulting distribution is called the *empirical probability distribution*, and the mass function is called the *empirical probability mass function.* We denote the empirical probability mass function of a random variable X by $\hat{p}(x)$.

It is possible and often useful to simulate the random variables $X_1, X_2, \ldots, X_n$ that make up a random sample. The focus here will be on simulating discrete random variables. This process is similar to the simulation of events on the computer as discussed in Section 1.3.

To simulate a discrete random variable, the unit interval is broken up into subintervals whose lengths are equal to the values of the probability mass function. The subinterval into which a simulated unit random number falls determines the value of the random variable.

EXAMPLE 2.5-2 Let the random variable X have the following probability mass function and cumulative distribution function:

x	10	20	30	40
$p(x)$	.10	.25	.15	.50
$F(x)$	.10	.35	.50	1

To simulate X, break the unit interval into four subintervals, one for each possible value of X. This is shown in Figure 2.5-2. According to this breakdown of the unit interval, if a unit random number satisfies $0 \le u < .1$, then $X = 10$. If $.1 \le u < .35$, then $X = 20$, and so forth.

FIGURE 2.5-2 Partitioning of the Unit Interval for Random Variable Generation

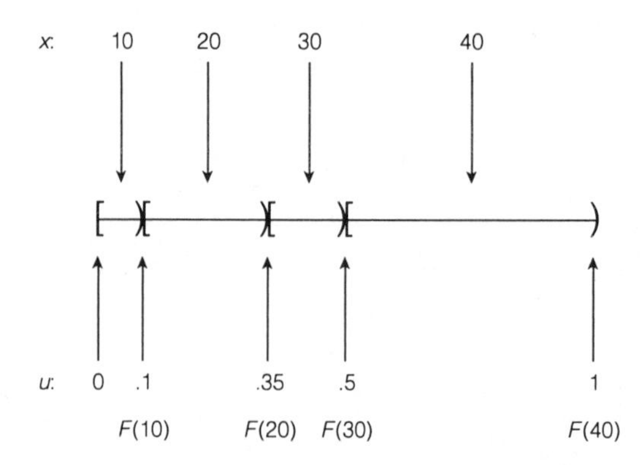

Notice that the breakpoints in the partitioning of the unit interval in Example 2.5-2 correspond to cumulative probabilities of X. In general, let $a_1 < a_2 < \cdots$ be the possible values of X, and let u be a unit random number. If $u < F(a_1)$, then assign $X = a_1$. If $F(a_{i-1}) \le u < F(a_i)$, for $i > 1$, then assign $X = a_i$. This method of random variable generation can be visualized by considering Figure 2.5-3, which is the step function graph of the cumulative distribution function of X from Example 2.5-2. If we let a unit random number correspond to a point on the vertical axis, then the corresponding value of X can be found by mapping this value to the horizontal axis, as shown in Figure 2.5-3. Repeating this process over and over gives us an observed random sample.

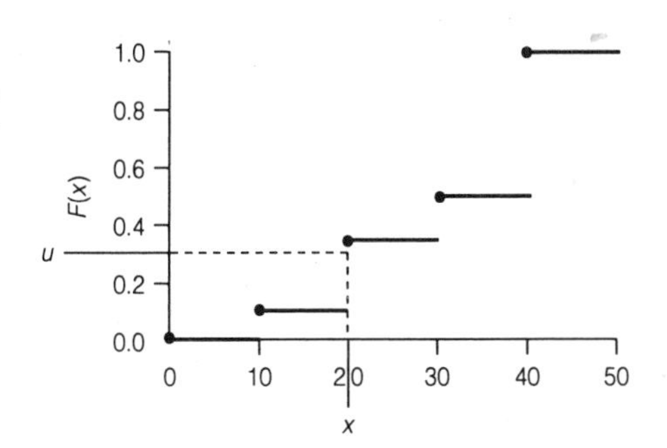

FIGURE 2.5-3 Generating a Random Number Using the Cumulative Distribution Function

EXAMPLE 2.5-3 Consider the random variable Y from Example 2.1-4, which denotes the lost revenue due to the number of vacant seats on a commuter airline flight. Recall that

y	\$0	\$70	\$140	\$210
$p(y)$	.5	.3	.15	.05
$F(y)$	.5	.8	.95	1

This random variable can be simulated according to the following algorithm.

1 Generate u, a unit random number.
2 If $u < .5$, then $Y = 0$;
else if $u < .8$, then $Y = 70$;
else if $u < .95$, then $Y = 140$;
else $Y = 210$.

Using this algorithm, a random sample of 1000 values of Y was simulated on a computer. A table of empirical probabilities is

y	$\hat{p}(y)$
0	.488
70	.310
140	.147
210	.055

As in the simulation of simple events, empirical probabilities are not necessarily equal to the true probabilities. If another 1000 values for Y were generated according to this same algorithm, we would not expect to get the same empirical probabilities as listed in the table. However, the more simulations the empirical probabilities are based on, the more likely they are to be close to the true probabilities.

EXAMPLE 2.5-4 The previous example showed a simulation of lost revenue on one flight. Suppose the airline has one flight each day of the work week, and the company wants to investigate its weekly lost revenue. Assume that each of the five flights functions independently, and the probability distribution of lost revenue for each flight is identical to the distribution for Y in Example 2.5-3. Let the random variable W denote the weekly lost revenue for the five flights. To simulate the random variable W, generate five independent values for the random variable Y from Example 2.5-3 and add them up. This is repeated 1000 times, giving us a random sample of 1000 values of W. From this sample, the empirical distribution of W is shown both in tabular form and as a histogram in Figure 2.5-4. From this distribution, the company can predict the likely range for lost revenue. For example, the empirical probability that the lost revenue will be between \$140 and \$350 a week is .726.

FIGURE 2.5-4 Empirical Distribution of Weekly Lost Revenue for Airline Flights

w	$\hat{p}(w)$
0	.021
70	.086
140	.170
210	.203
280	.195
350	.158
420	.084
490	.049
560	.016
630	.015
700	.002
770	.001
840	0
910	0
980	0
1050	0

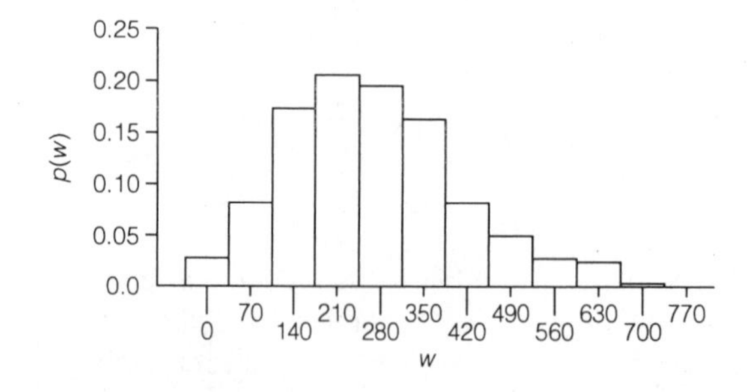

In Example 2.5-4, the true probability mass function for W could be determined by accounting for all possible outcomes that lead to each value of W. For example, the lost revenue will be \$70 if there is a lost revenue of \$70 on one of the five days and no lost revenue on the other days. The probability of this outcome is $5(.3)(.5)^4 = .094$. This method of enumeration of outcomes is tedious. The simulation of W, although only an approximation, gives a quick solution to the problem. In many problems, simulation is the only feasible solution.

Exercises 2.5

2.5-1 Using a table of random numbers, simulate the random variable Y in Example 2.5-3 25 times. Compute an empirical probability mass function from your simulation. Compare your results to the 1000 simulations given in Example 2.5-3.

2.5-2 Simulate the random variable Y in Example 2.5-3 1000 times and compute an empirical probability mass function. Compare your results to those in Example 2.5-3.

2.5-3 Write a program to do the simulation in Example 2.5-4. Simulate a random sample of size 1000 and obtain an empirical probability mass function for W. Compare these probabilities to the empirical probabilities given in that example.

2.5-4 Modify Example 2.5-4 to allow the distribution of lost revenues to be different on Wednesday. Let the distribution of lost revenue for Monday, Tuesday, Thursday, and Friday be the same as in Example 2.5-4. Let the distribution of lost revenue on Wednesday be

y	\$0	\$70	\$140	\$210
$p(y)$	.1	.2	.3	.4

Simulate the total lost revenue for the week. Repeat the simulation 1000 times and compute the empirical distribution for W.

2.5-5 Let the random variable Y denote the number of emergency calls received by a fire station in a day. Assume Y has the distribution

y	0	1	2	3	4
$p(y)$	.2	.2	.3	.2	.1

Find an empirical distribution for the number of calls in seven days. Assume that the calls from day to day are independent.

2.5-6 Let X and Y denote the number of barges leaving docks A and B, respectively, in a day. The distributions for X and Y are as follows. Assume X and Y are independent.

x	0	1	2
$p(x)$	.05	.4	.55

and

y	0	1	2	3
$p(y)$	.15	.2	.5	.15

a Simulate the total number of barges leaving the docks in a day. Find an empirical probability mass function for this total based on a random sample of size 1000.

b Find the true probability mass function for $X + Y$. Compare this mass function to your empirical distribution in part (a).

2.5-7 Show that the probability of lost revenue of \$140 in Example 2.5-4 is

$$\binom{5}{2}(.3)^2(.5)^2 + 5(.15)(.5)^4 = .1594.$$

SECTION

2.6 Sample Statistics

Sample statistics are quantities computed from a random sample. Given a random sample $X_1, X_2, \ldots, X_n$, we will define three important sample statistics: the sample mean, the sample variance, and the sample standard deviation.

Definition 2.6-1

The *sample mean,* denoted $\bar{X}$, is defined as

$$\bar{X} = \frac{1}{n}\sum_{i=1}^{n} X_i.$$

The *sample variance,* denoted S^2, is defined as

$$S^2 = \frac{1}{n}\sum_{i=1}^{n}(X_i - \bar{X})^2.$$

The *sample standard deviation,* denoted S, is $S = \sqrt{S^2}$.

An alternative definition of the sample variance is

$$S_{n-1}^2 = \frac{1}{n-1}\sum_{i=1}^{n}(X_i - \bar{X})^2.$$

The value of S_{n-1}^2 is slightly larger than S^2; however, for moderate to large values of n the difference is negligible.

EXAMPLE 2.6-1 Suppose an observed random sample of size $n = 20$ consists of the values 2, 0, 4, 2, 1, 2, 1, 5, 3, 3, 5, 1, 2, 3, 4, 4, 5, 1, 2, 3. Then

$$\begin{aligned}\bar{X} &= \frac{1}{20}(2+0+\cdots+3)\\ &= 2.65;\\ S^2 &= \frac{1}{20}[(2-2.65)^2+(0-2.65)^2+\cdots+(3-2.65)^2]\\ &= 2.13;\\ S &= \sqrt{2.13} = 1.46.\end{aligned}$$

The sample mean can alternatively be computed as the mean of the empirical probability mass function, and the sample variance can alternatively be computed as the variance of the empirical probability mass function. These results are illustrated in the next example.

EXAMPLE 2.6-2 The empirical probability mass function for the random sample in Example 2.6-1 is

x	0	1	2	3	4	5
$\hat{p}(x)$	1/20	4/20	5/20	4/20	3/20	3/20

The mean of this empirical probability mass function is

$$\begin{aligned}\sum_x x\hat{p}(x) &= 0\left(\frac{1}{20}\right)+1\left(\frac{4}{20}\right)+2\left(\frac{5}{20}\right)+3\left(\frac{4}{20}\right)+4\left(\frac{3}{20}\right)+5\left(\frac{3}{20}\right)\\ &= 2.65.\end{aligned}$$

The variance of the empirical probability mass function is

$$\begin{aligned}\sum_x (x-2.65)^2\hat{p}(x) &= (0-2.65)^2\left(\frac{1}{20}\right)+(1-2.65)^2\left(\frac{4}{20}\right)\\ &\quad+\cdots+(5-2.65)^2\left(\frac{3}{20}\right)\\ &= 2.13.\end{aligned}$$

Characteristics of a random variable, such as the various values of the probability mass function, the expected value, the variance, and the standard deviation, are called *parameters*. The empirical probabilities $\hat{p}(x)$ and the sample statistics $\bar{X}$, S^2, and S are called *estimates* of these parameters. An estimate can be thought of as an approximation of the parameter. In most cases of practical interest, estimates based on larger samples have higher probability of being "close" to the true parameter than estimates based on smaller samples. This is illustrated in Exercise 2.6-7.

Empirical distributions, sample means, and sample standard deviations are useful for summarizing the results of a simulation.

EXAMPLE **2.6-3** Suppose we were to ask people at random in which month they were born. Let the random variable X denote the number of people we would need to ask before we found two people born in the same month. The possible values for X are $2, 3, \ldots, 13$. That is, at least two people must be asked in order to have a match and no more than 13 need to be asked. With the simplifying assumption that every month is an equally likely candidate for a response, a computer simulation was used to estimate the probability mass function of X. The simulation generated birth months until a match was found. Based on 1000 repetitions of this experiment, the following empirical distribution and sample statistics were obtained:

x	2	3	4	5	6	7	8	9	10	11	12	13
$\hat{p}(x)$	.071	.143	.204	.183	.174	.121	.061	.021	.017	.005	.000	.000

The sample mean and sample standard deviation are 5.1 and 1.9, respectively. This simulation tells us that, on average, we would have to ask about five people to obtain a matching birth month.

Exercises 2.6

2.6-1 For the random sample $(11, 9, 3, 7, 8, 1, 5, 12, 3, 10)$, find the sample mean and sample standard deviation.

2.6-2 Compute the empirical probability mass function for the random sample $(1, 1, 3, 1, 3, 3, 1, 2, 2, 2, 3, 1, 3, 3, 2)$. Using the empirical probability mass function, find the sample mean and sample standard deviation.

2.6-3 The number of vacant rooms in a hotel was recorded for 15 randomly selected Saturdays. The following observed sample was obtained: $(5, 1, 2, 1, 2, 3, 6, 0, 4, 3, 7, 2, 1, 8, 9)$. Based on this information, estimate the expected value and standard deviation for the number of vacant rooms on a Saturday.

2.6-4 Write a program for computing the sample mean, sample variance, and sample standard deviation of the discrete random variable X, where the input of the program is an array containing the random sample $(X_1, X_2, \ldots, X_n)$. Apply your program to the observed random samples in Exercises 2.6-1 and 2.6-2.

2.6-5 Just as the variance of a random variable X can be expressed as

$$\text{VAR}(X) = E(X^2) - \mu^2,$$

the sample variance S^2 can be expressed as

$$S^2 = \frac{1}{n}\sum_{i=1}^{n} X_i^2 - (\bar{X})^2.$$

Verify that this holds for the sample in Exercise 2.6-1. This formula is sensitive to computer floating-point error in the case where the X_i's are large relative to the variance, and therefore it is not recommended as a general-purpose algorithm for the computation of the sample variance.

2.6-6 Let X denote the number of tosses it takes to obtain the first heads in tossing a fair coin repeatedly. Note that the possible values of X are infinite (i.e., $1, 2, 3, 4, \ldots$). Simulate the random variable X 1000 times. Compute the sample mean and sample standard deviation of X.

2.6-7 Let X denote the response of a randomly selected consumer to the question, "Please rate the taste of Crunchy Cereal on a scale of 1 to 5, where 1 is the least favorable score and 5 is the most favorable score." The probability mass function of X is

x	1	2	3	4	5
$p(x)$	.1	.2	.2	.3	.2

a Compute the expected value and standard deviation of X.

b Simulate a random sample of 10 values of X. Compute the sample mean and the sample standard deviation. Repeat this process 15 times.

c Simulate a random sample of 100 values of X. Compute the sample mean and sample standard deviation. Repeat this process 15 times.

d Based on these two sets of 15 sample means and sample standard deviations, in what sense is a sample of 100 better than a sample of 10?

2.6-8 The manufacturer of BURP beer has printed one of the four letters, B, U, R, and P, on the bottom of each bottle cap. The company offers a prize to anyone who collects four bottle caps that spell BURP. Suppose there is an equal probability that any of the four letters appears underneath a bottle cap. Simulate the number of bottles of BURP beer you must buy in order to win the prize. Repeat your simulation 500 times and compute the sample mean and the sample standard deviation for the number of bottles it takes to win the prize.

2.6-9 Modify the birth month problem from Example 2.6-3 so that the probability of a birth month is proportional to the number of days in the month. For example, the probability of the birth month being March would be 31/365. Estimate the mean and standard deviation based on 1000 simulations of the experiment.

2.6-10 Suppose we were to ask people at random on what day they were born. Let Y be the number of people we would need to ask to obtain a matching birthday. Assume that all 365 birthdays are equally likely to occur. Estimate the mean and standard deviation based on 1000 simulations of the experiment.

SECTION

2.7 Expected Values of Jointly Distributed Random Variables and the Law of Large Numbers

The concept of expected value for a single random variable carries over to jointly distributed random variables.

THEOREM 2.7-1 Let X and Y have joint probability mass function $p(x, y)$. Let $W = Q(X, Y)$. Then

$$E(W) = E(Q(X, Y)) = \sum_x \sum_y Q(x, y)\, p(x, y).$$

EXAMPLE 2.7-1 Let X and Y have joint probability mass function

	$p(x,y)$	x = 1	2	3
y	1	.25	.15	.10
	2	.05	.35	.10

Let $W = XY$. Then

$$\begin{aligned} E(W) &= (1)(1)(.25) + (2)(1)(.15) + (3)(1)(.10) \\ &\quad + (1)(2)(.05) + (2)(2)(.35) + (3)(2)(.10) \\ &= 2.95. \end{aligned}$$

Analogous to Theorem 2.3-2, we have the following theorem for the expected value of a linear combination of two random variables.

THEOREM 2.7-2 Let $W = aX + bY$. Then

$$E(W) = aE(X) + bE(Y).$$

Theorem 2.7-2 can be verified similarly to the verification of Theorem 2.3-2 (see Exercise 2.7-2).

EXAMPLE 2.7-2 Consider the joint probability mass function of X and Y in Example 2.7-1. The marginal probability mass functions for X and Y are

x	1	2	3
$p(x)$	.30	.50	.20

and

y	1	2
$p(y)$	.50	.50

Now, $E(X) = 1.9$ and $E(Y) = 1.5$. It follows that

$$E(aX + bY) = a(1.9) + b(1.5).$$

For example,

$$E(X - Y) = 1.9 - 1.5 = .4$$

and

$$E[(X + Y)/2] = (1.9 + 1.5)/2 = 1.7.$$

Theorem 2.7-2 extends immediately to any finite linear combination of random variables.

THEOREM 2.7-3 Let $W = a_1X_x + a_2X_2 + \cdots + a_nX_n$. Then

$$E(W) = a_1E(X_1) + a_2E(X_2) + \cdots + a_nE(X_n).$$

Verification of Theorem 2.7-3 is left to the reader (see Exercise 2.7-9).

EXAMPLE 2.7-3 Let Y denote the number of automobile accidents each week at an intersection. The probability mass function of Y is

y	0	1	2
$p(y)$	.5	.4	.1

We find $E(Y) = (0)(.5) + (1)(.4) + (2)(.1) = .6$. The total number of accidents in a year can be expressed as $W = \sum_{i=1}^{52} Y_i$, where each Y_i has the same probability distribution as the random variable Y above. Then

$$E(W) = \sum_{i=1}^{52} E(Y_i) = (52)(.6) = 31.2.$$

The following theorem applies to the product of two independent random variables.

THEOREM 2.7-4 If X and Y are independent random variables, then

$$E(XY) = E(X)\,E(Y).$$

The verification of Theorem 2.7-4 is as follows. Let $p(x, y)$ be the joint probability mass function of X and Y. If X and Y are independent, then $p(x, y) = p_X(x)\,p_Y(y)$. Therefore,

$$\begin{aligned} E(XY) &= \sum_x \sum_y xyp(x, y) \\ &= \sum_x \sum_y xyp_X(x)\,p_Y(y) \\ &= \sum_x xp_X(x) \sum_y yp_Y(y) \\ &= E(X)\,E(Y). \end{aligned}$$

The variance of a linear combination of two independent random variables is given in the next theorem.

THEOREM 2.7-5 Let $W = aX + bY$, where X and Y are independent random variables. Then,

$$\mathrm{VAR}(W) = a^2\mathrm{VAR}(X) + b^2\mathrm{VAR}(Y).$$

Theorem 2.7-5 is easily verified with the following algebra:

$$\begin{aligned}
\mathrm{VAR}(W) &= E[(W - E(W))^2] \\
&= E[(aX + bY - aE(X) - bE(Y))^2] \\
&= E[(a(X - E(X)) + b(Y - E(Y)))^2] \\
&= a^2E[(X - E(X))^2] + b^2E[(Y - E(Y))^2] \\
&\quad + 2abE[(X - E(X))(Y - E(Y))] \\
&= a^2\mathrm{VAR}(X) + b^2\mathrm{VAR}(Y) \\
&\quad + 2ab\{E(XY) - E(X)E(Y)\}.
\end{aligned}$$

Since X and Y are independent and $E(XY) = E(X)E(Y)$, the last term is zero and the result follows.

When dealing with the sum of independent random variables, it is important to note that it is the variances that are additive and not the standard deviations. That is, $\mathrm{STD}(X + Y) = \sqrt{\mathrm{Var}(X) + \mathrm{Var}(Y)} \neq \mathrm{STD}(X) + \mathrm{STD}(Y)$.

EXAMPLE 2.7-4 A company has two manufacturing plants for producing integrated circuits. Let X and Y denote the number of defective circuits produced at plants 1 and 2, respectively. Assume X and Y are independent. Suppose $E(X) = 23$, $\mathrm{STD}(X) = 5.5$, $E(Y) = 27$, and $\mathrm{STD}(Y) = 4.7$. The expected total number of defective circuits produced by the company in a day is $E(X + Y) = 23 + 27 = 50$, and

$$\begin{aligned}
\mathrm{VAR}(X + Y) &= \mathrm{VAR}(X) + \mathrm{VAR}(Y) \\
&= 5.5^2 + 4.7^2 \\
&= 52.34, \\
\mathrm{STD}(X + Y) &= \sqrt{52.34} = 7.23.
\end{aligned}$$

Note that a likely range for the daily total number of defective circuits is $50 \pm (2)(7.23)$, which is from 36 to 64 circuits when rounded to the nearest integer.

Theorem 2.7-5 extends immediately to any finite linear combination of random variables.

THEOREM 2.7-6 Let $W = a_1X_1 + a_2X_2 + \cdots + a_nX_n$, where $X_1, X_2, \ldots, X_n$ are independent random variables. Then

$$\text{VAR}(W) = a_1^2\,\text{VAR}(X_1) + a_2^2\,\text{VAR}(X_2) + \cdots + a_n^2\,\text{VAR}(X_n).$$

Verification of Theorem 2.7-6 is left as an exercise. Theorem 2.7-6 can be applied to find the expected value and variance of the sample mean $\bar{X}$.

THEOREM 2.7-7 Let the random variables $X_1, X_2, \ldots, X_n$ be a random sample with $E(X_i) = \mu$ and $\text{VAR}(X_i) = \sigma^2$, for $i = 1, 2, \ldots, n$. Then

$$E(\bar{X}) = \mu,$$
$$\text{VAR}(\bar{X}) = \frac{\sigma^2}{n},$$
$$\text{STD}(\bar{X}) = \frac{\sigma}{\sqrt{n}}.$$

Theorem 2.7-7 is a direct application of Theorems 2.7-3 and 2.7-6. First recall that if $X_1, X_2, \ldots, X_n$ is a random sample, then these random variables are independent and identically distributed. Therefore,

$$E(\bar{X}) = \frac{1}{n}\sum_{i=1}^{n} E(X_i) = \frac{1}{n} n\mu = \mu$$

and

$$\text{VAR}(\bar{X}) = \frac{1}{n^2}\sum_{i=1}^{n} \text{VAR}(X_i) = \frac{1}{n^2} n\sigma^2 = \frac{\sigma^2}{n}.$$

Notice that as the sample size increases, Var $(\bar{X})$ and hence STD $(\bar{X})$ get smaller. If we were to make a ± 2 STD$(\bar{X})$ interval about $E(\bar{X})$, (i.e., $\mu \pm 2\sigma/\sqrt{n}$), we would find that this interval gets narrower as n increases. This shows that the probability distribution of $\overline{X}$ piles up more closely around μ as n gets larger. Thus, if n is large, we have a good chance of obtaining an observed value of $\overline{X}$ that is close to μ. This is illustrated in Figure 2.7-1, which is a graph of $\overline{X}$ for random samples of size $n = 10$ to $n = 1000$. The samples are based on independent observations of a random variable X whose probability mass function is

x	0	1	2	3
$p(x)$	1/8	3/8	3/8	1/8

Figure 2.7-1 includes a graph of the interval $\mu \pm 2\sigma/\sqrt{n}$, where $\mu = 1.5$ and $\sigma = .87$. Clearly, we can see how $\overline{X}$ tends to get closer to μ as n increases.

FIGURE **2.7-1** Plots of $\overline{X}$ and $\mu \pm 2\sigma/\sqrt{n}$ for Values of n from 10 to 1000

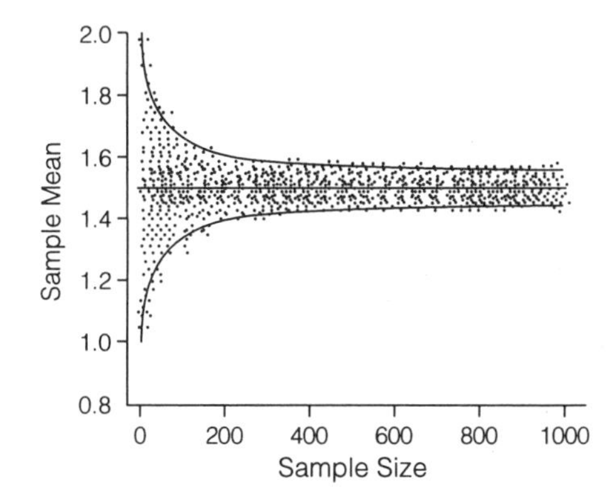

The notion that $\overline{X}$ gets close to μ as the size of the sample increases is the essence of the *Law of Large Numbers*. A version called the *Weak Law of Large Numbers* is stated in the following theorem.

THEOREM 2.7-8 Weak Law of Large Numbers: Let $X_1, X_2, \ldots, X_n$ be a random sample with $E(X_i) = \mu$ and $\mathrm{VAR}(X_i) = \sigma^2$, for $i = 1, 2, \ldots, n$. For all $\epsilon > 0$,

$$\lim_{n\to\infty} P(|\overline{X} - \mu| > \epsilon) = 0.$$

Theorem 2.7-8 can be verified by using Chebychev's inequality. Chebychev's inequality states

$$P\left(|\bar{X} - \mu| \le k\frac{\sigma}{\sqrt{n}} \right) \ge 1 - \frac{1}{k^2}.$$

Let $\epsilon = k\sigma/n$. Then, $k^2 = \epsilon^2 n/\sigma^2$,

$$P(|\bar{X} - \mu| \le \epsilon) \ge 1 - \frac{\sigma^2}{\epsilon^2 n},$$

and

$$P(|\bar{X} - \mu| > \epsilon) < \frac{\sigma^2}{\epsilon^2 n}.$$

Therefore, for any $\epsilon > 0$ as $n \to \infty$, $P(|\bar{X} - \mu| > \epsilon)$ goes to zero. In words, the probability that $\bar{X}$ is more than ϵ units away from μ goes to zero as n increases, regardless of the size of ϵ.

Exercises 2.7

2.7-1 The random variables X and Y have the following joint probability mass function:

	$p(x,y)$	x = 2	3	4
y	−1	.10	.20	.05
	0	.27	.10	.15
	1	0	.03	.10

a Find $E(X)$, $E(Y)$, $E(X - Y)$, and $E(2X - 3Y)$.

b Are X and Y independent?

c Show $E(XY) \neq E(X)E(Y)$.

2.7-2 Verify Theorem 2.7-2. Begin the proof by noting that

$$E(aX + bY) = \sum_x \sum_y (ax + by)\,p(x, y).$$

Then apply the distributive law for summations.

2.7-3 Let (X, Y) denote the face values that occur on the toss of a pair of fair dice. Note that X and Y are independent.

a Find $E(XY)$, using Theorem 2.7-1.

b Find $E(XY)$, using Theorem 2.7-4.

c Compute $E(X/Y)$. Does $E(X/Y) = E(X)/E(Y)$?

2.7-4 Let X and Y be independent random variables with $E(X) = 10$, $\text{STD}(X) = 2$, $E(Y) = 15$, and $\text{STD}(Y) = 4$. Find the expected value and standard deviation of the following random variables: $X + Y$, $X - Y$, $3X + 4Y$, and $3X - 4Y$.

2.7-5 Let (X, Y) denote the outcome of a toss of a pair of fair dice. Use Theorem 2.7-2 and Theorem 2.7-5 to find the expected value and standard deviation of the sum of the face values, $W = X + Y$.

2.7-6 The number of times a certain computer system crashes in a week is a random variable with probability mass function

y	0	1	2	3
$p(y)$	.50	.30	.15	.05

Assume the numbers of crashes from week to week are independent. Find the mean and standard deviation of the number of crashes in a year's time.

2.7-7 Let the random variable W be defined as in Example 2.5-4.

a Use the empirical distribution for W, the weekly lost revenue, from Example 2.5-4, to estimate the mean and standard deviation of W.

b Note that W may be expressed as $\sum_{i=1}^{5} Y_i$, where Y_i is the daily lost revenue for the commuter airline. We have assumed that the Y_i's are independent with identical probability mass functions

y	0	70	140	210
$p(y)$	.5	.3	.15	.05

Apply Theorems 2.7-2 and 2.7-5 to find $E(W)$ and $\text{STD}(W)$ Compare these values with the estimates.

2.7-8 Let X and Y be independent random variables. Let $g(X)$ and $h(Y)$ be functions of X and Y, respectively. Prove that $(E(g(X)h(Y)) = E(g(X))E(h(Y)))$.

2.7-9 Verify Theorems 2.7-3 and 2.7-6 by using mathematical induction.

2.7-10 Let the random variable X denote the sum of the face values on a pair of dice. Simulate a random sample of this random variable for samples of size $n = 10, 20, 30, 50, 75, 100, 200, 300, 400, 800$, and 1000. In generating the samples, use a new set of unit random numbers for each sample size. Obtain the sample means, $\bar{X}_n$, for each value of n. Plot these sample means versus n along with the theoretical intervals $\mu \pm 2\sigma/\sqrt{n}$ and $\mu \pm 3\sigma/\sqrt{n}$, where μ is $E(X)$ and σ is $\text{STD}(X)$. Observe how $\bar{X}$ gets close to the mean.

2.7-11 Repeat the simulation of the dice-tossing experiment of Exercise 2.7-10, only use the old samples as part of the new samples. That is, let the first sample of size 10 make up the first 10 outcomes of the sample of size 20, and let the sample of size 20 make up the first 20 outcomes of the sample of size 30, and so on. Plot the sample means versus the sample size. This graph is an illustration of the *Strong Law of Large Numbers,* which states that for any sequence of independent, identically distributed random variables, the sample mean is certain to converge to μ.

SECTION

2.8 Covariance and Correlation

When dealing with two random variables, we are often interested in studying how a change in one variable affects the other variable. Two functions that are useful for this purpose are *covariance* and *correlation.*

Definition 2.8-1

The *covariance* of a pair of random variables X and Y is

$$\mathrm{COV}(X, Y) = E[(X-E(X))(Y-E(Y))]$$

The *correlation* is

$$\mathrm{CORR}(X, Y) = \frac{\mathrm{COV}(X, Y)}{\mathrm{STD}(X)\,\mathrm{STD}(Y)}.$$

Correlation is a measure of the strength of the linear relationship between two random variables. The correlation can be shown to vary between -1 and 1. If the correlation is positive, then Y tends to increase as X increases. If the correlation is negative, the random variables tend to move in opposite directions. For example, if X and Y represent heights and weights of individuals, the correlation would be positive. If X represents outside temperature and Y represents an individual's home heating bill, we would expect X and Y to be negatively correlated. If the correlation equals -1 or 1, it can be shown that there is a perfect linear relationship between X and Y. If the correlation is 1, then $Y = a + bX$, where $b > 0$. If the correlation is -1, $Y = a + bX$, where $b < 0$. If the random variables are independent, the correlation and covariance are zero. To see this, note that

$$\begin{aligned}\mathrm{COV}(X, Y) &= E[(X-E(X))(Y-E(Y))]\\ &= E(XY) - E(X)\,E(Y).\end{aligned}$$

It follows directly from Theorem 2.7-4 that if X and Y are independent, $E(XY) = E(X)\,E(Y)$. Thus, $\mathrm{COV}(X, Y) = 0$ and $\mathrm{CORR}(X, Y) = 0$ when X and Y are independent.

EXAMPLE **2.8-1** Let X be the number of modules with programming errors in a piece of software, and let Y be the number of days it takes to debug the software. Suppose X and Y have the following joint probability mass function:

	$p(x,y)$	x = 0	1	2	3	4
y	0	.20	.08	.03	.02	.01
	1	0	.06	.09	.04	.01
	2	0	.04	.09	.06	.02
	3	0	.02	.06	.04	.03
	4	0	0	.03	.02	.02
	5	0	0	0	.02	.01

From this joint distribution, we find $E(XY) = 3.8$. The marginal probability mass function for X and Y is as follows.

x	0	1	2	3	4
$p(x)$	.20	.20	.30	.20	.10

with $E(X) = 1.8$ and $\text{STD}(X) = 1.25$;

y	0	1	2	3	4	5
$p(y)$	.34	.20	.21	.15	.07	.03

with $E(Y) = 1.5$ and $\text{STD}(Y) = 1.42$.
Thus,

$$\text{COV}(X, Y) = 3.8 - (1.8)(1.5) = 1.1,$$

$$\text{CORR}(X, Y) = \frac{1.1}{(1.25)(1.42)} = .62.$$

The next theorem gives the variance of a linear combination $W = aX + bY$ for any two random variables X and Y (e.g., not necessarily independent).

THEOREM 2.8-1 Let $W = aX + bY$. Then

$$\text{VAR}(W) = a^2\,\text{VAR}(X) + b^2\,\text{VAR}(Y) + 2ab\text{COV}(X, Y).$$

Theorem 2.8-1 can be verified with the following algebra:

$$\begin{aligned}\mathrm{VAR}(W) &= E[(W-E(W))^2]\\ &= E[(aX+bY-aE(X)-bE(Y))^2]\\ &= a^2E[(X-E(X))^2]+b^2E[(Y-E(Y))^2]\\ &\quad + 2abE[(X-E(X))(Y-E(Y))]\\ &= a^2\mathrm{VAR}(X)+b^2\mathrm{VAR}(Y)+2ab\mathrm{COV}(X,Y).\end{aligned}$$

Note that if X and Y are independent, $\mathrm{COV}(X,Y) = 0$ and Theorem 2.7-5 follows.

EXAMPLE 2.8-2 Suppose the cost of debugging the software in Example 2.8-1 is $1000 per defective module and $500 per day of debug time. This cost can be expressed as $W = 1000X + 500Y$. Therefore, the expected cost is $E(W) = 1000(1.8) + 500(1.5) =$ $2550. The variance is

$$\begin{aligned}\mathrm{VAR}(W) &= 1000^2(1.25)^2+500^2(1.42)^2+2(1000)(500)(1.1)\\ &= 3{,}166{,}600\\ \mathrm{STD}(W) &= \$1779.50.\end{aligned}$$

The notion of a random sample may be extended to pairs $(X_1, Y_1), \ldots, (X_n, Y_n)$. Just imagine lottery tickets with two numbers X and Y on each ticket. As we select tickets randomly, the pairs of numbers so obtained form a *bivariate random sample.* Heights and weights of randomly selected individuals would be a bivariate random sample. These pairs of random variables that make up the bivariate random sample are independent and identically distributed. We define it more formally.

Definition 2.8-2

Random pairs $(X_1, Y_1), (X_2, Y_2), \ldots, (X_n, Y_n)$ of discrete random variables are *mutually independent* if

$$\begin{aligned}&P[(X_1,Y_1)=(a_1,b_1),(X_2,Y_2)=(a_2,b_2),\ldots,(X_n,Y_n)=(a_n,b_n)]\\ &\quad = P[(X_1,Y_1)=(a_1,b_1)]P[(X_2,Y_2)=(a_2,b_2)]\cdots P[(X_n,Y_n)=(a_n,b_n)].\end{aligned}$$

Intuitively, the pairs are independent if the outcome of one pair does not affect the outcome of any other pair. The random pairs are identically distributed if they have the same joint probability mass function.

Definition 2.8-3

Random pairs $(X_1, Y_1), (X_2, Y_2), \ldots, (X_n, Y_n)$ are a *bivariate random sample* if they are independent and identically distributed.

Definition 2.8-4

Let $(X_1, Y_1), (X_2, Y_2), \ldots, (X_n, Y_n)$ be a bivariate random sample. Let $\bar{X}, \bar{Y}, S_X$, and S_Y be the sample means and sample standard deviations of the X_i's and Y_i's. The *sample covariance* is defined as

$$\operatorname{cov}(X, Y) = \frac{1}{n} \sum_{i=1}^{n} (X_i - \bar{X})(Y_i - \bar{Y}).$$

The *sample correlation* is defined as

$$\operatorname{corr}(X, Y) = \frac{\operatorname{cov}(X, Y)}{S_X S_Y}.$$

EXAMPLE **2.8-3** As in Example 2.8-1, let an experiment consist of observing a piece of software and recording the number of modules with errors (X) and the number of days it takes to debug the software (Y). Suppose nine pieces of software were observed, yielding the following observed random sample of (X, Y):

$$\{(1,2), (2,3), (0,0), (2,2), (0,0), (0,0), (3,2), (0,0), (3,4)\},$$

$$\bar{X} = 1.22, \quad \bar{Y} = 1.44, \quad S_X = 1.23, \quad S_Y = 1.42,$$

and

$$\begin{aligned} \operatorname{cov}(X, Y) &= \frac{1}{9}[(1 - 1.22)(2 - 1.44) + (2 - 1.22)(3 - 1.44) + \cdots \\ &\qquad + (3 - 1.33)(4 - 1.44)] \\ &= 1.57. \end{aligned}$$

Then

$$\begin{aligned} \operatorname{corr}(X, Y) &= \frac{1.57}{(1.23)(1.42)} \\ &= .90. \end{aligned}$$

Exercises 2.8

2.8-1 Let X denote the number of hotdogs and Y the number of beers consumed by an individual at a baseball game. Suppose X and Y have the following joint probability mass function:

	$p(x,y)$	x = 0	1	2	3
	0	.06	.15	.06	.03
y	1	.04	.20	.12	.04
	2	.02	.08	.06	.04
	3	.01	.03	.04	.02

a Find $E(X)$, $\text{STD}(X)$, $E(Y)$, and $\text{STD}(Y)$.

b Find $\text{COV}(X, Y)$ and $\text{CORR}(X, Y)$.

c Hotdogs cost \$3.00 each and beers cost \$3.50 each. Find an individual's expected total cost for beers and hotdogs at a baseball game. Find the standard deviation of this cost.

2.8-2 The hotdog and beer consumption for five individuals is $\{(1,0), (1,1), (2,1), (3,2), (3,3)\}$. Compute the sample correlation coefficient between hotdog and beer consumption for this sample.

2.8-3 Compute the sample covariance and correlation based on 100 simulations of the experiment in Example 2.8-1. To simulate this experiment, the pairs (X_i, Y_i) should be generated according to the joint distribution. For example, the pair $(0,0)$ should be generated so that it has a probability of .20 of occurring.

2.8-4 Let Y_1, Y_2, Y_3 be three random variables. Verify

$$\begin{aligned}\text{VAR}(aY_1 + bY_2 + cY_3) &= a^2\,\text{VAR}(Y_1) + b^2\,\text{VAR}(Y_2) + c^2\,\text{VAR}(Y_3)\\ &\quad + 2ab\,\text{COV}(Y_1, Y_2) + 2ac\,\text{COV}(Y_1, Y_3)\\ &\quad + 2bc\,\text{COV}(Y_2, Y_3).\end{aligned}$$

2.8-5 Extend the idea of a bivariate random sample to an arbitrary multivariate random sample. Give examples of random samples involving three variables (X, Y, Z).

2.8-6 Obtain the heights of 10 father–son pairs and 10 mother–daughter pairs from samples of acquaintances. Find the sample correlations for the father–son and mother–daughter pairs. To what extent do you think heights of sons can be predicted from heights of fathers, and heights of daughters can be predicted from heights of mothers?

SECTION

2.9 Conditional Expected Values

The *conditional expectation* of Y given $X = x$, denoted $E(Y \mid X = x)$, is found by computing the conditional probability mass function for Y given $X = x$ and taking the expected value of this distribution.

Definition 2.9-1

For discrete random variables X and Y the *conditional expectation* of Y given $X = x$ is

$$E(Y \mid X = x) = \sum_y yP(Y = y \mid X = x).$$

EXAMPLE **2.9-1** Let X and Y have the following joint probability mass function:

	$p(x,y)$	x = 2	3	4	$p_Y(y)$
y	0	.10	.20	.05	.35
	1	.30	.05	.10	.45
	2	.07	.13	0	.20
	$p_X(x)$	.47	.38	.15	

The conditional probability mass function of Y given $X = 2$ is

$$P(Y = 0 \mid X = 2) = \frac{.10}{.47} = .213,$$

$$P(Y = 1 \mid X = 2) = \frac{.30}{.47} = .638,$$

$$P(Y = 2 \mid X = 2) = \frac{.07}{.47} = .149.$$

Now,

$$E(Y \mid X = 2) = 0\,(.213) + 1\,(.638) + 2\,(.149) = .936.$$

In some problems, $E(Y \mid X = x)$ is easy to compute from intuitive considerations. These conditional expected values are functions of X and can be used to find $E(Y)$.

THEOREM 2.9-1 For random variables X and Y,

$$E(Y) = E(E(Y \mid X)).$$

Theorem 2.9-1 can be verified for discrete random variables as follows:

$$\begin{aligned} E(E(Y \mid X)) &= \sum_x E(Y \mid X = x)\, p_X(x) \\ &= \sum_x \sum_y yP(Y = y \mid X = x)\, p_X(x) \\ &= \sum_x \sum_y yp(x, y) \\ &= E(Y). \end{aligned}$$

EXAMPLE 2.9-2 A game consists of first tossing a die. If the face value on the die is X, then a coin is tossed X times. Let Y be the number of heads. It is intuitively clear that $E(Y \mid X = x) = x/2$. Also, $p_X(x) = 1/6$ for $x = 1, 2, \ldots, 6$. Thus,

$$E(Y) = \sum_{x=1}^{6} \left(\frac{x}{2}\right)\left(\frac{1}{6}\right) = 1.75.$$

The *conditional variance* of Y given $X = x$, $\mathrm{VAR}(Y \mid X = x)$, can also be found by using the conditional probability mass function.

Definition 2.9-2

For discrete random variables X and Y,

$$\begin{aligned}\mathrm{VAR}(Y \mid X=x) &= E[(Y-E(Y \mid X=x))^2 \mid X=x] \\ &= \sum_y (y-E(Y \mid X=x))^2 P(Y=y \mid X=x).\end{aligned}$$

Analogously to Theorem 2.4-3,

$$\mathrm{VAR}(Y \mid X=x) = E(Y^2 \mid X=x) - [E(Y \mid X=x)]^2.$$

EXAMPLE 2.9-3 Using the joint probability mass function from Example 2.9-1, we have $E(Y \mid X=2) = .936$,

$$\begin{aligned}\mathrm{VAR}(Y \mid X=2) &= \sum_{y=0}^{2} (y-.936)^2 P(Y=y \mid X=2) \\ &= (0-.936)^2(.213) + (1-.936)^2(.638) + (2-.936)^2(.149) \\ &= .358,\end{aligned}$$

and

$$\mathrm{STD}(Y \mid X=2) = \sqrt{.358} = .598.$$

It is frequently the case that the probability mass functions $P(Y=y \mid X=x)$ and $p_X(x)$ are more readily obtainable than the probability mass function of Y. Analogously to Theorem 2.9-1, these probability mass functions can be used to compute $\mathrm{VAR}(Y)$.

THEOREM 2.9-2 For random variables X and Y,

$$\mathrm{VAR}(Y) = E[\mathrm{VAR}(Y \mid X)] + \mathrm{VAR}[E(Y \mid X)].$$

Theorem 2.9-2 is verified for discrete random variables by noting that

$$\begin{aligned} E[\mathrm{VAR}(Y \mid X)] &= E[E(Y^2 \mid X) - \{E(Y|X)\}^2] \\ &= E(Y^2) - E[\{E(Y \mid X)\}^2] \\ &= E(Y^2) - \{E(Y)\}^2 - [E[\{E(Y \mid X)\}^2] - \{E(Y)\}^2] \\ &\qquad\qquad (\text{add and subtract } \{E(Y)\}^2) \\ &= \mathrm{VAR}(Y) - \mathrm{VAR}[E(Y \mid X)]. \end{aligned}$$

EXAMPLE 2.9-4 Let X denote the number of times a person goes fishing each month. The probability mass function for X is

x	0	1	2	3	4	5
$p(x)$	.25	.15	.20	.25	.10	.05

Note that $E(X) = 1.95$ and $\mathrm{VAR}(X) = 2.25$. The probability mass function for the number of fish caught each trip is

w	0	1	2	3
$p(w)$	.50	.30	.15	.05

For the random variable W, $E(W) = .75$ and $\mathrm{VAR}(W) = .79$. The number of fish caught each month can be expressed as $Y = \sum_{i=1}^{X} W_i$, where W_i is the number of fish caught on the ith trip. We will assume that the W_i are independent and identically distributed with probability mass function $p(w)$. We will now find $E(Y)$ and $\mathrm{STD}(Y)$. First,

$$E(Y \mid X = x) = E\left(\sum_{i=1}^{x} W_i\right) = \sum_{i=1}^{x} E(W_i) = \sum_{i=1}^{x} .75 = .75x.$$

Therefore,

$$E(Y) = E[E(Y \mid X)] = E(.75X) = .75\,E(X) = (.75)(1.95) = 1.46$$

and

$$\mathrm{VAR}[E(Y \mid X)] = \mathrm{VAR}(.75X) = (.75)^2\,\mathrm{VAR}(X) = (.75)^2(2.25) = 1.26.$$

Next we need to find $E[\mathrm{VAR}(Y \mid X)]$ to compute $\mathrm{VAR}(Y)$. Now,

$$\mathrm{VAR}(Y \mid X = x) = \mathrm{VAR}\left(\sum_{i=1}^{x} W_i\right) = \sum_{i=1}^{x} \mathrm{VAR}(W_i) = \sum_{i=1}^{x} .79 = .79x$$

and

$$E[\mathrm{VAR}(Y \mid X)] = E(.79X) = .79E(X) = (.79)(1.95) = 1.54.$$

Finally,

$$\mathrm{VAR}(Y) = 1.26 + 1.54 = 2.8$$

and

$$\mathrm{STD}(Y) = \sqrt{2.8} = 1.67.$$

Exercises 2.9

2.9-1 Let X and Y have the following joint probability mass function:

	$p(x,y)$	x = 2	3	4
y	−1	.10	.20	.05
	0	.27	.10	.15
	1	0	.03	.10

Find $E(Y \mid X = 2)$ and $E(Y \mid X = 0)$.

2.9-2 Let X be the number of heads in two tosses of a fair coin. Suppose a die is rolled $X + 1$ times after a value of X has been determined from the coin tosses. Let Y denote the total of the face values of the $X + 1$ rolls of the die. Find $E(Y \mid X = x)$ and $\mathrm{VAR}(Y \mid X = x)$ for $x = 0, 1, 2$. Use these conditional expected values to find $E(Y)$ and $\mathrm{VAR}(Y)$.

2.9-3 Let X denote the daily number of trucks arriving at a loading dock. The probability mass function for X is

x	0	1	2	3	4
$p(x)$	.10	.15	.25	.30	.20

Let T denote the tonnage carried by each truck. The probability mass function for T is

t	1	2	3	4	5
$p(t)$	.10	.20	.30	.25	.15

Let Y denote the total daily tonnage for the trucks at this loading dock. Find $E(Y)$ and $\mathrm{STD}(Y)$.

3

Special Discrete Random Variables

SECTION

3.1 Binomial Random Variable

An experiment that has only two possible outcomes is called a *Bernoulli trial.* An example is a single coin toss. We are interested in studying experiments that consist of a sequence of independent and identical Bernoulli trials. An example is repeatedly tossing a coin and observing the sequence of heads and tails that occurs. It is common with a Bernoulli trial to assign a value of 1 to one of the two outcomes and 0 to the other outcome. Such a random variable is called a *Bernoulli random variable.* It is also common to call the outcomes assigned to 1 a success and the outcome assigned to 0 a failure. Examples 3.1-1 through 3.1-5 illustrate experiments consisting of a sequence of independent and identical Bernoulli trials.

EXAMPLE **3.1-1** A true–false exam consists of 10 questions. If a student guesses at each question, then the questions are independent Bernoulli trials. The probability of a correct answer on each trial is 1/2.

EXAMPLE **3.1-2** A machine manufactures electronic components. An occasional "glitch" in the process can cause a component to be defective. Each manufactured part represents a

Bernoulli trial where the outcome is either a good component or a defective component. Provided the glitches occur independently and the probability of a glitch remains constant, this experiment can be regarded as a sequence of independent and identical Bernoulli trials.

EXAMPLE 3.1-3 A taste test is conducted to determine whether there is a difference between low-sodium crackers and regular crackers. Each judge is given three crackers; two are same and the third is different. The judge is asked to choose the cracker that is different from the other two. The response of each judge represents a Bernoulli trial where the outcome is either correct or incorrect. Different judges represent independent trials. If the crackers cannot be distinguished, the probability of a correct response is 1/3.

EXAMPLE 3.1-4 Suppose that a large population of items has each one labeled either 1 or 0 according to some characteristic of the item. Suppose that we sample randomly from this population as if we were drawing items in a lottery fashion from a large basket of items. Let X_i denote the numerical value (either 1 or 0) associated with the *i*th draw. As discussed in Section 2.5, the X_i's can be regarded as independent and identically distributed random variables provided the population size is large relative to the sample size. In this situation, $X_1, X_2, \ldots, X_n$ can be treated as independent and identical Bernoulli trials. As an example, suppose we randomly sample 1000 individuals from Los Angeles and ask them whether they favor a certain tax proposal. The outcomes of this survey can be regarded as independent and identical Bernoulli trials.

For an experiment consisting of a sequence of independent and identical Bernoulli trials, a random variable of interest is the number of successes.

Definition 3.1-1

A *binomial random variable* is the number of successes in n independent and identical Bernoulli trials.

EXAMPLE 3.1-5 Consider the experiment of tossing a fair coin n times and assigning the outcome success to a head. The random variable Y that denotes the total number of heads in the n tosses is a binomial random variable.

The probability mass function of a binomial random variable is called a *binomial distribution*. We derive this distribution for a special case in the next example. Following this example we give the general formula for the binomial distribution.

EXAMPLE 3.1-6 An experiment consists of inspecting five computers. The probability of finding a defective computer is $p = 1/6$. Assuming the computers fail independently of each other, the number of defective computers, Y, is a binomial random variable. The possible values for Y are 0, 1, 2, 3, 4, and 5. The probability mass function of Y is derived below.

It will be useful for this derivation to list the sequence of outcomes for the Bernoulli trials as a five-digit number, $c_1c_2c_3c_4c_5$, where c_i is either 0 or 1, depending on whether the ith computer is good or defective. For example, 01100 indicates that computers 1, 4, and 5 are good and computers 2 and 3 are defective. First consider $p(0)$ and $p(1)$.

$$\begin{aligned} p(0) &= P(Y=0) \\ &= P(00000) \\ &= \left(\frac{5}{6}\right)^5, \end{aligned}$$

$$\begin{aligned} p(1) &= P(Y=1) \\ &= P(10000) + P(01000) + P(00100) + P(00010) + P(00001) \\ &= \left(\frac{1}{6}\right)\left(\frac{5}{6}\right)^4 + \left(\frac{1}{6}\right)\left(\frac{5}{6}\right)^4 + \left(\frac{1}{6}\right)\left(\frac{5}{6}\right)^4 + \left(\frac{1}{6}\right)\left(\frac{5}{6}\right)^4 + \left(\frac{1}{6}\right)\left(\frac{5}{6}\right)^4 \\ &= 5\left(\frac{1}{6}\right)\left(\frac{5}{6}\right)^4. \end{aligned}$$

To derive $p(y)$ for $y = 2$, first consider two specific cases. One sequence of outcomes for which the two computers are defective is 00011. The probability of observing this sequence is $(5/6)^3(1/6)^2$. Another sequence for which two computers are defective is 00101. The probability for this sequence is also $(5/6)^3(1/6)^2$. It is easy to see that all sequences corresponding to two defective computers have probability $(5/6)^3(1/6)^2$. Thus, to determine the probability that there are two defective computers, we must count the number of sequences involving three 0's and two 1's. This can be done by using the counting technique in Section 1.4. Imagine writing the two 1's and three 0's in a row. The number of ways of selecting the two places for the 1's out of the five places available is $\binom{5}{2} = 10$, the other places automatically being assigned to the 0's. Thus,

$$\begin{aligned} p(2) &= \binom{5}{2}\left(\frac{1}{6}\right)^2\left(\frac{5}{6}\right)^3 \\ &= 10\left(\frac{1}{6}\right)^2\left(\frac{5}{6}\right)^3. \end{aligned}$$

Similarly,

$$\begin{aligned} p(3) &= \binom{5}{3}\left(\frac{1}{6}\right)^3\left(\frac{5}{6}\right)^2 \\ &= 10\left(\frac{1}{6}\right)^3\left(\frac{5}{6}\right)^2, \end{aligned}$$

$$\begin{aligned} p(4) &= \binom{5}{4}\left(\frac{1}{6}\right)^4\left(\frac{5}{6}\right)^1 \\ &= 5\left(\frac{1}{6}\right)^4\left(\frac{5}{6}\right), \end{aligned}$$

$$\begin{aligned} p(5) &= P(11111) \\ &= \left(\frac{1}{6}\right)^5. \end{aligned}$$

Note that $p(y)$ has the general form $\binom{5}{y}(1/6)^y(5/6)^{5-y}$, which includes $p(0)$ and $p(5)$. For instance,

$$p(0) = \binom{5}{0}\left(\frac{1}{6}\right)^0\left(\frac{5}{6}\right)^5 = \left(\frac{5}{6}\right)^5$$

and

$$p(5) = \binom{5}{5}\left(\frac{1}{6}\right)^5\left(\frac{5}{6}\right)^0 = \left(\frac{1}{6}\right)^5.$$

A general equation of the binomial distribution is given next.

THEOREM 3.1-1 Let Y be a binomial random variable based on n independent and identical Bernoulli trials with probability of a success p. Then

$$p(y) = \binom{n}{y} p^y (1-p)^{n-y}, \quad y = 0, 1, 2, \ldots, n.$$

Theorem 3.1-1 can be verified by considering the outcomes of the sequence of Bernoulli trials as an n-digit number consisting of 0's and 1's as in Example 3.1-6. There are $\binom{n}{y}$ of these n-digit numbers that have y 1's and $(n-y)$ 0's. Each of these has probability $p^y(1-p)^{n-y}$. The result follows immediately.

EXAMPLE 3.1-7 The probability that a randomly selected consumer will choose Cola A over Cola B is .4. Let Y denote the total number out of five randomly selected consumers that chose Cola A. The probability that exactly four consumers choose Cola A is

$$P(Y=4) = \binom{5}{4}(.4)^4(.6) = .0768.$$

The probability that a majority of the consumers choose Cola A is

$$P(Y \geq 3) = \binom{5}{3}(.4)^3(.6)^2 + \binom{5}{4}(.4)^4(.6) + \binom{5}{5}(.4)^5(.6)^0$$
$$= .2304 + .0768 + .0102 = .3174.$$

The quantities n and p are parameters of the binomial distribution. The mean and variance of a binomial random variable can be expressed in terms of n and p.

THEOREM 3.1-2 The mean and variance of a binomial random variable Y with parameters n and p are

$$E(Y) = np$$

and

$$\mathrm{VAR}(Y) = np(1-p).$$

Theorem 3.1-2 can be verified by expressing Y as a sum of n independent Bernoulli random variables. That is, $Y = \sum_{i=1}^{n} X_i$, where the X_i's are independent with $P(X_i = 1) = p$ and $P(X_i = 0) = 1 - p$. It is easy to show that $E(X_i) = p$ and $\mathrm{VAR}(X_i) = p(1-p)$ (see Exercise 3.1-8). From Theorems 2.7-3 and 2.7-6,

$$E(Y) = \sum_{i=1}^{n} E(X_i) = np$$

and

$$\mathrm{VAR}(Y) = \sum_{i=1}^{n} \mathrm{VAR}(X_i) = np(1-p).$$

EXAMPLE 3.1-8 Let Y denote the number of heads in 100 tosses of a fair coin. Then $E(Y) = 100(1/2) = 50$. That is, we would observe 50 heads, on average, in 100 tosses. In addition, $\mathrm{VAR}(Y) = 100(1/2)(1-1/2) = 25$ and $\mathrm{STD}(Y) = 5$. A likely range of values based on a two-standard-deviation interval about the mean is 40 to 60 heads.

The shape of the binomial distribution depends on the values of the parameters n and p. If $p = 1/2$, then the distribution is symmetric about $n/2$. For $p \neq 1/2$, the distribution is asymmetric. The degree of asymmetry increases as p approaches 0 or 1 and decreases as n increases. The histograms in Figure 3.1-1 illustrate the effects n and p have on the shape of the distribution. A range of likely binomial random variable values based on a two-standard-deviation interval is indicated on each histogram in Figure 3.1-1. Generally speaking, around 95% of the probability of a binomial random variable will fall within two standard deviations of its mean.

FIGURE 3.1-1 Binomial Random Variable Histograms

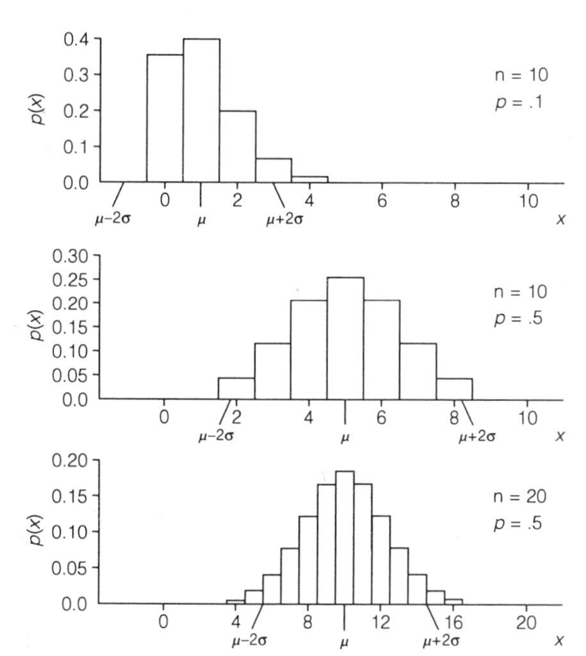

The cumulative probabilities of the binomial random variable for various values of n and p are given in Table 2 in Appendix A. For example, let Y be a binomial random variable with $n = 10$ and $p = .4$. Reading from the table we find $P(Y \leq 3) = .3823$.

EXAMPLE **3.1-9** Fifteen percent of the VCRs sold by a particular store will be returned for warranty service. Suppose the store sells 20 VCRs. Let Y denote the number requiring warranty service. If the VCR returns are independent of each other, then Y is a binomial random variable with $n = 20$ and $p = .15$. The expected number of VCRs that will be returned is $20(.15) = 3$ with a standard deviation of $\sqrt{20(.15)(.85)} = 1.60$. If more than five VCRs are returned, the store will lose money. The probability that the store loses money on the warranty service can be computed by using Table 2 in the Appendix:

$$\begin{aligned} P(Y > 5) &= 1 - P(Y \leq 5) \\ &= 1 - .9327 \\ &= .0673. \end{aligned}$$

EXAMPLE **3.1-10** A fireworks buyer makes a decision to accept or reject a large shipment of rockets based on drawing a random sample of 10 rockets from the shipment and testing them. If at least eight of these work, then the shipment will be accepted; otherwise it will be rejected. Suppose that the buyer receives a bad shipment in which the probability that a rocket in the shipment works is only .60. Let Y denote the number of working rockets in the sample. The probability that the shipment will be accepted may be found by referring to the binomial table with $n = 10$ and $p = .6$.

$$\begin{aligned} P(\text{shipment accepted}) &= P(Y \geq 8) \\ &= 1 - P(Y \leq 7) \\ &= 1 - .8327 \\ &= .1673. \end{aligned}$$

Suppose the acceptance rule is changed so that the shipment will be accepted provided at least nine of the rockets work. Now,

$$\begin{aligned} P(\text{shipment accepted}) &= P(Y \geq 9) \\ &= 1 - .9536 \\ &= .0464. \end{aligned}$$

In practice, an acceptance criterion is established so that there is only a small chance that a poor shipment is accepted. At the same time, the criterion should not be so stringent that good shipments are rejected. Suppose in our example that the probability of a working rocket is .95 and our acceptance criterion requires at least nine rockets to work. Then

$$
\begin{aligned}
P(\text{shipment accepted}) &= P(Y \geq 9) \\
&= 1 - .0861 \\
&= .9139.
\end{aligned}
$$

An easy way to simulate a binomial random variable is to simulate n independent Bernoulli trials and sum them.

EXAMPLE **3.1-11** Consider the experiment of observing five computers to determine how many are defective, as described in Example 3.1-6. Table 3.1-1 gives the empirical probabilities for the random variable Y based on 1000 simulations of this experiment. Each of the 1000 simulated values of the random variable Y required simulating five Bernoulli trials. The following algorithm was used.

Initialize $Y = 0$ and repeat the following steps five times.

1 Generate u, a unit random number.

2 If $u < 1/6$, then increment $Y = Y + 1$.

The empirical probabilities are relatively close to the true probabilities, as indicated in Table 3.1-1. Similarly, the sample mean and variance are about the same as $E(Y)$ and $\text{VAR}(Y)$. That is, $E(Y) = .833$ and $\text{STD}(Y) = .833$, while $\bar{Y} = .834$ and $S = .867$.

TABLE **3.1-1**
Empirical Probability Based on 1000 Simulated Binomial Variables

y	Empirical Probabilities	True Probabilities
0	.418	.4019
1	.380	.4019
2	.155	.1608
3	.045	.0322
4	.001	.0032
5	.001	.0001

Sometimes rather than the total number of successes in n independent and identical Bernoulli trials, the proportion of the n trials that result in a success is of interest. For example, suppose we poll a large population and record whether a respondent favors a particular candidate. It is customary and more informative to express the outcome of the poll in terms of the proportion of respondents who favor the candidate rather than the number.

Let Y be binomial random variable based on n independent and identical Bernoulli trials. The proportion of successes is

$$\hat{P} = \frac{Y}{n}.$$

The mean, variance, and probability mass function for $\hat{P}$ are easily derived from the same quantities about Y. Applying Theorem 3.1-2, we find

$$\begin{aligned} E(\hat{P}) &= E\left(\frac{Y}{n}\right) \\ &= \frac{1}{n}E(Y) \\ &= \frac{1}{n}np \\ &= p, \end{aligned}$$

$$\begin{aligned} \mathrm{VAR}(\hat{P}) &= \mathrm{VAR}\left(\frac{Y}{n}\right) \\ &= \frac{1}{n^2}\mathrm{VAR}(Y) \\ &= \frac{1}{n^2}np(1-p) \\ &= \frac{p(1-p)}{n} \end{aligned}$$

and

$$\begin{aligned} P\left(\hat{P} = \frac{i}{n}\right) &= P(Y = i) \\ &= \binom{n}{i} p^i (1-p)^{n-i}. \end{aligned}$$

In the context of observing outcomes of a random event whose probability is p, the observed value of $\hat{P}$ is nothing more than the empirical probability of a success.

EXAMPLE **3.1-12** Suppose we toss a coin 100 times. Let heads denote a success. The standard deviation for the proportion, $\hat{P}$, of heads in 100 tosses is $\sqrt{.5(1-.5)/100} = .05$. A likely range of values for $\hat{P}$ is $.5 \pm 2(.05)$, which is from .4 to .6. Suppose the coin is tossed 1600 times. In this case, the standard deviation for $\hat{P}$ is .0125, and a likely range of values for $\hat{P}$ is $.5 \pm 2(.0125)$, which is from .475 to .525. Thus, as the number of tosses increases, the likely range of values for $\hat{P}$ shrinks closer to the true value of $p = .5$.

By computing the two-standard-deviation interval, we can get some idea about how close an observed binomial proportion is likely to be to the true probability p. The quantity $p(1-p)$ is maximized when $p = .5$. Regardless of the value of p,

$$2\sqrt{\frac{p(1-p)}{n}} \le 2\sqrt{\frac{.5(1-.5)}{n}} = \frac{1}{\sqrt{n}}.$$

Suppose a simulation is used to estimate the probability p of some event. If the estimate is based on 100 simulations, then it is likely that the empirical probability of a success would differ from the true probability p by no more than $\pm 1/\sqrt{100} = \pm .1$. If the estimate is based on 1600 simulations, then it is likely that this empirical probability would differ from p by no more than $\pm .025$. In most of our examples, the empirical probabilities have been based on 1000 repetitions of the experiment. Thus, our empirical probabilities are typically within $\pm .03$ of the true probabilities.

EXAMPLE **3.1-13** The preceding discussion has implications for opinion polling. Suppose an opinion poll is based on a random sample of 1000, which is typical for opinion polling. Assuming that the two-standard-deviation interval is used to measure the margin of error, the margin of error would be $\pm .03$. For instance, if the opinion poll showed that 54% of the sample favors candidate A, we would be reasonably certain that between 51% and 57% of the population favors candidate A.

Exercises 3.1

3.1-1 Use your calculator to find $P(Y = 0)$, $P(Y = 1)$, $P(2 \le Y \le 4)$, and $P(Y > 2)$ for a binomial random variable Y with $n = 8$ and $p = .25$. Compute these probabilities by using Table 2 of Appendix A and compare your results.

3.1-2 Dr. Dribble has a probability of .8 of making a free throw each time he shoots. Assume that shots are independent of one another.

a Find the probability of making at least 8 out of 10 free throws.

b Find the probability of making fewer than 5 free throws in 10 shots.

3.1-3 Each question on a multiple choice test has four possible answers, one of which is correct. If there are five questions on the test and the student guesses on each one, what is the probability that he or she will get at least four of the five correct?

3.1-4 Inspectors on an assembly line find that an average of 1 in 10 parts is defective. If the defective parts occur independently of one another, what is the probability that the inspectors will find fewer than 3 defectives in 20 parts?

3.1-5 A fair die is to be tossed 600 times. Let Y denote the number of times a 1 occurs.

a Find the expected value and standard deviation of Y.

b Within what range would the number of 1's likely fall?

3.1-6 An experiment consists of throwing 25 darts at a dart board. The probability of hitting the bull's-eye is .4. Assume the throws are independent.

a Find the probability of exactly five bull's-eyes.

b Within what range would you expect the number of bull's-eyes to fall?

3.1-7 Suppose that the probability of a customer buying a hamburger at Burger Barn is .60. If there are 15 customers in line and 10 hamburgers already prepared, what is the probability that a customer will have to wait for a hamburger?

3.1-8 Let X be Bernoulli random variable with probability $P(X = 1) = p$ and $P(X = 0) = 1 - p$. Show that $E(X) = p$, $E(X^2) = p$, and $\text{VAR}(X) = p(1-p)$.

3.1-9 Suppose that 55% of a large population favors Candidate A.

a If a pollster selects a random sample of 100 from this population, what are the expected value and standard deviation of the percent in the sample that favor Candidate A?

b Compute a likely range for the percent in a sample of 100 that favor Candidate A.

c Repeat parts (a) and (b) for sample sizes 400 and 1600.

3.1-10 It is known that 75% of mice inoculated with a serum are protected from a certain disease. Suppose a shipment of 20 mice that were supposed to have been inoculated arrive at your lab.

a In shipments of size 20, how many mice on average would you expect to contract the disease?

b In your shipment, 10 mice die from the disease. Do you think the mice were inoculated?

3.1-11 For $0 < p < 1$, show that the maximum value of $p(1-p)$ occurs when $p = .5$. Thus, show

$$2\,\text{STD}(\hat{P}) \le \frac{1}{\sqrt{n}}.$$

3.1-12 An experiment consists of tossing a die 200 times. Compute the ±2-standard-deviation interval for the proportion of die tosses that result in a 3. Simulate 200 die tosses and compute the empirical probability of a 3. Does this empirical probability fall in the two-standard-deviation interval?

3.1-13 Another method for generating binomial random variables is to use directly the cumulative distribution function of the binomial random variable. Compute the cumulative distribution function for a binomial random variable with $n = 8$ and $p = .35$. Use an algorithm based on this function to simulate 100 binomial random variables. Use the binomial random variable simulation method given in the test to simulate 100 of these binomial random variables. Which method would you recommend be used in general binomial random variable simulation? Why?

3.1-14 Suspicions of illegal discrimination may be raised when hiring decisions result in outcomes that do not reflect the applicant pool. Suppose that an applicant pool has 50% females and 50% males, and suppose that the number of qualified applicants is equally divided among the two groups. If hiring does not result in an equal number of men and women being hired, to what extent might the discrepancy be attributed to chance? Discuss the role that the binomial distribution could play in answering this question.

SECTION

3.2 Geometric and Negative Binomial Random Variables

The *geometric* and *negative binomial* random variables are based on a sequence of independent and identical Bernoulli trials.

Definition 3.2-1

The *geometric random variable* is the number of independent and identical Bernoulli trials it takes to obtain the first success. Given an integer $k > 1$, a *negative binomial random variable* is the number of independent and identical Bernoulli trials it takes to obtain the kth success.

THEOREM 3.2-1 The probability mass function of the geometric random variable X, with probability p of a success, is

$$p(x) = p(1-p)^{x-1}, \quad x = 1, 2, \ldots.$$

Theorem 3.2-1 is easy to verify. In order for the first success to occur on the xth trial, there must have been $x - 1$ trials resulting in a failure each with probability $1 - p$ followed by one trial resulting in a success with probability p.

EXAMPLE 3.2-1 A gambler has a probability of .4 of picking a winning team. The probability that the gambler will pick a winning team for the first time on the third attempt is

$$p(3) = .4(.6)^2 = .144.$$

The probability that it takes the gambler three or more tries to choose a winning team is

$$P(X \geq 3) = \sum_{x=3}^{\infty} .4(.6)^{x-1}$$

$$= .4\,(.6)^2\left(\frac{1}{1-.6}\right)$$

$$= .36.$$

THEOREM 3.2-2 The mean and variance of a geometric random variable, with probability p of a success, are

$$\mu = \frac{1}{p}$$

and

$$\sigma^2 = \frac{1-p}{p^2}.$$

Verification of Theorem 3.2-2 is given in Section 3.6 and Exercise 3.2-8.

EXAMPLE 3.2-2 Suppose that the probability of winning a prize in a raffle is .1. The expected number of raffles a person would need to enter to win for the first time is 10 with a standard deviation of $\sqrt{(1-.1)/.1^2} = 9.5$.

EXAMPLE 3.2-3 Suppose a computer program has an error in it. To locate the error, investigators conduct a series of independent tests. Each test is based on a set of random inputs for the program. If the probability that a random test detects the error is .16, the expected number of tests that will need to be conducted to detect the error is $\mu = 1/.16 = 6.25$. The standard deviation is $\sigma = \sqrt{(1-.16)/.16^2} = 5.73$.

THEOREM 3.2-3 The probability mass function of a negative binomial random variable Y, with probability p of a success and $k > 1$, is

$$p(y) = \binom{y-1}{k-1} p^k (1-p)^{y-k}, \quad y = k, k+1, \ldots .$$

Theorem 3.2-3 can be verified as follows. If the kth success occurs on the yth trial, then $k-1$ successes and $y-k$ failures occurred during the first $y-1$ trials. From the binomial distribution, the probability of this happening is

$$\binom{y-1}{k-1} p^{k-1} (1-p)^{y-k}.$$

The probability of a success occurring on the yth trial is p, and the result follows.

EXAMPLE 3.2-4 Consider the random testing problem of Example 3.2-3. Suppose that the error is not removed after it is detected but that testing of the program with random inputs continues. Let Y denote the number of tests it takes to encounter the error three times. The probability that it takes exactly 18 tests to encounter the error three times is

$$\begin{aligned} P(Y = 18) &= \binom{18-1}{3-1} (.16)^3 (1-.16)^{18-3} \\ &= .041. \end{aligned}$$

THEOREM 3.2-4 The mean and variance of a negative binomial random variable, with probability of success p and $k > 1$, are

$$\mu = \frac{k}{p}$$

and

$$\sigma^2 = \frac{k(1-p)}{p^2}.$$

To verify Theorem 3.2-4, express the negative binomial random variable as the sum of k independent geometric random variables each with mean $1/p$ and variance

$(1-p)/p^2$. Let X_1 denote the number of trials to the first success, X_2 the number of trials from the first to the second success, ... X_k the number of trials from the $(k-1)$th to the kth success. Then $Y = \Sigma_{i=1}^{k} X_i$. Applying Theorems 2.7-3 and 2.7-6 gives

$$E(Y) = \sum_{i=1}^{k} E(X_i) = \frac{k}{p},$$

$$\text{VAR}(Y) = \sum_{i=1}^{k} \text{VAR}(X_i)$$

$$= \frac{k(1-p)}{p^2}.$$

EXAMPLE 3.2-5 For the random testing of a computer program in Example 3.2-4, the expected number of tests it takes to encounter the error three times is $3/.16 = 18.75$, and the standard deviation is $\sqrt{3(1-.16)/.16^2} = 9.92$.

Geometric and negative binomial random variables can be easily simulated by generating a sequence of Bernoulli trials and counting the number of trials that occur when the first or kth success is found. An alternative method that can be used for generating these random variables is given next.

The cumulative distribution function of a geometric random variable is

$$\begin{aligned} F(x) &= P(X \le x) \\ &= 1 - P(\text{all of the first } x \text{ trials are failures}) \\ &= 1 - (1-p)^x. \end{aligned}$$

From the techniques of Section 2.5, if u is a unit random number such that

$$1 - (1-p)^{x-1} \le u < 1 - (1-p)^x,$$

then x is the observed value of a geometric random variable X. This inequality is equivalent to

$$x - 1 \le \frac{\ln(1-u)}{\ln(1-p)} < x.$$

Thus, the observed random variable is the first integer larger than

$$\ln(1-u)/\ln(1-p).$$

For instance, if $p = .25$ and $u = .87901$, then $\ln(1 - .87901)/\ln(1 - .25) = 7.34$ and $x = 8$.

To generate a negative binomial random variable, we generate k geometric random variables $X_1, X_2, \ldots, X_k$. Then

$$Y = \sum_{i=1}^{k} X_i$$

is a negative binomial random variable.

Exercises 3.2

3.2-1 In a game of billiards, a player shoots until a miss occurs. A particular player misses on any given shot with probability 1/3. Let the random variable X denote the number of shots taken to obtain the first miss.

- **a** Give the probability mass function for X.
- **b** What is the probability that the player will take exactly five shots?
- **c** What is the probability that the player will take at least four shots?
- **d** What is the average number of shots this player will take?

3.2-2 Each time customers visit a restaurant they are given a game card. Suppose the probability of winning a prize with the game card is .2.

- **a** What is the probability that a customer will win a prize for the first time on the fourth visit?
- **b** What is the probability that it will take at most four visits to win a prize?
- **c** On average, how many visits would a customer need to make to win a prize? What is the standard deviation?
- **d** What is the probability that a customer will win the fourth prize on the tenth visit?
- **e** What is the average number of visits necessary to win four prizes? What is the standard deviation?

3.2-3 Suppose the probability of male birth is .5. A couple desires to have three boys. They have decided to have children until they have exactly three boys.

- **a** What is the probability that the family will have exactly three children? exactly four? exactly five?
- **b** What do you think is a likely range for the number of children the family would have?

3.2-4 A roulette wheel has 38 divisions. Eighteen are red, 18 are black, and 2 are green. A player may bet on red, black, or both. If the wheel stops on the color the player bets, the player wins.

- **a** If a player bets on red, how many spins in a row, on average, can the player be expected to win?
- **b** If a player bets on both red and black, how many spins in a row, on average, can the player be expected to win?

3.2-5 Refer to Exercise 3.2-4. With a random number table, simulate 20 games of roulette for a player who bets on red. What was the longest winning streak? What was the longest losing streak? What was the average length of the winning streaks? Are the answers consistent with what you know about the expected value and standard deviation of a geometric random variable?

3.2-6 Suppose the arrivals of planes at an airport are independent and identical Bernoulli trials with a probability of .4 of an arrival each minute. Let X denote the time (rounded to the nearest minute) from one arrival to the next. Compute $E(X)$ and $\mathrm{STD}(X)$.

3.2-7 Refer to Exercise 3.2-6. Let Y denote the time (rounded to the nearest minute) it takes for three planes to arrive. Simulate 100 values of Y. Compute the sample mean and sample standard deviation of these 100 values. Compare these results to $E(Y)$ and $\mathrm{STD}(Y)$ for a negative binomial random variable.

3.2-8 Derive the expected value and variance of a geometric random variable X, using the following steps. First, note that $\Sigma_{i=1}^{\infty} r^i = r/(1-r)$. Differentiate both sides of the equation with respect to r and show that $\Sigma_{i=1}^{\infty} ir^{i-1} = 1/(1-r)^2$. Apply this result to

$$E(X) = \sum_{i=1}^{\infty} xp(1-p)^{x-1}.$$

Differentiate again to find $E[X(X-1)]$ from which $E(X^2)$ may be obtained. Then apply Theorem 2.4-3.

3.2-9 Generate 1000 values for a geometric random variable with $p = .3$, using two different techniques discussed in this section. Compare these methods, discussing both speed and number of unit random numbers needed. Try other values of p, such as $p = .003$ and $p = .95$.

3.2-10 Simulate 100 values of a negative binomial random variable Y with $k = 4$. Compute the sample mean and sample standard deviation and compare with $E(Y)$ and $\mathrm{STD}(Y)$.

SECTION

3.3 Hypergeometric Random Variable

The *hypergeometric random variable* is the number of successes that arise from sampling without replacement. We begin with an example.

EXAMPLE 3.3-1 A box contains 12 poker chips of which 7 are green and 5 are blue. Five chips are selected at random without replacement from this box. Let the random variable X denote the number of green chips selected. To compute $P(X = 2)$, for instance, we first find the number of ways in which we can select two green and three blue chips from the box. This is $\binom{7}{2}\binom{5}{3}$. The total number ways of selecting 5 chips from the 12 is $\binom{12}{5}$. Therefore,

$$p(2) = \frac{\binom{7}{2}\binom{5}{3}}{\binom{12}{5}}.$$

In general, the probability mass function of X is

$$p(x) = \frac{\binom{7}{x}\binom{5}{5-x}}{\binom{12}{5}}, \quad x = 0, 1, \ldots, 5.$$

Consider what happens in the experiment if we select eight chips instead of five. In this case, it is impossible to select eight chips that are all blue. Therefore, $P(X = 0) = 0$. Also, $P(X = 1) = 0$, $P(X = 2) = 0$, and $P(X = 8) = 0$. In this case, the probability mass function is

$$p(x) = \frac{\binom{7}{x}\binom{5}{8-x}}{\binom{12}{8}}, \quad x = 3, 4, \ldots, 7.$$

Definition 3.3.-1

A set contains N items of which r have characteristic C and $N - r$ do not have characteristic C. A hypergeometric random variable denotes the number of items with characteristic C in a sample of n items selected at random without replacement from this set. The probability mass function of a hypergeometric random variable is

$$p(x) = \frac{\binom{r}{x}\binom{N-r}{n-x}}{\binom{N}{n}}, \quad x = \max(0, n - (N-r)), \ldots, \min(n, r).$$

EXAMPLE 3.3-2 An applicant pool for a software engineering position contains 4 females and 10 males, all equally qualified. The company plans to hire five people from this pool. Let X denote the number of females hired. If the company chooses at random from this pool, then

$$p(x) = \frac{\binom{4}{x}\binom{10}{5-x}}{\binom{14}{5}}, \quad x = 0, 1, 2, 3, 4.$$

The probability that the company hires at most one female is

$$P(X \le 1) = p(0) + p(1) = .126 + .420 = .546.$$

The following theorem is stated without verification.

THEOREM 3.3-1 The mean and variance of the hypergeometric random variable are

$$\mu = n\left(\frac{r}{N}\right)$$

and

$$\sigma^2 = n\left(\frac{r}{N}\right)\left(1 - \frac{r}{N}\right)\left(\frac{N-n}{N-1}\right).$$

EXAMPLE 3.3-3 The mean, variance, and standard deviation of X in Example 3.3-1 are

$$\mu = 5\left(\frac{7}{12}\right) = 2.92,$$

$$\sigma^2 = 5\left(\frac{7}{12}\right)\left(\frac{5}{12}\right)\left(\frac{7}{11}\right) = .773,$$

$$\sigma = .879.$$

EXAMPLE 3.3-4 Suppose a box contain 12,000 poker chips of which 7000 are green and 5000 are blue. A sample of 100 chips is selected without replacement. Let the random variable X denote the number of green chips in the sample. Based on the hypergeometric distribution,

$$\mu = 100\left(\frac{7000}{12{,}000}\right) = 58.33,$$

and

$$\sigma^2 = 100\left(\frac{7000}{12,000}\right)\left(\frac{5000}{12,000}\right)\left(\frac{11,900}{11,999}\right)$$
$$= 24.11,$$

and

$$\sigma = 4.91.$$

Note that if the selected poker chip was replaced each time, this could be treated as 100 independent and identical Bernoulli trials, each with probability $p = 7000/(12,000) = .5833$ of selecting a green chip. Then

$$\mu = 100(.5833) = 58.33$$
$$\sigma^2 = 100(.5833)(1 - .5833) = 24.31,$$

and

$$\sigma = 4.93.$$

As discussed in Example 3.3-4, Theorem 3.3-1 reinforces the notion that sampling without replacement can be treated as sampling with replacement when n is small relative to N. Let $p = r/N$. If n is small relative to N, then the ratio

$$\frac{N-n}{N-1} \approx 1,$$

and the mean and variance of the hypergeometric random variable coincide with the mean and variance of a binomial random variable. The distributions of these two random variables are also essentially the same when n is small relative to N.

Exercises 3.3

3.3-1 Let X be a hypergeometric random variable with $N = 20$, $r = 9$, and $n = 12$.

a What are the possible values for X?

b Find $E(X)$ and $\text{VAR}(X)$.

c Repeat parts (a) and (b) with $r = 5$.

3.3-2 A grocery display contains 25 apples of which 11 are Jonathans and 14 are McIntosh.

a If a customer randomly selects six apples, what is the probability that three will be Jonathans and three will be McIntosh?

b Let the random variable X denote the number of Jonathan apples in the sample of six randomly selected apples. Find the probability mass function of X.

c Find $E(X)$ and $\text{STD}(X)$.

3.3-3 A scientific expedition has captured, tagged, and released eight sea turtles in a particular region. The expedition assumes that the population size in this region is 35, which would mean that 8 are tagged and 27 are not tagged. The expedition will now capture 10 turtles and note how many in this new sample are tagged.

a If the expedition's assumption about the population size is correct, what is the probability of at most three tagged turtles in this new sample?

b The expedition found five tagged turtles in the new sample. Is this evidence that they have overestimated the population size? Explain by using the hypergeometric distribution.

3.3-4 Simulate a capture–tag–recapture experiment similar to Exercise 3.3-3. Assume 20 endangered animals have been captured and tagged.

a Obtain through simulation the percent of tagged animals in samples of size 25 for true population sizes of 35, 50, 75, 100, and 500.

b What values would you expect to see for the percents in part (a)? Use the mean and standard deviation of the hypergeometric distribution to answer this question.

3.3-5 A small company has 20 accounts of which 10 are delinquent. An auditor inspects eight accounts and finds one delinquent. Is this consistent with random selection of the accounts? Answer by computing the probability of finding at most one delinquent account.

3.3-6 Compare hypergeometric probabilities for $N = 30$, $r = 18$, and $n = 4$ to binomial probabilities with $n = 4$ and $p = 18/30 = .60$.

SECTION

3.4 Multinomial Random Variables

Suppose an experiment consists of n independent and identical trials where the outcome of each trial falls into exactly one of k mutually exclusive classes. For example, in a game of roulette, each spin of the wheel will result in the ball landing on one of three colors, typically red, green, or black. We will be interested in the number of times each of the k classes occurs in the n trials.

EXAMPLE 3.4-1 A box contains four blue marbles, three green marbles, and five red marbles. The marbles will be selected at random, one after another with replacement. Let the random variables Y_1, Y_2, and Y_3 correspond to the number of blue, green, and red marbles in 10 draws. On each draw, $p_1 = 4/12$, $p_2 = 3/12$, and $p_3 = 5/12$. The joint probability of five blue, two green, and three red marbles can be derived by first determining the number of ways in which this arrangement of marbles can occur. There are $\binom{10}{5}$ ways of selecting the blue marbles. Then, there are $\binom{5}{2}$ ways of selecting the green marbles for the five remaining draws, and $\binom{3}{3} = 1$ way of selecting the red marbles. Using the Multiplication Rule to account for any arrangements, we obtain

$$\binom{10}{5}\binom{5}{2}\binom{3}{3} = \frac{10!}{5!2!3!}.$$

Since the selections are done with replacement, the probability of each of these outcomes is

$$\left(\frac{4}{12}\right)^5\left(\frac{3}{12}\right)^2\left(\frac{5}{12}\right)^3.$$

Finally,

$$P(Y_1 = 5, Y_2 = 2, Y_3 = 3) = \frac{10!}{5!2!3!}\left(\frac{4}{12}\right)^5\left(\frac{3}{12}\right)^2\left(\frac{5}{12}\right)^3.$$

THEOREM 3.4-1 Let the random variables $Y_1, Y_2, \ldots, Y_k$ denote the number of times that the mutually exclusive and exhaustive outcomes $C_1, C_2, \ldots, C_k$ occur in n independent and identical trials. Let $p_i = P(C_i)$, for $i = 1, 2, \ldots, k$. Then the *multinomial probability mass function* is

$$P(Y_1 = y_1, Y_2 = y_2, \ldots, Y_k = y_k) = \frac{n!}{y_1!y_2!\cdots y_k!}p_1^{y_1}p_2^{y_2}\cdots p_k^{y_k},$$

where $\sum_{i=1}^{k} y_i = n$ and $\sum_{i=1}^{k} p_i = 1$.

EXAMPLE 3.4-2 Consider the experiment of Example 3.4-1. A general expression of the joint probability mass function is

$$p(y_1, y_2, y_3) = \frac{10!}{y_1!y_2!y_3!}\left(\frac{4}{12}\right)^{y_1}\left(\frac{3}{12}\right)^{y_2}\left(\frac{5}{12}\right)^{y_3}.$$

Each random variable in Theorem 3.4-1 can be expressed as a sum of the outcomes of independent and identical Bernoulli trials. That is, Y_i is equal to the number of times in n trials that outcome C_i occurs. Therefore, Y_i is a binomial random variable. In particular, $E(Y_i) = np_i$ and $\mathrm{VAR}(Y_i) = np_i(1 - p_i)$.

EXAMPLE **3.4-3** The table gives the national percentages for income, broken down into categories, in a certain community.

Income	Percentages
More than \$100,000	2
\$75,000 to \$100,000	15
\$50,000 to \$74,999	14
\$25,000 to \$49,999	20
\$15,000 to \$24,999	18
\$5,000 to \$14,999	19
Below \$5,000	12

In a sample of 50 people we would expect, on average, $50(.02) = 1$ person with an income over \$100,000 and $50(.12) = 6$ people with an income below \$5000. The standard deviations are $\sqrt{50(.02)(.98)} = .99$ for income over \$100,000 and $\sqrt{50(.12)(.88)} = 2.30$ for income below \$5000.

The random variables $Y_1, Y_2, \ldots, Y_k$ from Theorem 3.4-1 are clearly not independent since $\sum_{i=1}^{k} Y_i = n$. The covariance for any two of these random variables is given in the next theorem.

THEOREM **3.4-2** Let $Y_1, Y_2, \ldots, Y_k$ have a multinomial probability distribution. Then

$$\mathrm{COV}(Y_s, Y_t) = -np_s p_t, \quad s \neq t.$$

The covariance is negative, as we would expect since a large number of trials resulting in outcome C_s forces a small number of trials resulting in outcome C_t.

EXAMPLE **3.4-4** The community has a special tax of \$120 per year for those whose income is over \$100,000 and \$80 per year for those whose income is between \$75,000 and \$100,000 per year. The rest are not assessed the special tax. Let Y_1 and Y_2 denote the number of individuals in a sample of size 50 who fall into the over \$100,000 and the \$75,000–\$100,000 income brackets, respectively. The special tax paid by the individuals is

$$T = 120Y_1 + 80Y_2.$$

We find

$$\begin{aligned} E(T) &= 120E(Y_1) + 80E(Y_2) \\ &= 120(50)(.02) + 80(50)(.15) \\ &= \$770 \end{aligned}$$

and

$$\begin{aligned} \mathrm{VAR}(T) &= (120)^2\,\mathrm{VAR}(Y_1) + (80)^2\,\mathrm{VAR}(Y_2) + 2(120)(80)\,\mathrm{COV}(Y_1, Y_2) \\ &= 14{,}400(50)(.02)(.98) + 6400(50)(.15)(.85) \\ &\quad + (19{,}200)(-50)(.02)(.15) \\ &= 52{,}032, \end{aligned}$$

$$\begin{aligned} \mathrm{STD}(T) &= \sqrt{52{,}032} \\ &= \$228. \end{aligned}$$

Exercises 3.4

3.4-1 Bolts of fabric manufactured at a particular company have color flaws, weaving flaws, both color and weaving flaws, or no flaws with probabilities .15, .07, .03, .75, respectively. If 30 bolts of fabric are randomly selected from this company, what is the probability that 5 will have color flaws, 4 will have weaving flaws, 1 will have both weaving and color flaws, and 20 will be flawless?

3.4-2 A rat's maze has four exits labeled A, B, C, and D. Each time the rat runs the maze it has a probability of .25 of choosing any of the exits.

a If the rat runs the maze 10 times, what is the probability it chooses exit A four times, exit B three times, exit C twice, and exit D once?

b If the rat runs the maze four times, what is the probability it chooses each exit exactly once?

3.4-3 Computer chips manufactured by a particular company have a probability of .22 of one defect, .13 of more than one defect, and .65 of no defects. The cost of repairing the chips is \$10.00 for one defect and \$30.00 for more than one defect. For a random sample of 50 chips, what is the expected repair cost and the standard deviation of this cost?

3.4-4 The adjusted census figures on adults over the age of 18 are given as follows:

Age	Population Proportion
18–24	.18
25–34	.23
35–44	.16
45–64	.27
65 and over	.16

The ages from a random sample of 100 residents over the age of 18 in Manhattan, Kansas, are

Age	Sample Proportion
18–24	.30
25–34	.23
35–44	.20
45–64	.15
65 and over	.12

a Does the sample from Manhattan appear to follow the national figures? Explain by using properties of the multinomial probability distribution.

b Simulate a sample of 100 ages by using the national age distribution. Compute the empirical probabilities for each age category. Repeat this simulation 10 times. Does this simulation reinforce your answer to part (a)?

SECTION

3.5 Poisson Random Variable

A useful distribution for many random phenomena is the *Poisson distribution*, so named for the French mathematician Simeon D. Poisson, who studied its properties. Typical examples of Poisson random variables include the number of misprints in a newspaper, the number of automobile accidents per week at a busy intersection, the number of particles emitted per unit of time by a radioactive source, the number of customers arriving for service at a drive-in banking facility during some interval of time, and the number of telephone calls per minute coming into an office.

Definition 3.5-1

The Poisson probability mass function is defined on the non-negative integers and has functional form

$$p(x) = \frac{e^{-\mu}\mu^x}{x!}, \quad x = 0, 1, 2, \ldots .$$

THEOREM 3.5-1 Let X denote a Poisson random variable with probability mass function given in Definition 3.5-1. Then

$$E(X) = \mu$$

and

$$\text{VAR}(X) = \mu.$$

Verification of Theorem 3.5-1 is given in Section 3.6 and Exercises 3.5-8 and 3.5-9. Table 3 in Appendix A is a table of the cumulative distribution function of the Poisson random variable for selected values of μ.

There is an interesting relationship between the binomial distribution and the Poisson distribution. This relationship will help us identify phenomena for which the Poisson distribution is appropriate.

THEOREM 3.5-2 Let X be a binomial random variable based on a sequence of n independent and identical Bernoulli trials with probability p of success on each trial. If $n \to \infty$ and $p \to 0$ in such a way that the mean $\mu = np$ is constant, then

$$\lim_{\substack{n\to\infty,\, p\to 0 \\ \mu = np}} P(X = x) = \lim_{\substack{n\to\infty,\, p\to 0 \\ \mu = np}} \binom{n}{x} p^x (1-p)^{n-x}$$

$$= \frac{e^{-\mu}\mu^x}{x!}.$$

Theorem 3.5-2 is verified with the following calculations. If $\mu = np$ or $p = \mu/n$, then $n \to \infty$ implies $p \to 0$. We will express the binomial probabilities in terms of n and μ and let $n \to \infty$.

$$P(X = x) = \frac{n!}{x!(n-x)!}\left(\frac{\mu}{n}\right)^x\left(1 - \frac{\mu}{n}\right)^{n-x}$$

$$= \frac{\mu^x}{x!}\left(1 - \frac{\mu}{n}\right)^n\left[\frac{n!}{n^x(n-x)!}\right]\left(1 - \frac{\mu}{n}\right)^{-x}.$$

In the limit,

$$\lim_{n\to\infty}\left(1 - \frac{\mu}{n}\right)^n = e^{-\mu},$$

$$\lim_{n\to\infty}\left[\frac{n!}{n^x(n-x)!}\right] = \lim_{n\to\infty}\frac{n}{n}\left(\frac{n-1}{n}\right)\left(\frac{n-2}{n}\right), \ldots, \left(\frac{n-x+1}{n}\right) = 1,$$

$$\lim_{n\to\infty}\left(1 - \frac{\mu}{n}\right)^{-x} = 1.$$

Therefore,

$$\lim_{\substack{n\to\infty,\, p\to 0\\ \mu=np}} P(X = x) = \frac{\mu^x}{x!}e^{-\mu}, \quad x = 0, 1, 2, \ldots.$$

Theorem 3.5-2 tells us that binomial probabilities can be approximated by Poisson probabilities when n is large and p is small. As a rule, the approximation will be adequate when n is 20 or greater and p is .05 or less. Table 3.5-1 is a comparison of the binomial and Poisson probabilities for $n = 20$ and $p = .05$.

TABLE **3.5-1**
Poisson and Binomial Comparisons

Values of X	**0**	**1**	**2**	**3**	**4**	**>4**
Binomial	.36	.38	.19	.06	.01	.00
Poisson ($\mu = 1$)	.37	.37	.18	.06	.02	.00

Theorem 3.5-2 also suggests possible applications of the Poisson distribution. If there are many independent opportunities for an outcome of interest to occur, but a small constant probability that it will occur at each opportunity, then the Poisson distribution will be a good model for the number of times the outcome of interest occurs. For instance, there are many words in a newspaper, but there is a small chance for a misprint in each word. Thus, the Poisson distribution is likely to be a

good model for the number of misprints in the newspaper. Similarly, at a busy intersection, there is an opportunity for an accident each time a car enters the intersection, but there is a small probability that the accident will occur. The Poisson distribution is often appropriate for the number of accidents per unit of time.

EXAMPLE **3.5-1** Let the random variable X be the number of calls per minute coming into a switchboard. Assume X is a Poisson random variable with mean of .9 call per minute. The probability of no calls in a minute is

$$P(X = 0) = p(0) = e^{-.9} = .4066.$$

The probability of three or more calls in a minute is

$$\begin{aligned} P(X \geq 3) &= 1 - P(X \leq 2) \\ &= 1 - \sum_{x=0}^{2} \frac{(.9)^x e^{-.9}}{x!} \\ &= 1 - .9371 \\ &= .0629. \end{aligned}$$

EXAMPLE **3.5-2** The number of customers requesting automobile repair at a garage is a Poisson random variable with a mean of 10 per day. Suppose that the garage can service at most 15 cars in a day. The probability that the garage will not be able to handle all the requests for service in a given day can be computed by using Table 3 in the Appendix to be

$$\begin{aligned} P(X > 15) &= 1 - P(X \leq 15) \\ &= 1 - .9513 \\ &= .0487. \end{aligned}$$

EXAMPLE **3.5-3** The probability that a microchip is defective is .01. If we can assume that the defective chips occur independently of one another, then the number of defective chips in 30 has a binomial distribution with $n = 30$ and $p = .01$. Theorem 3.5-2 can be used to find the probability that there is 1 or more defective chips in 30. We use a Poisson approximation to the binomial with $\mu = np = .3$. If X denotes the number of defective chips, then

$$\begin{aligned} P(X \geq 1) &= 1 - P(X = 0) \\ &= 1 - e^{-.3} \\ &= 1 - .7408 \\ &= .2592. \end{aligned}$$

It is unlikely that there would be more than two defective chips since

$$\begin{aligned} P(X > 2) &= 1 - P(X \leq 2) \\ &= 1 - .9964 \\ &= .0036. \end{aligned}$$

Exercises 3.5

3.5-1 Use your calculator to find $P(X = 0)$, $P(X = 1)$, $P(2 \leq X \leq 4)$, and $P(X > 2)$ for the Poisson distribution with $\mu = 2.0$. Compute these probabilities, using Table 3 of Appendix A and compare your results.

3.5-2 The number of jobs submitted per minute to a supercomputer center is a Poisson random variable with $\mu = 8$.

- **a** Find the probability that there will be no jobs submitted in any one minute.
- **b** Find the probability that there will be two or fewer jobs submitted in a minute.
- **c** If there are more than 10 jobs submitted in a minute, there will be a delay. What is the probability of this happening?

3.5-3 The probability that a given line of computer code has a mistake in it is .02. Suppose there are 20 such lines of code and that the mistakes occur independently from line to line. Use the Poisson approximation to the binomial distribution to find

- **a** the probability of no mistakes,
- **b** the probability of two or more mistakes.

3.5-4 An engineer claims that a process should produce no more than 1 in 1000 defective items. We randomly select 100 of these items and find three defects. Is this consistent with the engineer's claim? Explain by using the Poisson approximation to the binomial.

3.5-5 During any year the number of reported sightings of UFOs has a Poisson distribution with $\mu = 30$. Find a likely range for the number of sightings based on a two-standard-deviation interval about the mean.

3.5-6 The number of defects in a sheet of glass is a Poisson random variable with a mean of .2. If a builder buys a sheet of glass, what is the probability that it is defect free?

3.5-7 The number of burglaries committed in a small city is a Poisson random variable with a mean of nine per month. Suppose 13 burglaries are reported in a given month. Does this indicate that burglaries are on the increase?

3.5-8 Show that $E(X) = \mu$ for a Poisson random variable as follows: show that

$$\sum_{x=0}^{\infty} \frac{xe^{-\mu}\mu^x}{x!} = \mu \sum_{x=1}^{\infty} \frac{e^{-\mu}\mu^{x-1}}{(x-1)!} = \mu.$$

3.5-9 Show that $\mathrm{VAR}(X) = \mu$ for a Poisson random variable as follows: show that

$$E[X(X-1)] = \mu^2 \sum_{x=2}^{\infty} \frac{e^{-\mu}\mu^{x-2}}{(x-2)!}.$$

Then show that

$$\mathrm{VAR}(X) = E[X(X-1)] + E(X) - (E(X))^2.$$

3.5-10 The Poisson distribution describes many random outcomes that we observe in everyday life. For instance, the number of coughs that occur among the students in a large lecture class could have a Poisson distribution. Give four other such examples not mentioned in the text.

SECTION

3.6 Moments and Moment-Generating Functions

The purpose of this section is to introduce a useful mathematical tool called the *moment-generating function*. It has many applications in the theory of probability. We will use it to derive the mean and variance of several important distributions introduced previously in this chapter. We begin with the following definition.

Definition 3.6-1

The *kth moment* of a random variable X is defined to be $E(X^k)$.

The first moment is obviously the mean of X. The relationship between the second moment and the variance was established in Theorem 2.4-3. That is,

$$\mathrm{VAR}(X) = E(X^2) - \mu^2.$$

For a discrete random variable X with probability mass function $p(x)$, the kth moment may be computed as

$$E(X^k) = \sum_x x^k p(x).$$

The following example illustrates this computation.

EXAMPLE 3.6-1 Let the random variable X have the following probability mass function.

x	0	1	2	3	4
$p(x)$	.1	.3	.4	.1	.1

The third moment of X is

$$\begin{aligned} E(X^3) &= 0^3(.1) + 1^3(.3) + 2^3(.4) + 3^3(.1) + 4^3(.1) \\ &= 12.6. \end{aligned}$$

Definition 3.6-2

The *moment-generating function* of a random variable X is defined by

$$M_X(t) = E(e^{Xt}).$$

The moment-generating function is said to exist if there is some positive constant b such that $M_X(t)$ is finite for $-b < t < b$.

THEOREM 3.6-1 For any positive integer k, the kth derivative of the moment-generating function with respect to t, evaluated at $t = 0$, is the kth moment of the random variable X.

Theorem 3.6-1 can be verified for a discrete random variable X as follows. The moment-generating function is

$$M_X(t) = E(e^{tX}) = \sum_x e^{xt} p(x).$$

The kth derivative of $M_X(t)$, with respect to t, is

$$\begin{aligned} M_X^{(k)}(t) &= \frac{d^k M_X(t)}{dt^k} \\ &= \sum_x x^k e^{xt} p(x). \end{aligned}$$

Evaluating the kth derivative at $t = 0$, we find

$$\begin{aligned} M_X^{(k)}(0) &= \sum_x x^k p(x) \\ &= E(X^k). \end{aligned}$$

The mean and variance of a random variable X may be expressed in terms of the moment-generating function as

$$\mu = M^{(1)}(0), \quad \sigma^2 = M^{(2)}(0) - [M^{(1)}(0)]^2.$$

The advantage of the moment-generating function as a tool for computing moments, and in particular for computing means and variances, occurs when the moment-generating function has a "nice" form. This is the case of the binomial, geometric, and Poisson distributions.

EXAMPLE 3.6-2 (binomial)

If X has a binomial distribution, then

$$\begin{aligned} M_X(t) &= \sum_{x=0}^{n} e^{xt} \binom{n}{x} p^x (1-p)^{n-x} \\ &= \sum_{x=0}^{n} \binom{n}{x} (pe^t)^x (1-p)^{n-x}. \end{aligned}$$

This summation is of the form

$$\sum_{x=0}^{n} \binom{n}{x} (a)^x (b)^{n-x} = (a+b)^n,$$

where $a = pe^t$ and $b = 1 - p$. Thus,

$$M_X(t) = (pe^t + 1 - p)^n.$$

The first and second derivatives of this moment-generating function are

$$\begin{aligned} M_X^{(1)}(t) &= npe^t (pe^t + 1 - p)^{n-1}, \\ M_X^{(2)}(t) &= n(n-1) p^2 e^{2t} (pe^t + 1 - p)^{n-2} + npe^t (pe^t + 1 - p)^{n-1}. \end{aligned}$$

Evaluating these derivatives at $t = 0$, we find

$$M_X^{(1)}(0) = np,$$
$$M_X^{(2)}(0) = n(n-1)p^2 + np.$$

Therefore, $\mu = np$ and $\sigma^2 = n(n-1)p^2 + np - (np)^2 = np(1-p)$.

EXAMPLE **3.6-3**
(geometric)

If X has a geometric distribution, then

$$\begin{aligned} M_X(t) &= \sum_{x=1}^{\infty} e^{xt} p(1-p)^{x-1} \\ &= pe^t \sum_{x=1}^{\infty} [e^t(1-p)]^{x-1} \\ &= \frac{pe^t}{1-e^t(1-p)}. \end{aligned}$$

Note that the summation above is finite provided $e^t(1-p) < 1$ or $t < -\ln(1-p)$. The summation is infinite for other values of t. Letting $q = 1-p$, we obtain

$$M_X^{(1)}(t) = \frac{pe^t}{(1-qe^t)^2},$$
$$M_X^{(2)}(t) = \frac{pe^t(1-qe^t) + 2pqe^{2t}}{(1-qe^t)^3}.$$

Evaluating these derivatives at $t = 0$, we have

$$\begin{aligned} M_X^{(1)}(0) &= \frac{1}{p}, \\ M_X^{(2)}(0) &= \frac{p(1-q) + 2pq}{p^3} \\ &= \frac{2-p}{p^2}. \end{aligned}$$

Thus,

$$\sigma^2 = \frac{2-p}{p^2} - \left(\frac{1}{p}\right)^2 = \frac{1-p}{p^2}.$$

EXAMPLE 3.6-4 (Poisson) If X has a Poisson distribution, then

$$M_X(t) = \sum_{x=0}^{\infty} e^{xt}\left(\frac{e^{-\mu}\mu^x}{x!}\right) = e^{-\mu}\sum_{x=0}^{\infty}\frac{(\mu e^t)^x}{x!} = e^{-\mu}e^{\mu e^t} = e^{\mu(e^t-1)}.$$

The first two derivatives are

$$M_X^{(1)}(t) = \mu e^t e^{\mu(e^t-1)},$$
$$M_X^{(2)}(t) = (\mu e^t)^2 e^{\mu(e^t-1)} + \mu e^t e^{\mu(e^t-1)}.$$

Evaluating the derivatives at $t = 0$, we find

$$M_X^{(1)}(0) = \mu, \quad M_X^{(2)}(0) = \mu^2 + \mu.$$

Therefore,

$$\sigma^2 = \mu^2 + \mu - \mu^2 = \mu.$$

It is possible that some of the moments of a random variable will not exist. For example, if X is a random variable with probability mass function of the form

$$p(x) = \frac{c}{x^6}, \quad x = 1, 2, 3, \ldots,$$

then the series

$$E(X^k) = \sum_{x=1}^{\infty} x^k p(x)$$
$$= \sum_{x=1}^{\infty} c x^{k-6}$$

is finite provided $k < 5$. The moments of the random variable for $k \geq 5$ do not exist. In this example, the moment-generating function is infinite for $t > 0$. Hence, by Definition 3.6-2, the moment-generating function does not exist and cannot be used to generate the moments.

The moment-generating function, when it exists, is unique in the sense that only one distribution can give rise to a moment-generating function of a specific form. For instance, if we are given that a random variable X has the moment-generating function

$$M_X(t) = e^{3(e^t - 1)},$$

then we know the random variable has a Poisson distribution with a mean of 3.

Exercises 3.6

3.6-1 Let X be a Bernoulli random variable with $P(X = 1) = p$ and $P(X = 0) = 1 - p$. Find the moment-generating function of X.

- **a** Use the moment-generating function to find $E(X)$ and $\mathrm{VAR}(X)$.
- **b** Use the moment-generating function to find $E(X^k)$, for all integer values of k.

3.6-2 Let X have the following probability mass function:

x	-1	0	1
$p(x)$	.2	.3	.5

Find the moment-generating function of X and use it to find the first three moments of X.

3.6-3 Let X and Y be discrete, independent random variables with moment-generating functions $M_X(t)$ and $M_Y(t)$, respectively. Prove that the moment-generating function of $Z = X + Y$

is $M_Z(t) = M_X(t)\,M_Y(t)$. (*Hint*: $M_Z(t) = E(e^{(X+Y)t}) = E(e^{Xt}e^{Yt})$. Apply Theorem 2.7-4 to the random variables e^{Xt} and e^{Yt}.)

3.6-4 Suppose that X and Y are independent Poisson random variables with means μ_X and μ_Y, respectively. Use Exercise 3.6-3 to find the moment-generating function of $Z = X + Y$. Recognize that $M_Z(t)$ is the moment-generating function of a Poisson random variable with mean $\mu_X + \mu_Y$. By the uniqueness of the moment-generating function, conclude that Z has a Poisson distribution with mean $\mu_Z = \mu_X + \mu_Y$.

4

Markov Chains

1–6

SECTION

4.1 Introduction: Modeling a Simple Queuing System

A queue is a waiting line. We encounter queues every day in dealing with systems that deliver services to customers. Examples include waiting to be seated at a restaurant, waiting to take the driving test at the motor vehicle office, and waiting on hold when making a telephone call to a busy office. Such systems are called *queuing systems*. Our discussion will begin with computers that break down and are repaired. We model this queuing system by using the simulation tools developed in the previous chapters. Later in this chapter we analyze the system by using the analytical tools developed in this chapter.

The Little Green Computing Machine Company (LGCM) has assigned a technician to handle service for five computers. Each LGCM computer has a probability of .2 of failing during any day in which it is in operation. The failures are independent of one another. The technician can fix one computer a day. When a computer breaks down, it will be fixed the day it fails provided there is no backlog of requests for service. If there is a backlog, the computer will join a service queue and wait until the technician fixes those ahead of it.

We simulate the behavior of this system over five days of operation. The system begins on Monday with no backlog of service and ends on Friday evening. We observe two characteristics of the system:

1 The number of computers waiting for service at the end of each day (the backlog)

2 The number of days in the week the technician is idle

The following algorithm describes the simulation of this system. The steps of the algorithm are illustrated in Table 4.1-1.

Define the variables as follows:

DAY	Day of the week labelled 1,2,...,5
BACKLOG(·)	Array consisting of the backlogs at the end of each day
GOOD	Number of good computers at the beginning of the day
FAIL	Number of computers that fail during the day
IDLEDAYS	Number of days in the week the technician is idle

Initialize the variables as follows:

BACKLOG(0) = 0	No backlog at the beginning of the week
GOOD = 5	All computers are working at the start
FAIL = 0	No failed computers at the start of the day
IDLEDAYS = 0	Count of idle days begins at zero

Repeat the following steps for DAY = 1 to 5.

1 Determine the number of computers that FAIL during the day. This is done by generating a unit random number for each GOOD computer and determining how many of these numbers are less than .2, the probability of failure. Each unit random number less than .2 represents a computer failure.

2 Determine the backlog at the end of the day. This is done as follows: If BACKLOG(DAY – 1) = 0 and FAIL = 0, then BACKLOG(DAY) = 0; else BACKLOG(DAY) = BACKLOG(DAY – 1) + FAIL – 1. (Note that if yesterday's backlog is zero and there is no failure today, then today's backlog is zero. Otherwise, since one computer can be fixed a day, today's backlog will be one less than the sum of yesterday's backlog and the number that fail today).

3 Increment IDLEDAYS as follows: If BACKLOG(DAY – 1) = 0 and FAIL = 0, then IDLEDAYS = IDLEDAYS + 1. (Note that the number of days idle is incremented by 1 only if there is no work to be done, and this occurs only if yesterday's backlog is zero and there are no failures today).

4 Update the number of GOOD computers as follows: GOOD = 5 – BACKLOG(DAY)

An important idea in simulation, and indeed an important idea in all experimental work, is that of *replication*. By replication, we mean repeating the experiment more than once. Table 4.1-1 represents a single experiment. If we were to do the simulation a second time, starting out the same way, the results probably would not be exactly the same as the first. The backlogs and days idle generally differ from one set of five days to the next. To see how these variables could change, we must do the simulation several times. If we do the simulation enough times, the patterns of behavior of the various variables of interest will become clear.

TABLE **4.1-1** Simulation of the Simple Queuing System

DAY	Unit Random Numbers					Status at end of day FAIL	BACKLOG	IDLEDAYS	GOOD
1	*.10480*	.22368	.24130	.42167	.37570	1	0	0	5
2	.77921	.99562	.96301	.89579	.85475	0	0	1	5
3	.29818	.63553	*.09428*	*.10365*	*.07119*	3	2	1	3
4	.51085	*.02368*	*.01011*			2	3	1	2
5	.48663	.54164				0	2	1	3

Note: Italic unit random numbers correspond to failed computers.

To show the idea of replication, we repeated the simulation of this queuing system 1000 times. This yielded 1000 values of BACKLOG(·) for each day of the week and 1000 values of IDLEDAYS. Using these values, we constructed empirical probability distributions for the daily backlogs and for the idle days. The results are shown in Table 4.1-2.

We can learn much about the behavior of the system by studying the empirical distributions. To illustrate, consider the distribution of backlogs on Wednesday. We see that the empirical probability of zero is .584. This tells us that there was no backlog on Wednesday evening in 584 of the 1000 repetitions of the experiment. Similarly, in 266 of the 1000 repetitions of the experiment there was a backlog of one computer on Wednesday evening. The other empirical probabilities are also interpreted in this manner. The sample mean of the backlog obtained from the empirical distribution for Wednesday is .59 with a sample standard deviation of .80. Thus, there was an average backlog of less than one computer a day on Wednesday evening.

TABLE **4.1-2** Empirical Distributions for Simulation of Simple Queuing System

DAY	BACKLOG	Empirical Probabilities 0	1	2	3	4	$\bar{X}$	S
Mon.	BACKLOG(1)	.741	.205	.050	.004	.000	.32	.58
Tue.	BACKLOG(2)	.616	.273	.091	.019	.001	.52	.75
Wed.	BACKLOG(3)	.584	.266	.125	.025	.000	.59	.80
Thr.	BACKLOG(4)	.541	.295	.132	.030	.002	.66	.83
Fri.	BACKLOG(5)	.509	.312	.143	.034	.002	.71	.85

IDLEDAYS	Empirical Probabilities 0	1	2	3	4	5	$\bar{X}$	S
	.343	.316	.220	.098	.019	.004	1.15	1.08

Notice the trends in the empirical probabilities of the number of computers waiting for service. The distributions change quite a bit in going from Monday to Tuesday. For example, the probability of finding no backlog drops from .741 to .616. The empirical distributions, sample means, and sample standard deviations change less dramatically after that. The system is "settling down" to a *steady-state* condition in which changes in the probability distributions across time become negligible. This type of behavior is common in many queuing systems and is discussed in detail in this chapter.

Again looking at the backlog in the system, we see that on every day there is over an 80% chance that the backlog will be either 0 or 1, and the chances are relatively small that there are three or four computers waiting to be repaired. If at most a day's delay can be tolerated by the computer users, then the LGCM Company is providing adequate service. On the other hand, if it is important not to have any backlog, then the LGCM is not doing so well. For instance, the chance for no backlog on Friday is only around 50%. If fast service is essential, LGCM needs to hire more technicians.

Balanced against service to the customer is the cost of the service. One measure of cost is the amount of time the technician is idle. We see that in 343 weeks out of 1000 the technician was busy every day. It is much more likely that the technician is idle at least one day of the week. Thus, in order to provide the level of service to the customer that would keep backlogs low, the company must allow for the possibility that the technician will spend at least part of the week idle. Having quantitative information on such things as idle time and backlog can help the LGCM Company evaluate the costs and quality of service.

It is often the case in simulation studies that we play the "what if" game. What if the probability of breakdown could be reduced from .2 to .1? How would that affect the behavior of the system? What if another technician were added? Could we be reasonably assured that there would be no backlog? What if the backlog could not be handled over the weekend but had to be carried forward to the following week? Would this seriously affect the quality of service? Questions like this can be answered by simulation simply by making changes in the computer program to reflect changing circumstances.

The distribution of backlogs in this simple system can be obtained theoretically— that is, without the aid of simulation. We will do so later. However, it is not hard to make changes in the system so that it is not amenable to theoretical analysis. For example, the probability of a breakdown could vary over time and the computers could fail at different rates. In such cases, theoretical solutions can become prohibitive. On the other hand, with just the rudiments of probability theory one can model such systems through simulation.

Exercises 4.1

4.1-1 . Write a program to produce results analogous to those in Table 4.1-2. Simulate 1000 weeks of operation of the system, assuming that each Monday begins with no backlog from the previous Friday. Although your empirical probabilities will not be the same as those in Table 4.1-2, they should be generally within ±.05 of those values. Change the probability of a breakdown from .2 to .1 and rerun your program. Compare your results with those given in the text.

4.1-2 Change the queuing example of this section so that there are two technicians instead of one. Assume each can fix one computer per day. Using a table of random numbers, simulate the behavior of the system for one week. Then repeat for a second and third week. Do you think there is much chance of having a backlog on any given day?

4.1-3 Write a computer program to simulate the system described in Exercise 4.1-2. Repeat the experiment 1000 times, and obtain empirical probability distributions for the backlogs on Monday through Friday.

4.1-4 Modify the queuing example of this section so that the backlog from Friday is carried over to Monday. Obtain the backlog at the end of the second week. Write a program to repeat this experiment 1000 times. Find an empirical distribution for the backlog at the end of the second week. Compare to the distribution of the backlog at the end of the first week.

4.1-5 Consider a queuing system with 10 computers and two technicians. Let the assumptions for failures and repairs be the same as those in the example of this section. Write a program to obtain empirical distributions of the backlogs on Monday through Friday.

4.1-6 Modify the queuing example of this section so that the technician is not certain to fix a computer on a given day. Assume that at most one computer can be fixed per day, and if there is a computer to be fixed the technician has a probability of .7 of completing service on the computer.

- **a** Use a table of random numbers to simulate this system for a week. Then repeat for a second and third week.
- **b** Write a computer program for modeling this system. Obtain empirical distributions for the backlogs on Monday through Friday based on 1000 repetitions of the experiment.

SECTION

4.2 The Markov Property

Random events often cause conditions of a system to change with time. The random breakdowns of the computers in the queuing system in Section 4.1 cause the backlogs of broken computers to fluctuate from day to day. Randomness in the demand for automobiles causes a car dealer's inventory to vary from week to week. Random electrical disturbances in the environment cause radio signals to fluctuate, producing static in the receiver. Processes that fluctuate with time as a result of random events acting upon a system are called *stochastic processes*. A stochastic process is a collection of random variables denoted $\{X(n), n \in N\}$. In practice, the index set N usually refers to time, and the values of $X(n)$ represent measurements or observations on a system at time n. The possible values of $X(n)$ are called the *states of the process*.

EXAMPLE **4.2-1** Consider the queuing system in Section 4.1. If the backlogs are the quantities observed, then the states of this process are 0, 1, 2, 3, 4.

A stochastic process may be put into one of four broad categories. These categories are determined by how time is measured and by how the states of the process are classified. Time can be measured on a discrete scale or on a continuous scale, and the states can be discrete or continuous. For example, if time is measured in days and states are the number of computers waiting for service, the process is a discrete-time and discrete-state process. On the other hand, atmospheric temperature, which changes continuously with time, is an example of a continuous-time and continuous-state process. The number of calls arriving at a switchboard during the day would be a continuous-time and discrete-state process provided that the calls can be received at any instant during the day. The distance traveled by a truck driver each day is a discrete-time (days), continuous-state (distance) process.

We limit our attention to discrete-time, discrete-state processes. This restriction is not as severe as it might seem. It is often possible to approximate the behavior of a process having continuous time or continuous states by one having discrete time and discrete states. For example, atmospheric temperature may be approximated with a discrete-time, discrete-state process by taking readings every minute and measuring temperature to the nearest degree. Thus, the techniques developed here have wide applicability.

Depending on the nature of the process, the states may be nonnumerical or numerical quantities. In queuing systems, for example, the states are often taken to be the number of customers waiting for service. If we are modeling the movement of a package in an airline delivery system, the states could be the possible locations of the package at any given time. We arbitrarily assign numerical values to nonnumerical states so that the state of the system at any time can be thought of as a random variable.

The random variables representing the states of a discrete-time stochastic process are denoted as $X(n)$, for times $n = 1, 2, \ldots$. The starting value of a process is called the *initial state* and is denoted $X(0)$. For instance, if $X(n)$ is the price of a stock on the nth day after it was purchased, then $X(0)$ would be the purchase price of the stock.

We are interested in modeling the relationships among the $X(n)$'s as the process moves through time. In the simplest systems, the $X(n)$'s are independent. That is, the outcome at one time is not affected by the outcomes at other times. Repeated tosses of a fair die are like that. In other systems, the outcome at one time may be affected by all previous outcomes. For example, suppose we draw numbers randomly, one after the other, without replacement from a box containing the integers $1, 2, \ldots, 1000$. Clearly, the numbers that may be drawn in the future depend on all numbers that have been previously drawn.

We study processes that possess the *Markov property*, named for a Russian mathematician. In a stochastic process having the Markov property, each outcome depends only on the one immediately preceding it. This concept will be defined more precisely.

Definition 4.2-1

For times $n = 0, 1, 2, \ldots$ let $\{X(n)\}$ denote a stochastic process and let $\{s(n)\}$ denote any collection of states of the process. The process is said to satisfy the Markov property if

$$P[X(n+1) = s(n+1) \mid X(n) = s(n), X(n-1) = s(n-1), \ldots, X(0) = s(0)]$$
$$= P[X(n+1) = s(n+1) \mid X(n) = s(n)]$$

for $n = 0, 1, 2, \ldots$.

A stochastic process that satisfies the Markov property is called a *Markov process*. Such processes are also called *Markov chains*. This term refers to the fact that each outcome is linked to the one immediately preceding it.

EXAMPLE 4.2-2 Many board games have the Markov property. That is, the next position of a token on the board depends only on the present position and the roll of the dice.

EXAMPLE 4.2-3 Consider a simple game of chance in which a fair coin is tossed repeatedly. We win a dollar each time heads occurs and lose a dollar each time tails occurs. Let $X(n)$ be our accumulated winnings after the nth toss of the coin, where a negative amount indicates a net loss and $X(0) = 0$. In this example the $X(n)$'s have the Markov property. To illustrate, consider a special case. Suppose that we have accumulated \$6 after 10 tosses of the coin. It makes no difference what happened on the previous trials to get us to this point. Our accumulated winnings after the 11th trial will either be \$7 or \$5 with probability $1/2$ each. That is, our accumulated winnings after the 11th trial depend only on the accumulated winnings after the 10th trial. Since this is clearly the case for all other trials, the process satisfies the Markov property.

EXAMPLE 4.2-4 This example has the same probability structure as the previous example, but it is put in a different context. A fair coin is tossed repeatedly. Each time heads occurs, a person steps one unit to the right, and each time tails occurs the person steps one unit to the left. Let $X(n)$ be the person's position after the nth toss of the coin. As in the previous example, $X(n)$ satisfies the Markov property. That is, the person's next position will depend only on the present position and not on any other past positions. This particular Markov process is called a *random walk*.

A random walk can be used as a model to approximate the one-dimensional movement of a particle suspended in a fluid. As the particle collides with other particles, it moves randomly back and forth much as the person would move back and forth according to the toss of the coin. Of course, these movements occur rapidly and are measured on a microscopic scale.

Pictorial representations of the possible changes of state, or transitions, that a Markov chain can make are sometimes useful in describing a system. A *state diagram* shows the possible states of the process and the transitions that may be made from each state in one period of time. The state diagram of a random walk is shown in Figure 4.2-1.

FIGURE **4.2-1** State Diagram for the Random Walk

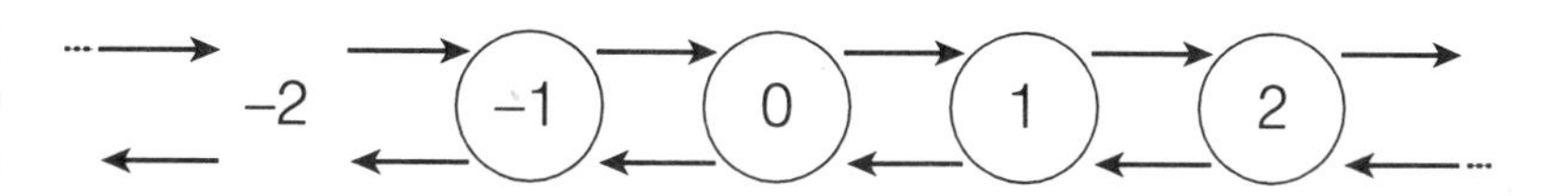

EXAMPLE **4.2-5** In the queuing problem considered in Section 4.1, the number of computers waiting for service at the end of the day has the Markov property. Tomorrow's backlog depends only on today's backlog and the number that happen to fail tomorrow. Figure 4.2-2 contains a partial state diagram representing the transitions that can be made into and out of state 1, one computer waiting for service.

FIGURE **4.2-2** Partial State Diagram for Transitions into and out of State 1 for Example 4.2-5

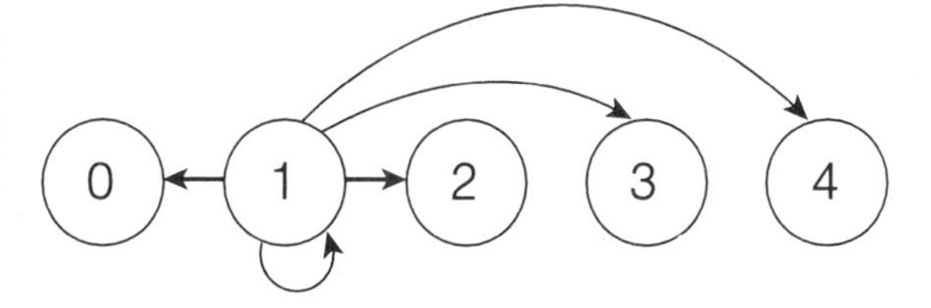

EXAMPLE **4.2-6** This is an example of a stochastic process for which the Markov property would seem unlikely to hold. Consider an individual who is thinking about buying a new car. Whether this person buys a new car will surely depend on when he or she last purchased a car. Suppose $\{X(n)\}$, $n = 0, 1, 2, 3,$ denotes an individual's car buying history for four years, where $X(n) = 1$ if a car was bought in year n and $X(n) = 0$ otherwise. If $\{X(n)\}$ were a Markov process, then

$$\begin{aligned} P[X(3) = 1 \mid X(2) = 0] &= P[X(3) = 1 \mid X(2) = 0, X(1) = 1, X(0) = 0] \\ &= P[X(3) = 1 \mid X(2) = 0, X(1) = 0, X(0) = 1]. \end{aligned}$$

These equalities would imply that a person who bought a car two years ago would be as likely to buy a new car this year as someone who bought a new car three years ago. For most car buyers, this seems like an unreasonable assumption. Therefore, it is unlikely that the preceding equalities hold, and thus unlikely that $\{X(n)\}$ would satisfy the Markov property.

In some stochastic processes it may be unclear whether the Markov property holds. For example, some would argue that the changes in stock prices have the Markov property. That is, if we know the present level of the stock, past levels will give us no additional knowledge for predicting tomorrow's prices. Do you believe this? It is beyond the scope of this text to give methods for inferring whether such systems satisfy the Markov property. In some modeling problems, the Markov property is an assumption rather than a verifiable fact.

Definition 4.2-2

The *one-step transition probability* is the conditional probability defined as

$$P_n(i \to j) = P[X(n+1) = j \mid X(n) = i].$$

In words, the one-step transition probability is the probability of going from state i to state j in one time period beginning at time n. The one-step transition probability plays a key role in the theory and application of Markov chains.

EXAMPLE 4.2-7 Consider the one-step transition probabilities for the game of chance in Example 4.2-3. If the amount of our winnings at time n is i, then the amount of our winnings at time $n + 1$ will be either $j = i + 1$ or $j = i - 1$, each with probability 1/2. Thus,

$$P_n(i \to j) = \begin{cases} \frac{1}{2}, & j = i+1 \text{ or } j = i-1, \\ 0, & \text{otherwise.} \end{cases}$$

In the example, the transition probabilities do not depend on time n. A process with this property is called a *time-homogeneous process*. In such cases, the subscript n is suppressed, and the one-step transition probabilities are denoted $P(i \to j)$. If the transition probabilities depend on time, the process is said to be *nonhomogeneous*.

EXAMPLE 4.2-8 A modification of the game of chance in Example 4.2-3 will be used to illustrate a nonhomogeneous process. In this game, one coin is tossed at time $n = 0$, two coins are tossed at time $n = 1$, and so forth. We win \$1 if all coins are heads and lose \$1 otherwise. The one-step transition probabilities are

$$P_n(i \to i+1) = \left(\frac{1}{2}\right)^{n+1} \quad \text{(all heads appear)},$$

$$P_n(i \to i-1) = 1 - \left(\frac{1}{2}\right)^{n+1} \quad \text{(at least one tail appears)},$$

$$P_n(i \to j) = 0, \quad j \neq i+1 \text{ or } i-1.$$

This is a nonhomogeneous process because the nth step transition probabilities depend on n.

EXAMPLE 4.2-9 During any minute of time, the number of calls coming into the switchboard at the statistics office is a random variable having one of three possible values: 0,1, or 2. The respective probabilities are .80, .15, and .05. Assume the number of calls during one minute is independent of the number of calls during any other minute. Let $X(n)$ be the total number of calls after n minutes of time. The possible states of the process are $0, 1, 2, 3, \ldots$. Suppose that i calls have come into the switchboard in n minutes of time. The only possibilities for $X(n+1)$ are i, $i+1$, and $i+2$, corresponding to no calls, one call, or two calls coming into the switchboard during the next minute. This is a time-homogeneous process with one-step transition probabilities

$$\begin{aligned} P(i \to i) &= .80, \\ P(i \to i+1) &= .15, \\ P(i \to i+2) &= .05, \\ P(i \to j) &= 0, \quad \text{for } j \neq i,\ i+1, \text{ or } i+2. \end{aligned}$$

Typically the rate at which calls arrive at a switchboard will vary with time of day. For instance, the probabilities of 0, 1, or 2 calls might apply to the morning hours but change to .60, .30, and .10 in the afternoon hours. In this case, the transition probabilities of .80, .15, and .05 would be replaced by .60, .30, and .10 for afternoon hours. This change would make the process nonhomogenous.

EXAMPLE 4.2-10 A sequential search file consists of K items stored in K locations, one item in each location. When a request is made for an item, a sequential search is conducted, beginning with the first location and continuing until the item is found. It is advantageous for the most frequently requested items to be in the first locations of the file in order to reduce the amount of time necessary for searching. A problem occurs when it is not known which items will be the most frequently requested. One solution to the problem is to move a requested item to the first location and to move those previously ahead of it down one location in the file. With this method, those items most frequently requested should remain near the top of the file while those least frequently requested should remain near the bottom of the file.

The movement of items among the locations of the file can be put into the framework of a Markov chain. To illustrate, suppose that there are just three items, A, B, and C, stored in locations 1, 2, and 3. Each request for an item is made independently of other such requests. Let $P(A)$, $P(B)$, and $P(C)$ denote the probabilities of requests for A, B, and C, respectively. The states of the system will be the six possible arrangements of the items A, B, and C in the three locations as listed in the table.

	Location		
State	1	2	3
1	A	B	C
2	A	C	B
3	B	A	C
4	B	C	A
5	C	A	B
6	C	B	A

There are 36 possible transitions to consider, but not all can occur. For example, it would be impossible to make a transition from state 1 to state 2 because item C is not moved to the front of the list. There are, in fact, only three transitions that can be made from state 1. These are

$$\begin{aligned} P(1 \to 1) &= P(ABC \to ABC) = P(A \text{ is requested item}) = P(A), \\ P(1 \to 3) &= P(ABC \to BAC) = P(B \text{ is requested item}) = P(B), \\ P(1 \to 5) &= P(ABC \to CAB) = P(C \text{ is requested item}) = P(C). \end{aligned}$$

The transition probabilities from the other states are left as an exercise.

A useful way to represent the transition probabilities of a time-homogenous Markov chain is with a *one-step transition matrix*. The one-step transition matrix is a matrix whose elements correspond to the one-step transition probabilities.

EXAMPLE 4.2-11 Let P denote the one-step transition matrix for a three-state Markov process. Suppose

$$P = \begin{array}{c} \\ 1 \\ 2 \\ 3 \end{array} \begin{array}{c} \begin{array}{ccc} 1 & 2 & 3 \end{array} \\ \begin{bmatrix} .30 & .10 & .60 \\ .75 & .20 & .05 \\ 0 & .55 & .45 \end{bmatrix} \end{array}.$$

The row index represents the present state, and the column index represents the next state. For example, the probability that the process moves from state 3 to state 2 in one step is $P(3 \rightarrow 2) = .55$, whereas the probability of moving from state 3 to state 1 in one step is zero.

EXAMPLE 4.2-12 A transition matrix can have infinite dimensions if there are infinitely many states in the process. Consider the transition matrix for the number of phone calls in Example 4.2-9. It can be expressed as

$$P = \begin{array}{c} \\ 0 \\ 1 \\ 2 \\ 3 \\ \vdots \end{array} \begin{array}{c} \begin{array}{cccccc} 0 & 1 & 2 & 3 & 4 & \cdots \end{array} \\ \begin{bmatrix} .80 & .15 & .05 & 0 & 0 & \cdots \\ 0 & .80 & .15 & .05 & 0 & \cdots \\ 0 & 0 & .80 & .15 & .05 & \cdots \\ 0 & 0 & 0 & .80 & .15 & \cdots \\ \vdots & \vdots & \vdots & \vdots & \vdots & \vdots \end{bmatrix} \end{array}.$$

Exercises 4.2

4.2-1 An insect is put in a circular box with six compartments as shown. Each time the insect makes a move, it moves to an adjacent compartment. Assume it has a probability of 1/2 of moving either direction. Let $X(n)$ be the compartment occupied by the insect after n moves. Give the one-step transition matrix for $X(n)$.

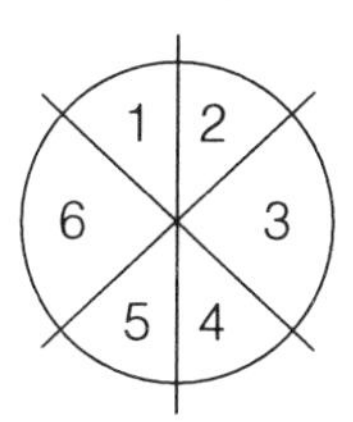

4.2-2 Four balls, numbered 1 to 4, are distributed among two compartments in a box as shown. An integer from 1 to 4 is selected at random, and the ball corresponding to that integer is moved from the compartment it is in to the other compartment. This procedure is repeated again and again. Let $X(n)$ be the number of balls in compartment 1 after the nth ball has been moved. Give the transition matrix for $X(n)$.

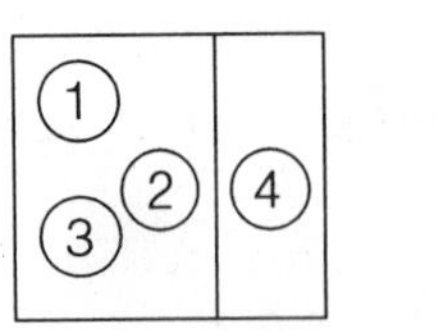

4.2-3 Modify Exercise 4.2-2 so that there are K balls instead of four. Find the one-step transition probabilities when there are i balls in compartment 1. This movement of balls between compartments is a simplified model for the random movement of molecules between adjacent cells across a permeable membrane.

4.2-4 Consider a game of random ladder climbing of the type that might be part of a video game. A person begins on the bottom rung of a ladder. Each time heads appears in a toss of a fair coin, the person moves up one rung. Each time tails is tossed, the person moves to the bottom rung and starts over. Once at the top, the person stays there. Suppose there are five rungs on the ladder. Let $X(n)$ be the person's position after the nth toss of the coin. Give the one-step transition matrix for $X(n)$.

4.2-5 In going from one generation to the next, a particular type of cell will either divide and produce two new cells or die. The probability of a successful division is .4.

- **a** If there are now two such cells, give the number of possible cells that may be in the next generation, and give the possible transition probabilities.
- **b** If there are now i such cells, give the number of possible cells that may be in the next generation, and give the possible transition probabilities.

4.2-6 Give the one-step transition probability matrix for Example 4.2-10 and draw a state diagram.

4.2-7 Suppose Example 4.2-10 is modified so that the following rule is used to interchange items. Each time an item is selected, its position is interchanged with the one immediately ahead of it instead of being moved to the top. Give the one-step transition probabilities for this process.

4.2-8 A state diagram of a Markov chain is shown. Such a diagram might represent the modules through which data could move during the execution of a program or the flow of paperwork through an office. Assume that each path leaving a state has an equal chance of being selected. Find the one-step transition matrix of the chain.

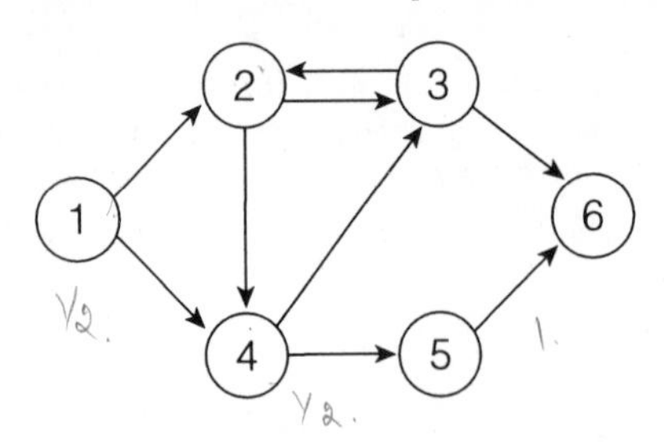

4.2-9. Suppose a box contains the numbers 10, 20, 30, 40, and 50. Numbers are drawn from the box one after the other without replacement. Let $X(n)$ be the number selected on the nth draw. Verify that $\{X(n)\}$, $n = 1, 2, \ldots, 5$, is not a Markov process.

SECTION

4.3 Computing Probabilities for Markov Chains

In our discussion we assume that the Markov chains are time homogeneous. Consider computing the probability that a Markov chain visits states $s(1), s(2), \ldots, s(n)$ at times $1, 2, \ldots, n$ given that the chain begins in state $s(0)$. A sequence of states through which a process may move is called a *path* of the process. It follows from the Markov property that the probability of a path is just the product of one-step transition probabilities. That is,

$$P[X(1) = s(1), X(2) = s(2), X(n) = s(n) \mid X(0) = s(0)]$$
$$= P[s(0) \to s(1)]\, P[s(1) \to s(2)] \cdots P[s(n-1) \to s(n)].$$

To demonstrate the validity of this result, consider the case $n = 2$. From the Multiplication Rule of conditional probability, if we have events A, B, and C then

$$\begin{aligned} P(A \cap B \cap C) &= P(C \mid A \cap B)\, P(A \cap B) \\ &= P(C \mid A \cap B)\, P(B \mid A)\, P(A). \end{aligned}$$

Applying this to the case $n = 2$, we have

$$P[X(2) = s(2), X(1) = s(1), X(0) = s(0)]$$
$$= P[X(2) = s(2) \mid X(1) = s(1), X(0) = s(0)]$$
$$\times P[X(1) = s(1) \mid X(0) = S(0)]\, P[X(0) = s(0)]$$

Since the process is a Markov chain,

$$P[X(2) = s(2) \mid X(1) = s(1), X(0) = s(0)] = P[s(1) \to s(2)].$$

By definition, $P[X(1) = s(1) \mid X(0) = s(0)] = P[s(0) \to s(1)]$. Therefore,

$$P[X(2) = s(2), X(1) = s(1), X(0) = s(0)]$$
$$= P[s(1) \to s(2)]\, P[s(0) \to s(1)]\, P[X(0) = s(0)],$$

and

$$P[X(2) = s(2), X(1) = s(1) \mid X(0) = s(0)]$$
$$= \frac{P[X(2) = s(2), X(1) = s(1), X(0) = s(0)]}{P[X(0) = s(0)]}$$
$$= P[s(0) \to s(1)]\, P[s(1) \to s(2)].$$

A similar argument can be applied in the general case.

EXAMPLE 4.3-1 A rat moves through the maze shown in Figure 4.3-1. Each time the rat makes a move from one compartment to another, it chooses a passageway at random. Let $X(n)$ be the compartment occupied by the rat after n moves. Time in this case is measured by the moves of the rat rather than by the clock.

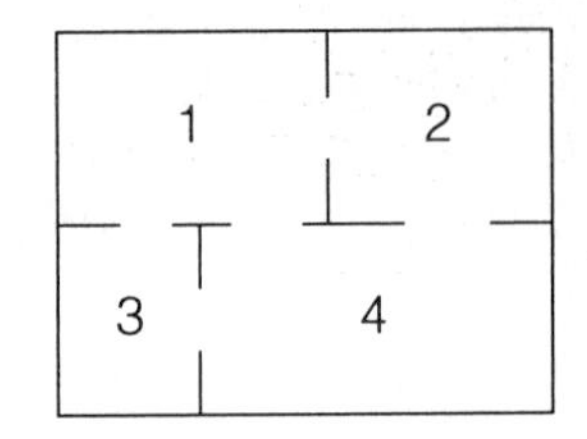

FIGURE 4.3-1 Maze for Example 4.3-1

The one-step transition matrix is

$$P = \begin{array}{c} \\ 1 \\ 2 \\ 3 \\ 4 \end{array} \begin{array}{c} \begin{array}{cccc} 1 & 2 & 3 & 4 \end{array} \\ \begin{bmatrix} 0 & 1/3 & 1/3 & 1/3 \\ 1/2 & 0 & 0 & 1/2 \\ 1/2 & 0 & 0 & 1/2 \\ 1/3 & 1/3 & 1/3 & 0 \end{bmatrix} \end{array}.$$

If the rat begins in compartment 1, the probability that it first goes to 2, and then to 4, and then back to 2 is

$$P(1 \to 2)\,P(2 \to 4)\,P(4 \to 2) = \left(\frac{1}{3}\right)\left(\frac{1}{2}\right)\left(\frac{1}{3}\right) = \frac{1}{18}.$$

If it begins in compartment 2, the probability that it goes to 1, then to 3, and then to 4 is

$$P(2 \to 1)\,P(1 \to 3)\,P(3 \to 4) = \left(\frac{1}{2}\right)\left(\frac{1}{3}\right)\left(\frac{1}{2}\right) = \frac{1}{12}.$$

In some cases, the initial state is not constant but is a random variable. The probability mass function of the initial state $X(0)$, which is called the *initial distribution* of the process, would then figure into the computation of the probabilities of various paths.

EXAMPLE 4.3-2 Suppose that the rat in Example 4.3-1 has a probability of 1/4 of starting in any of the compartments. That is, the initial distribution of the starting point of the rat assigns probability 1/4 to all the compartments. The probability that the rat begins in compartment 1 and traverses the path $1 \to 2 \to 4 \to 2$ is

$$\frac{1}{4}[P(1 \to 2)\,P(2 \to 4)\,P(4 \to 2)] = \frac{1}{4}\left[\left(\frac{1}{3}\right)\left(\frac{1}{2}\right)\left(\frac{1}{3}\right)\right] = \frac{1}{72}.$$

Instead of the probability of a particular path, we might be interested in the probability that the process will be in state j a fixed number of time periods after it was in state i.

Definition 4.3-1

The *k-step transition probability* is defined to be

$$P^{(k)}(i \to j) = P[X(n+k) = j \mid X(n) = i].$$

This transition probability is the probability of being in the jth state k time periods after being in the ith state. Since we are dealing with time-homogenous processes, the k-step transition probability does not depend on time n. The next example illustrates the computation of two-step transition probabilities.

EXAMPLE 4.3-3 Referring to Example 4.3-1, consider the probability that the rat ends up in compartment 1 two moves after it starts from compartment 1. There are three possible paths the rat can follow in going from 1 to 1 in two moves. These paths are $1 \to 2 \to 1$, $1 \to 3 \to 1$, and $1 \to 4 \to 1$. Thus,

$$\begin{aligned} p^{(2)}(1 \to 1) &= P(1 \to 2)P(2 \to 1) + P(1 \to 3)P(3 \to 1) + P(1 \to 4)\,P(4 \to 1) \\ &= \left(\frac{1}{3}\right)\left(\frac{1}{2}\right) + \left(\frac{1}{3}\right)\left(\frac{1}{2}\right) + \left(\frac{1}{3}\right)\left(\frac{1}{3}\right) \\ &= \frac{4}{9}. \end{aligned}$$

There is an interesting way to obtain a general expression for two-step transition probabilities. Imagine making a list of the possible paths that the process can follow in going from i to j in two steps. These paths can be expressed in the form $i \to s \to j$, where s is the state of the process after one step. Compute the probability of all such paths and sum the results to obtain $P^{(2)}(i \to j)$. Thus,

$$P^{(2)}(i \to j) = \sum_{s} P(i \to s) P(s \to j),$$

where the summation is taken over the possible states of the process. What this expression shows is that $P^{(2)}(i \to j)$ is the i,jth element of the square of the one-step transition matrix. Thus, squaring the one-step transition matrix is a convenient way to determine the two-step transition probabilities. This matrix is called the *two-step transition matrix*.

EXAMPLE 4.3-4 The square of the one-step transition matrix in Example 4.3-1 is

$$P^2 = \begin{array}{c} \\ 1 \\ 2 \\ 3 \\ 4 \end{array} \begin{array}{c} \begin{array}{cccc} 1 & 2 & 3 & 4 \end{array} \\ \begin{bmatrix} 4/9 & 1/9 & 1/9 & 1/3 \\ 1/6 & 1/3 & 1/3 & 1/6 \\ 1/6 & 1/3 & 1/3 & 1/6 \\ 1/3 & 1/9 & 1/9 & 4/9 \end{bmatrix} \end{array}$$

The element in the $(1,1)$ position is the probability that a transition is made from compartment 1 to 1 in two steps. This probability is $4/9$, which agrees with the computations in Example 4.3-3. Likewise, the probabilities of being in states 2, 3, and 4 two moves after beginning in state 1 are $1/9, 1/9$, and $1/3$, respectively. If we read across rows 2, 3, and 4, we find the two-step probabilities starting in states 2, 3, and 4, respectively.

THEOREM 4.3-1 The k-step transition probabilities are obtained by raising the one-step transition matrix to the kth power.

The matrix P^k is called the *kth-step transition matrix*. Theorem 4.3-1 is illustrated in the following example.

EXAMPLE 4.3-5 The fourth power of the transition matrix in Example 4.3-1 is

$$P^4 = \begin{array}{c} \\ 1 \\ 2 \\ 3 \\ 4 \end{array} \begin{array}{c} \begin{array}{cccc} 1 & 2 & 3 & 4 \end{array} \\ \begin{bmatrix} .347 & .160 & .160 & .333 \\ .241 & .259 & .259 & .241 \\ .241 & .259 & .259 & .241 \\ .333 & .160 & .160 & .346 \end{bmatrix} \end{array}.$$

If the rat begins in compartment 1, the elements of the first row of this matrix tell us that it has probabilities .347, .160, .160, and .333 of being in compartments 1, 2, 3, or 4 in four moves, respectively. The other rows of this matrix are interpreted similarly.

EXAMPLE 4.3-6 A signal, either 0 or 1, is sent through a series of five relay stations after it leaves the source. Errors in transmission can cause the signal to change when it is sent from one place to the next. The changes behave as a Markov chain with one-step transition matrix

$$P = \begin{array}{c} \\ 0 \\ 1 \end{array} \begin{array}{c} \begin{array}{cc} 0 & 1 \end{array} \\ \begin{bmatrix} .95 & .05 \\ .01 & .99 \end{bmatrix} \end{array}.$$

That is, in going from transmitter to receiver, the probability that a 0 is received given that a 0 was transmitted is .95, and the probability that a 1 is received given that a 1 was transmitted is .99. The two-step through five-step transition matrices are

$$P^2 = \begin{array}{c} \\ 0 \\ 1 \end{array} \begin{array}{c} \begin{array}{cc} 0 & 1 \end{array} \\ \begin{bmatrix} .90 & .10 \\ .02 & .98 \end{bmatrix} \end{array},$$

$$P^3 = \begin{array}{c} \\ 0 \\ 1 \end{array} \begin{array}{c} \begin{array}{cc} 0 & 1 \end{array} \\ \begin{bmatrix} .86 & .14 \\ .03 & .97 \end{bmatrix} \end{array},$$

$$P^4 = \begin{array}{c} \\ 0 \\ 1 \end{array} \begin{array}{c} \begin{array}{cc} 0 & 1 \end{array} \\ \begin{bmatrix} .82 & .18 \\ .04 & .96 \end{bmatrix} \end{array},$$

$$P^5 = \begin{matrix} & \begin{matrix} 0 & 1 \end{matrix} \\ \begin{matrix} 0 \\ 1 \end{matrix} & \begin{bmatrix} .78 & .22 \\ .04 & .96 \end{bmatrix} \end{matrix}.$$

These transition matrices contain the probabilities that 0 and 1 are received at stations 1 through 5, respectively, given that either a 0 or a 1 was initially transmitted. Consider the five-step transition matrix, for instance. If a 0 was transmitted initially, the signal has probability .78 of being received as a 0 at the fifth station. Similarly, if the original signal was a 1, it has probability .96 of being received as a 1 after five transmissions.

Suppose that 80% of the time a 0 is initially transmitted and 20% of the time a 1 is initially transmitted. The probability of a correct signal being received at station 5 is

$$\begin{aligned} P[X(0) = 0]\,P^5(0 \to 0) + P[X(0) = 1]\,P^5(1 \to 1) &= .8\,(.78) + .2\,(.96) \\ &= .816. \end{aligned}$$

In using a k-step transition matrix, we assume we know the initial state. Matrix computations can also be used to find probabilities when the initial state is a random variable. For example, to find the probability that the chain is in state i after one step, we add up the probabilities of all paths that lead from the initial state to state i. Thus,

$$P[X(1) = i] = \sum_s P[X(0) = s]P(s \to i),$$

where the summation is over the possible states s in the process. Let $\pi(n)$ be a row vector whose entries consist of the probabilities of the form $P[X(n) = s]$ for all possible states. The foregoing computation shows

$$\pi(1) = \pi(0)P,$$

where P is the one-step transition matrix. Likewise, the probabilities for the chain after k steps can be computed as

$$\pi(k) = \pi(0)P^k.$$

EXAMPLE **4.3-7** Consider the signal transmission problem in Example 4.3-6 with $P[X(0) = 0] = .8$. Then the probability that a 0 and 1 are received at station 5 can be found by multiplying the initial probability vector (.80, .20) on the right by the five-step transition matrix.

That is,

$$\pi(0)P^5 = (.80, .20)\begin{bmatrix} .78 & .22 \\ .04 & .96 \end{bmatrix}$$
$$= (.632, .368).$$

Thus, station 5 has probability .632 of receiving a 0 and .368 of receiving a 1.

Exercises 4.3

4.3-1 A machine can be in one of two states: up (1) or down (0). The machine changes state from one hour to the next according to a Markov chain with one-step transition matrix

$$P = \begin{array}{c} \\ 0 \\ 1 \end{array}\begin{array}{c} \begin{array}{cc} 0 & 1 \end{array} \\ \begin{bmatrix} .5 & .5 \\ .8 & .2 \end{bmatrix} \end{array}.$$

a Suppose the machine is up this hour. Find the probability that it is up the next 3 hours and down 2 hours in a row after that.

b If the machine is up this hour, what is the probability that it is down 2 hours later? If the machine is down this hour, what is the probability it is up 2 hours later?

4.3-2 Suppose that the pattern of sunny days and cloudy days follows a Markov chain with one-step transition matrix

$$P = \begin{array}{c} \\ \text{Sunny} \\ \text{Cloudy} \end{array}\begin{array}{c} \begin{array}{cc} \text{Sunny} & \text{Cloudy} \end{array} \\ \begin{bmatrix} .7 & .3 \\ .5 & .5 \end{bmatrix} \end{array}.$$

If it is sunny today, what is the probability that it will be sunny three days from now?

4.3-3 Suppose the rat in the maze in Example 4.3-1 has a probability of 1/4 of starting in any compartment.

a Find the probability that it is in each of the four compartment after one move.

b Find the probability that it is in each of the four compartments after two moves.

4.3-4 A taxicab moves between the airport, Hotel A, and Hotel B according to a Markov chain with transition matrix

$$P = \begin{array}{c} \\ \text{Airport} \\ \text{Hotel A} \\ \text{Hotel B} \end{array}\begin{array}{c} \begin{array}{ccc} \text{Airport} & \text{Hotel A} & \text{Hotel B} \end{array} \\ \begin{bmatrix} 0 & .75 & .25 \\ .90 & 0 & .10 \\ .80 & .20 & 0 \end{bmatrix} \end{array}.$$

a If the taxicab begins at the airport, what is the probability that it will be at Hotel A three moves later?

b Suppose the taxicab starts at the airport with probability .5 and starts at Hotel A and Hotel B with probability .25 each. What is the probability that the taxicab will be at Hotel A three moves later?

4.3-5 Referring to Exercise 4.2-2, find the probability that there are two balls in compartment 1 after three draws.

4.3-6 Referring to Exercise 4.2-4, find the probability that the person is either at the top of the ladder or at the bottom of the ladder after four tosses of the coin.

4.3-7 Students at a certain college take a sequence of three mathematics courses. The mathematical knowledge that the student has at the beginning of the sequence is measured by the letter grades A, B, or C. As students progress through the sequence of courses, their grades may stay the same or change according to a Markov chain with transition matrix

$$P = \begin{array}{c} \\ A \\ B \\ C \end{array} \begin{array}{c} \begin{array}{ccc} A & B & C \end{array} \\ \begin{bmatrix} .7 & .2 & .1 \\ .3 & .4 & .3 \\ .1 & .3 & .6 \end{bmatrix} \end{array}$$

a What is the probability that a student has an A at the end of the third course given that he or she has a B at the beginning of the first course?

b The probability distribution of grades at the beginning is given as $P(A) = .3$, $P(B) = .5$, $P(C) = .2$. Find the probability distribution of grades at the end of the three-course sequence.

4.3-8 An instructor places five copies of an exam on file in the library. Students check out copies of the exam for short periods of time and then return them. Let $X(n)$ denote the number of copies of the exam available for checkout at time n. Assume $\{X(n)\}$ is a Markov chain with transition matrix

$$P = \begin{array}{c} \\ 0 \\ 1 \\ 2 \\ 3 \\ 4 \\ 5 \end{array} \begin{array}{c} \begin{array}{cccccc} 0 & 1 & 2 & 3 & 4 & 5 \end{array} \\ \begin{bmatrix} .9 & .1 & 0 & 0 & 0 & 0 \\ .1 & .8 & .1 & 0 & 0 & 0 \\ 0 & .1 & .8 & .1 & 0 & 0 \\ 0 & 0 & .1 & .8 & .1 & 0 \\ 0 & 0 & 0 & .1 & .8 & .1 \\ 0 & 0 & 0 & 0 & .1 & .9 \end{bmatrix} \end{array}.$$

a If there are five copies on file now, what is the probability that there will be zero copies on file five time periods later?

b If there are five copies on file now, what is the probability that there will be at least one copy available three time periods later?

SECTION

4.4 The Simple Queuing System Revisited

In the queuing example of Section 4.1, we analyzed the behavior of the number of computers waiting for service at the end of each day, using simulation. Now we analyze the behavior of that system by using what we have learned about Markov chains.

First we find the one-step transition matrix. The transitions that can be made from state 0, which represents no backlog of computers waiting for service, are

$0 \to 0$	None or one of the five computers breaks down (remember that one can be fixed each day)
$0 \to 1$	Two of the five computers break down
$0 \to 2$	Three of the five computers break down
$0 \to 3$	Four of the five computers break down
$0 \to 4$	All five computers break down

Since the computers fail independently of one another with probability .2, the number of failures each day is a binomial random variable. Thus, the desired transition probabilities are determined from the binomial distribution. For the one-step transitions from no backlogs,

$$\begin{aligned}
P(0 \to 0) &= \binom{5}{0}(.2)^0(.8)^5 + \binom{5}{1}(.2)(.8)^4 \\
&= .7373, \\
P(0 \to 1) &= \binom{5}{2}(.2)^2(.8)^3 \\
&= .2048, \\
P(0 \to 2) &= \binom{5}{3}(.2)^3(.8)^2 \\
&= .0512, \\
P(0 \to 3) &= \binom{5}{4}(.2)^4(.8) \\
&= .0064, \\
P(0 \to 4) &= \binom{5}{5}(.2)^5(.8)^0 \\
&= .0003.
\end{aligned}$$

If there is a backlog of one computer waiting for service at the end of today, the possible transitions are

$1 \to 0$	None of the four computers breaks down
$1 \to 1$	One of the four computers breaks down
$1 \to 2$	Two of the four computers break down
$1 \to 3$	Three of the four computers break down
$1 \to 4$	All four of the computers break down

Again, the transition probabilities can be expressed in terms of the binomial distribution. In this case, the probabilities are based on four, instead of five, Bernoulli trials. Thus,

$$\begin{aligned}
P(1 \to 0) &= \binom{4}{0}(.2)^0(.8)^4 \\
&= .4096, \\
P(1 \to 1) &= \binom{4}{1}(.2)(.8)^3 \\
&= .4096, \\
P(1 \to 2) &= \binom{4}{2}(.2)^2(.8)^2 \\
&= .1536, \\
P(1 \to 3) &= \binom{4}{3}(.2)^3(.8) \\
&= .0256, \\
P(1 \to 4) &= \binom{4}{4}(.2)^4(.8)^0 \\
&= .0016.
\end{aligned}$$

Continuing in this way, we can obtain all the elements of the one-step transition matrix. The two-step through five-step transition matrices can be obtained by raising the one-step transition matrix to the appropriate powers.

The one-step through five-step transition matrices for this example are given in Table 4.4-1. The entries in row i of the kth-step transition matrix are the probabilities of no, one, two, three, or four computers waiting for service at the end of the kth day of the week given that the week begins with i computers waiting for service. For example, if we begin the week with no computers waiting for service, the probability that there are three computers waiting for service at the end of Wednesday is the $(0, 3)$ entry of the three-step transition matrix, which is .0261.

TABLE 4.4-1
K-step Transition Matrices for the Simple Queuing System

One-Step (Monday's Backlog)

$P =$

	0	1	2	3	4
0	.7373	.2048	.0512	.0064	.0003
1	.4096	.4096	.1536	.0256	.0016
2	.0000	.5120	.3840	.0960	.0080
3	.0000	.0000	.6400	.3200	.0400
4	.0000	.0000	.0000	.8000	.2000

Two-Step (Tuesday's Backlog)

$P^2 =$

	0	1	2	3	4
0	.6275	.2611	.0930	.0172	.0013
1	.4698	.3303	.1593	.0373	.0034
2	.2097	.4063	.2875	.0871	.0093
3	.0000	.3277	.4506	.1958	.0259
4	.0000	.0000	.5120	.4160	.0720

Three-Step (Wednesday's Backlog)

$P^3 =$

	0	1	2	3	4
0	.5696	.2831	.1189	.0261	.0023
1	.4817	.3130	.1598	.0414	.0041
2	.3211	.3566	.2393	.0747	.0084
3	.1342	.3649	.3487	.1350	.0171
4	.0000	.2621	.4628	.2399	.0351

Four-Step (Thursday's Backlog)

$P^4 =$

	0	1	2	3	4
0	.5359	.2935	.1350	.0325	.0031
1	.4833	.3087	.1606	.0430	.0044
2	.3828	.3343	.2109	.0647	.0072
3	.2484	.3555	.2832	.1006	.0122
4	.1074	.3444	.3715	.1560	.0207

Five-Step (Friday's Backlog)

$P^5 =$

	0	1	2	3	4
0	.5153	.2991	.1452	.0368	.0036
1	.4828	.3077	.1613	.0437	.0045
2	.4192	.3233	.1934	.0578	.0064
3	.3288	.3415	.2405	.0799	.0094
4	.2202	.3533	.3009	.1117	.0139

It is interesting to compare the values in the k-step transition matrices to those obtained in the simulation study in Section 4.1. In the simulation study, we began with no backlog of computers. Therefore the empirical probabilities should be close to the probabilities in the zeroth rows of the kth-step transition matrices. Indeed this is the case. For example, we have the following empirical and actual probabilities for the backlog at the end of Wednesday.

Backlog	0	1	2	3	4
Empirical probabilities (from simulation study)	.584	.266	.125	.025	.000
Actual probabilities (from three-step matrix)	.5696	.2831	.1189	.0261	.0023

The probabilities obtained from the Markov chain analysis are preferred because they are exact values. However, additional information on idle time can be easily obtained by simulation but cannot be obtained directly from the transition matrices of this Markov chain. Thus, there is a benefit to the simulation approach as well as the theoretical approach in this simple example.

Exercises 4.4

4.4-1 Modify the simple queuing system of Section 4.1 so that there are two computers and one technician.

a Find the one-step transition matrix.

b Find the probability distributions of backlogs for Tuesday through Friday, assuming that there are no backlogs at the beginning of the week.

4.4-2 Modify the simple queuing system of Section 4.1 so that there are five computers and two technicians. Assume that each technician can fix one computer a day.

a Find the one-step transition matrix.

b Find the probability distributions of the number of backlogs on Tuesday through Friday, assuming that there is a backlog of two computers to be fixed beginning Monday morning.

4.4-3 Find the sixth power of the transition matrix of the simple queuing system of Section 4.1. Interpret the elements of this matrix.

SECTION

4.5 Simulating the Behavior of a Markov Chain

Given a starting value and the one-step transition matrix, it is possible to simulate paths of a Markov chain. To illustrate, consider a Markov chain with the following one-step transition matrix:

$$P = \begin{array}{c} \\ 1 \\ 2 \\ 3 \end{array} \begin{array}{c} \begin{array}{ccc} 1 & 2 & 3 \end{array} \\ \begin{bmatrix} .7 & .2 & .1 \\ .5 & .3 & .2 \\ .3 & 0 & .7 \end{bmatrix} \end{array}.$$

Suppose that the chain begins in state 1. The possible transitions are to states 1, 2, and 3 with probabilities .7, .2, and .1, respectively. To simulate the transition, select a unit random number u. If $u < .7$, make the transition to state 1; if $.7 \le u < .9$, make the transition to state 2; otherwise make the transition to state 3.

Suppose that $u = .94518$ so that the transition is made to state 3. Now that we are in state 3, we refer to transition probabilities in row 3 of the one-step transition matrix to simulate the next transition. Select another unit random number u. If $u < .3$ make the transition to state 1; otherwise make the transition to state 3. If $u = .56781$, for instance, the transition would be made to state 3, and the simulation would continue using the transition probabilities for state 3. Table 4.5-1 contains a sequence of 10 transitions from this Markov chain.

TABLE **4.5-1**
Ten Transitions of Markov Chain

n	$X(n)$	u	$X(n+1)$
0	1	.95413	3
1	3	.56781	3
2	3	.29880	1
3	1	.06115	1
4	1	.20655	1
5	1	.09922	1
6	1	.56873	1
7	1	.66969	1
8	1	.87589	2
9	2	.94970	3
10	3	.11398	1

Since there are many theoretical results for Markov chains, one might wonder what value there is in simulating their behavior. Without diminishing the value of theoretical results, there are some good reasons for doing simulations.

Why Simulate Markov Chains?

1. Simulation may be the most expedient way to study the behavior of a Markov chain because it is easy to do.
2. Certain characteristics of Markov chains are difficult to study theoretically. In these cases, simulation is the only feasible method of obtaining information about these characteristics.
3. The Markov chain may be only one part of a larger, more complex, system. Simulation of the Markov chain then becomes a part of the study of a larger system.

EXAMPLE 4.5-1 Four people (A, B, C, D) are having a conversation. When one person finishes speaking, it is equally likely that any of the other three begins. Under this assumption, the transitions from one speaker to the next follow the rules of a Markov chain with one-step transition matrix

$$P = \begin{array}{c} \\ A \\ B \\ C \\ D \end{array} \begin{array}{c} \begin{array}{cccc} A & B & C & D \end{array} \\ \begin{bmatrix} 0 & 1/3 & 1/3 & 1/3 \\ 1/3 & 0 & 1/3 & 1/3 \\ 1/3 & 1/3 & 0 & 1/3 \\ 1/3 & 1/3 & 1/3 & 0 \end{bmatrix} \end{array}.$$

The random variable of interest is the number of transitions that it takes for all persons in the conversation to speak. We assume that the conversation begins with person A. For example, if the conversation goes $(A \to B \to A \to C \to D)$, then it takes four transitions for all to speak.

The minimum number of transitions that it can take is three. There are six paths, beginning with A, in which this can occur. These paths are

$$\{(A \to B \to C \to D),\ (A \to B \to D \to C),\ (A \to C \to B \to D),$$
$$(A \to C \to D \to B),\ (A \to D \to B \to C),\ (A \to D \to C \to B)\}.$$

Each path has probability $(1/3)^3$ of occurring, so the probability that it takes three transitions for all to speak is $6(1/3)^3 = .22$. The probability that it takes four, five, or more transitions for all to speak is more difficult to compute. However, an empirical distribution can be obtained by simulation. As we simulate the paths of the chain, we keep track of the number of distinct speakers that have spoken and the number of transitions that have been made. When all four have spoken, we record the total number of transitions. An algorithm for carrying this out is as follows.

Variables:

SPEAKER	The state of the chain, or person speaking, at any point in time.
IA, IB, IC, ID	Variables coded 0 or 1, which indicate whether persons A, B, C, and D have ever spoken in the conversation. For instance, IB = 0 indicates person B has not spoken, whereas IB = 1 indicates that person B has spoken at some point in the conversation.
NT	Number of transitions among speakers in the conversation.

Initialize:

SPEAKER = A, IA = 1	(Conversation begins with A)
IB = 0, IC = 0, ID = 0	
NT = 0	

Simulation:

1 Generate a transition of the Markov chain to obtain a new SPEAKER.

2 Increment NT = NT + 1.

3 Set IB, IC, or ID equal to 1 according to the value of the variable SPEAKER. For instance, if SPEAKER = C, then set IC = 1. Note that once an indicator is set to 1, it stays there.

4 If IA + IB + IC + ID < 4 (not all persons have spoken), then repeat steps 1, 2, and 3;
else exit with the value of NT.

This procedure was repeated 1000 times, and the results are shown in Figure 4.5-1. The most striking feature is the asymmetrical appearance of the empirical probability mass function with its occasional large values. For instance, in 106 of the 1000 cases, it took 10 or more transitions to allow all persons to speak. The mean number of transitions in this simulation was 5.8 with a standard deviation of 3.1.

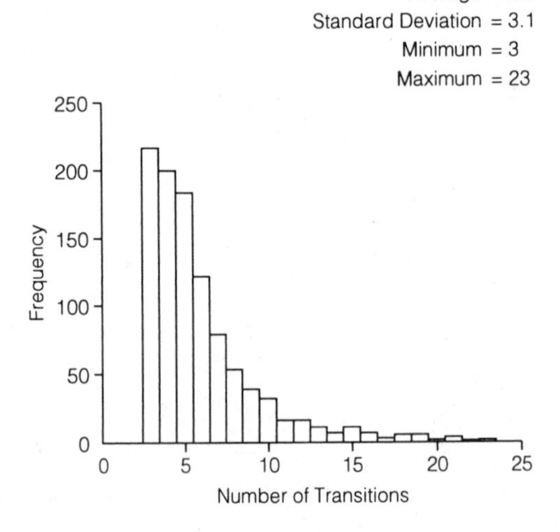

FIGURE 4.5-1
Transitions Required for All to Join a Conversation

EXAMPLE 4.5-2 A quantity frequently of interest in Markov chains is the amount of time it takes the chain to go from one state to another. We illustrate this idea with a model of the cumulative damage sustained by a system over time. Let the system be an automobile, and suppose that the auto begins in excellent condition, which we denote as 1. As the auto is driven, its condition deteriorates due to mechanical wear-out and such environmental factors as road bumps and weather. The automobile will pass through states 2, 3, and 4 (good, fair, and poor) on its way to the junk pile. Suppose that the yearly transition from one state to another is a Markov chain with one-step transition matrix

$$P = \begin{array}{c} \\ 1 \\ 2 \\ 3 \\ 4 \end{array} \begin{array}{c} \begin{array}{cccc} 1 & 2 & 3 & 4 \end{array} \\ \begin{bmatrix} .7 & .3 & 0 & 0 \\ 0 & .6 & .4 & 0 \\ 0 & 0 & .5 & .5 \\ 0 & 0 & 0 & 1 \end{bmatrix} \end{array}$$

The quantity of interest here is the amount of time it will take an auto to reach state 4 (poor condition). For instance, if the path is $(1 \to 1 \to 2 \to 2 \to 3 \to 4)$, then it has taken five transitions, or five years, for the automobile to reach the poor condition. Note that the chain will stay in state 4 once it reaches it.

The desired probabilities may be easily estimated by simulation. We count the number of transitions as the chain moves from one state to the next and end the count when the chain reaches state 4. The empirical distribution of time for an automobile to reach the poor state, based on the simulated performance of 1000 autos, is given in Figure 4.5-2. The empirical distribution has a mean of 7.7 years and a standard deviation of 3.5 years. The minimum of the 1000 simulated times is 3 years, and the maximum is 26 years. Approximately half the autos of this type reach poor condition in about 7 years, and about 90% are in poor condition by the 12th year. This chain is treated theoretically in Section 4.7.

FIGURE **4.5-2**
Time for an Automobile to Reach Poor Condition

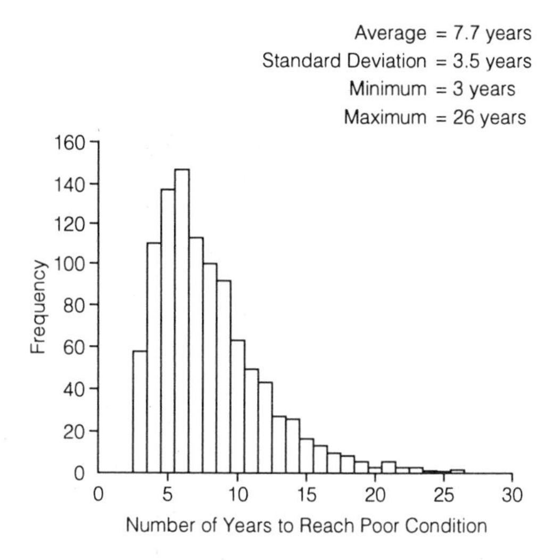

EXAMPLE **4.5-3**

We illustrate simulating a Markov chain as part of a larger problem of determining the operating costs of a machine. A machine can be in one of two states, either up (1) or down (0). The machine changes states from one week to the next in accordance with the rules of a Markov chain. The one-step transition matrix is

$$P = \begin{array}{c} \\ 0 \\ 1 \end{array} \begin{array}{c} \begin{array}{cc} 0 & 1 \end{array} \\ \begin{bmatrix} .5 & .5 \\ .1 & .9 \end{bmatrix} \end{array}.$$

The weekly cost of maintaining or repairing the machine, measured in hundreds of dollars, is 1 if the machine is up and 5 if the machine is down. The quantity of interest is the total cost over 52 weeks. The week before accounting begins, the machine is up. To illustrate the cost computation, if the machine follows the path ($1 \to 0 \to 0 \to 1$) in the first three weeks, then the three-week cost is $5 + 5 + 1 = 11$ hundred dollars. An empirical distribution of the 52-week costs is given in Figure 4.5-3. This distribution is based on repeating the 52-week period 1000 times. The simulation involves the following three simple steps:

1 Generate the state of the system each week,
2 Compute the cost based on whether the state is up or down.
3 Accumulate the cost as the machine goes through 52 weeks of operation.

The average yearly operating cost of the machine, based on the results of this simulation, is 86.3 hundred dollars with a standard deviation of 15.7 hundred dollars. Yearly costs have approximately a 90% chance of being between 64 and 112 hundred dollars. The variability among costs is as important a feature as the average cost. Knowledge of what variability to expect can allow a company to provide adequate reserves in case costs go up.

FIGURE 4.5-3
Costs of Operating a Machine

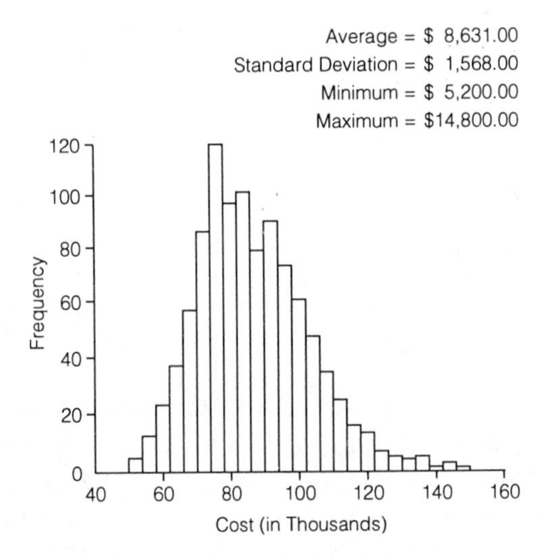

Exercises 4.5

4.5-1 A gambler begins with $500. Each game, he may win $100 with probability 1/2 or lose $100 with probability 1/2. He will play until he either doubles his money or loses it all. Use simulation to determine how long he can expect to play the game.

4.5-2 In Exercise 4.5-1, suppose the gambler has a probability of .7 of winning and .3 of losing each game. Use simulation to determine how long he can expect to play the game, and estimate the probability that he doubles his money.

4.5-3 Refer to Exercise 4.2-4. Use simulation to obtain a sample mean and sample standard deviation for the amount of time it takes the climber to get to the top rung of the ladder.

4.5-4 A machine can move among three states (good, worn, broken) as it goes from one week to the next. The one step transition matrix is

$$P = \begin{array}{l} \text{Good} \\ \text{Worn} \\ \text{Broken} \end{array} \begin{bmatrix} \text{Good} & \text{Worn} & \text{Broken} \\ .7 & .2 & .1 \\ 0 & .6 & .4 \\ .8 & 0 & .2 \end{bmatrix}.$$

The number of units that can be manufactured by the machine depends on its state. If the machine is good, it produces 1000 units a week; if it is worn, it produces 800 units a week; if it is broken, it produces no units per week. Use simulation to obtain a sample mean, sample standard deviation, and empirical probability mass function for the number of units produced by this machine in 20 weeks. Assume the machine starts in the good state.

4.5-5 Refer to Exercise 4.5-4. Use simulation to obtain an estimated mean, standard deviation, and probability mass function for the number of weeks it takes the machine to reach the broken state from the good state. For instance, if the path is good → good → worn → broken, then it takes three weeks to go from good to broken.

4.5-6 Refer to Example 4.5-1. Suppose that the transition probabilities among the speakers are not equal. Speakers B, C, and D each have probability .5 of turning the conversation over to A, whereas they have probability .25 of turning it over to each of the other two participants. The transition probabilities for A remain at 1/3. Repeat the procedures outlined in Example 4.5-1, and note the effects of having person A as the dominant speaker in the conversation.

4.5-7 Refer to Example 4.5-1. Assuming that the conversation begins with person A, let T be the number of transitions it takes for the conversation to return to A. Use simulation to estimate the mean, standard deviation, and probability mass function of T.

4.5-8 A token is placed at point A on the 3×3 grid. The token may move vertically or horizontally to an adjacent point. The movement is determined randomly with each of the possible directions having an equally likely chance of being selected. Simulate the number of transitions it takes the token to get to point B. Repeat the simulation 1000 times and obtain an estimated mean and standard deviation for the time in question.

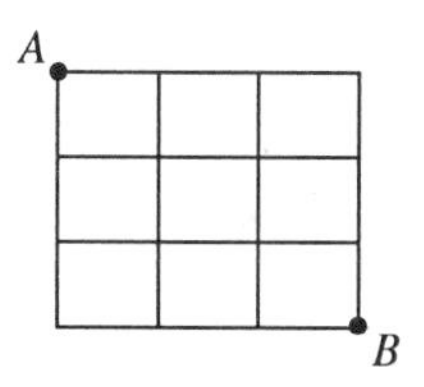

4.5-9 Many believe that a basketball player has a greater probability of making the next shot after having made the present shot than after having missed the present shot. The argument goes that the player gains confidence by making a shot and therefore shoots better the next time. Skeptics believe that the confidence factor is negligible, so the probability of making the next shot is essentially independent of what happened on previous shots. What do you believe is the case? What differences would you expect to see in the pattern of made and missed shots if future makes and misses do or do not depend on previous makes and misses? Simulate both the situation in which the confidence gained by making a shot increases the player's chance of making the next shot and the situation in which this does not occur. Compare the pattern of makes and misses for these two situations.

4.5-10 In a Markov chain $\{X(n)\}$, $n = 0, 1, 2, \ldots$ with two states 0 and 1. Suppose

$$P[X(n+1) = 1 \mid X(n) = 0] = P[X(n+1) = 1 \mid X(n) = 1], \quad n = 0, 1, 2, \ldots.$$

Show that $X(1), X(2), \ldots,$ are independent and identically distributed Bernoulli random variables.

SECTION

4.6 Steady-State Probabilities

An interesting thing can happen to the k-step transition matrix when k becomes large. Consider the 1st-, 2nd-, 4th-, 8th- and 16th-step transition matrices for the rat in the maze in Example 4.3-1. These matrices are

$$P = \begin{array}{c} \\ 1 \\ 2 \\ 3 \\ 4 \end{array} \begin{array}{c} \begin{array}{cccc} 1 & 2 & 3 & 4 \end{array} \\ \begin{bmatrix} 0 & 1/3 & 1/3 & 1/3 \\ 1/2 & 0 & 0 & 1/2 \\ 1/2 & 0 & 0 & 1/2 \\ 1/3 & 1/3 & 1/3 & 0 \end{bmatrix} \end{array},$$

$$P^2 = \begin{array}{c} \\ 1 \\ 2 \\ 3 \\ 4 \end{array} \begin{array}{c} \begin{array}{cccc} 1 & 2 & 3 & 4 \end{array} \\ \begin{bmatrix} 4/9 & 1/9 & 1/9 & 1/3 \\ 1/6 & 1/3 & 1/3 & 1/6 \\ 1/6 & 1/3 & 1/3 & 1/6 \\ 1/3 & 1/9 & 1/9 & 4/9 \end{bmatrix} \end{array},$$

$$P^4 = \begin{array}{c} \\ 1 \\ 2 \\ 3 \\ 4 \end{array} \begin{array}{c} \begin{array}{cccc} 1 & 2 & 3 & 4 \end{array} \\ \begin{bmatrix} .347 & .160 & .160 & .333 \\ .241 & .259 & .259 & .241 \\ .241 & .259 & .259 & .241 \\ .333 & .160 & .160 & .347 \end{bmatrix} \end{array},$$

$$P^8 = \begin{array}{c} \\ 1 \\ 2 \\ 3 \\ 4 \end{array} \begin{array}{c} \begin{array}{cccc} 1 & 2 & 3 & 4 \end{array} \\ \begin{bmatrix} .308 & .192 & .192 & .308 \\ .288 & .212 & .212 & .288 \\ .288 & .212 & .212 & .288 \\ .302 & .192 & .192 & .302 \end{bmatrix} \end{array},$$

$$P^{16} = \begin{matrix} & 1 & 2 & 3 & 4 \\ 1 \\ 2 \\ 3 \\ 4 \end{matrix} \begin{bmatrix} .300 & .200 & .200 & .300 \\ .300 & .200 & .200 & .300 \\ .300 & .200 & .200 & .300 \\ .300 & .200 & .200 & .300 \end{bmatrix}.$$

The thing to note is that all the rows of the 16th-step transition matrix are the same, at least to three decimals. If we were to continue to raise the one-step transition matrix to higher powers, we would find that the entries would not change beyond this point. That is, the k-step transition matrix for values of $k > 16$ will have the same entries as P^{16} at least up to the first three decimal places. Thus, regardless of the starting compartment, after about 16 moves, the rat has probabilities .300, .200, .200, and .300 of being in states 1, 2, 3, or 4, respectively. Even at eight moves, the probabilities are close to these values. In this situation, we say that the system attains the *steady-state* condition.

We restrict our discussion in this section to Markov chains with a finite number of states. Under this condition we can define a class of chains in which the steady-state condition can be attained.

Definition 4.6-1

A finite state Markov chain is said to be *regular* if the k-step transition matrix has all nonzero entries for some value of $k > 0$.

THEOREM 4.6-1 Let $X(n)$, $n = 0, 1, 2, \ldots$, be a regular Markov chain with one-step transition matrix P. Then there exists a matrix Π having identical rows with nonzero entries such that

$$\lim_{k \to \infty} P^k = \Pi.$$

Theorem 4.6-1 states that if the k-step transition matrix of a finite-state Markov chain has all nonzero entries at some point in time, then the chain will eventually attain the steady-state condition. For a proof of this result, refer to Bhat (1984).

Let each of the identical rows of the limiting matrix Π be denoted by row vector π. Since the elements of π constitute a probability distribution, π is called the *steady-state probability vector* and the entries of π are called *steady-state probabilities*.

EXAMPLE 4.6-1 A communication line is sampled once a minute to see whether it is busy (1) or not busy (2). Transition among the two states is a Markov process with one-step transition matrix

$$P = \begin{array}{c} \\ 1 \\ 2 \end{array} \begin{array}{c} \begin{array}{cc} 1 & 2 \end{array} \\ \begin{bmatrix} .6 & .4 \\ .1 & .9 \end{bmatrix} \end{array}.$$

Since all the elements of P are nonzero, according to Theorem 4.6-1 this Markov process will obtain a steady state. The steady-state probability vector can be found by raising P to a sufficiently high power so that the rows of P^k are the same. Consider the following:

$$P^2 = \begin{array}{c} \\ 1 \\ 2 \end{array} \begin{array}{c} \begin{array}{cc} 1 & 2 \end{array} \\ \begin{bmatrix} .40 & .60 \\ .15 & .85 \end{bmatrix} \end{array},$$

$$P^4 = \begin{array}{c} \\ 1 \\ 2 \end{array} \begin{array}{c} \begin{array}{cc} 1 & 2 \end{array} \\ \begin{bmatrix} .250 & .750 \\ .188 & .812 \end{bmatrix} \end{array},$$

$$P^8 = \begin{array}{c} \\ 1 \\ 2 \end{array} \begin{array}{c} \begin{array}{cc} 1 & 2 \end{array} \\ \begin{bmatrix} .203 & .797 \\ .199 & .801 \end{bmatrix} \end{array},$$

$$P^{12} = \begin{array}{c} \\ 1 \\ 2 \end{array} \begin{array}{c} \begin{array}{cc} 1 & 2 \end{array} \\ \begin{bmatrix} .200 & .800 \\ .200 & .800 \end{bmatrix} \end{array}.$$

The computation indicates that the steady-state probability vector is $\pi = (.2, .8)$. If the line is sampled after it has been operating for a long time, there is a probability of .2 of finding it busy and a probability of .8 of finding it not busy.

Consider the fraction of visits a Markov chain makes to a particular state in k

transitions. That is, let

$$V(j, k) = \frac{\text{Number of visits to state } j \text{ in } k \text{ transitions}}{k}.$$

For instance, referring to Example 4.6-1, if we begin in state j_0 (either 1 or 2) and if the states 1, 2, 1, 2, 2 are visited in the first five transitions, then $V(2, 5) = 3/5$ and $V(1, 5) = 2/5$. We are interested in $E(V(j, k))$, the expected fraction of visits a Markov chain makes to state j in k transitions.

THEOREM 4.6-2 Let $X(n)$, $n = 0, 1, 2, \ldots$ be a regular Markov chain. Let π_j be the steady state probability for state j. Then

$$\lim_{k \to \infty} E(V(j, k)) = \pi_j.$$

For verification of Theorem 4.6-2 see Exercises 4.6-13 and 4.6-14. This theorem tells us that the long-run fraction of visits a Markov chain makes to state j is the same as its steady-state probability. In Example 4.6-1, the steady-state probabilities for states 1 and 2 were .2 and .8, respectively. This tells us that in the long run the line is busy 20% of the time it is sampled, and it is not busy 80% of the time.

As indicated in the previous examples, the steady-state probabilities may become apparent by raising the one-step transition matrix to successively higher powers and seeing the elements of the k-step transition matrix converge to their limiting values. Matrix multiplication is an easy matter for the computer, and there are many specially designed software packages available for handling matrix operations. The only caution here is that there should be sufficient accuracy in the matrix multiplication procedure to avoid the cumulative effects of computer floating-point arithmetic error. The next theorem gives an alternative to matrix multiplication for finding the steady-state probabilities when they exist.

THEOREM 4.6-3 Let $X(n)$, $n = 0, 1, 2, \ldots$ be a regular Markov chain with one-step transition matrix P. Then the steady-state probability vector $\pi = (\pi_1, \pi_2, \ldots, \pi_s)$ may be found by solving the system of equations

$$\pi P = \pi, \quad \pi_1 + \pi_2 + \cdots + \pi_s = 1.$$

To see that the steady-state probability vector should satisfy the system of equations in Theorem 4.6-3, consider the following. By definition, $P^{k-1}P = P^k$. Since $\lim_{k\to\infty} P^{k-1} = \lim_{k\to\infty} P^k = \Pi$, then $\Pi P = \Pi$. Therefore, $\pi P = \pi$.

EXAMPLE 4.6-2 Refer to the one-step transition matrix for the rat in the maze in Example 4.3-1. The system of equations in Theorem 4.6-3 is

$$\begin{aligned}
0\pi_1 + (1/2)\pi_2 + (1/2)\pi_3 + (1/3)\pi_4 &= \pi_1,\\
(1/3)\pi_1 + 0\pi_2 + 0\pi_3 + (1/3)\pi_4 &= \pi_2,\\
(1/3)\pi_1 + 0\pi_2 + 0\pi_3 + (1/3)\pi_4 &= \pi_3,\\
(1/3)\pi_1 + (1/2)\pi_2 + (1/2)\pi_3 + 0\pi_4 &= \pi_4,\\
\pi_1 + \pi_2 + \pi_3 + \pi_4 &= 1.
\end{aligned}$$

By direct substitution it can be verified that $\pi_1 = .300$, $\pi_2 = .200$, $\pi_3 = .200$, $\pi_4 = .300$ satisfy this system of equations. The most direct method for finding this solution would be to substitute $\pi_4 = 1 - \pi_1 - \pi_2 - \pi_3$ into any three of the first four equations and solve for π_1, π_2, and π_3 using the usual algebraic methods or using a computer program designed to solve systems of linear equations. The first four equations do not have a unique solution. Use of the last equation is essential. The details of solving the equations are left to the reader.

EXAMPLE 4.6-3 Consider the Markov process of Example 4.6-1. The steady-state probability vector can be found directly by solving the system of equations $\pi P = \pi$. This is,

$$\begin{aligned}
.6\pi_1 + .1\pi_2 &= \pi_1,\\
.4\pi_1 + .9\pi_2 &= \pi_2,\\
\pi_1 + \pi_2 &= 1.
\end{aligned}$$

Substituting $\pi_2 = 1 - \pi_1$ from the third equation into the first equation, we find $\pi_1 = .2$; therefore, $\pi_2 = .8$.

EXAMPLE 4.6-4 The weekly demand for a certain commodity is classified as being high (1), moderate (2), or low (3). Assume that the transitions among these states behave as a

Markov chain with one-step transition matrix

$$P = \begin{array}{c} \\ 1 \\ 2 \\ 3 \end{array} \begin{array}{c} \begin{array}{ccc} 1 & 2 & 3 \end{array} \\ \begin{bmatrix} .80 & .15 & .05 \\ .10 & .80 & .10 \\ .10 & .30 & .60 \end{bmatrix} \end{array}.$$

The steady-state probabilities may be found by solving the system of

$$\begin{aligned} .80\pi_1 + .10\pi_2 + .10\pi_3 &= \pi_1, \\ .15\pi_1 + .80\pi_2 + .30\pi_3 &= \pi_2, \\ .05\pi_1 + 10\pi_2 + .60\pi_3 &= \pi_3, \\ \pi_1 + \pi_2 + \pi_3 &= 1. \end{aligned}$$

The solution may also be inferred by raising the one-step transition matrix to a sufficiently high power. For instance, the steady-state probabilities are apparent by the 20th step, where

$$P^{20} = \begin{array}{c} \\ 1 \\ 2 \\ 3 \end{array} \begin{array}{c} \begin{array}{ccc} 1 & 2 & 3 \end{array} \\ \begin{bmatrix} .333 & .500 & .167 \\ .333 & .500 & .167 \\ .333 & .500 & .167 \end{bmatrix} \end{array}.$$

Thus, $\pi_1 = 1/3$, $\pi_2 = 1/2$, $\pi_3 = 1/6$. In the long run, the demand will be high $1/3$ of the time, moderate $1/2$ of the time, and low $1/6$ of the time.

Next consider the amount of time it takes to return to a state j given that the chain is currently in state j. Suppose, for example, that a chain is currently in state 1 and the path $(1 \to 2 \to 3 \to 2 \to 1)$ is observed. In this case, the return time is four transitions. A simple relationship between the *expected return time* and the steady-state probabilities is expressed in the following theorem.

THEOREM 4.6-4 Let $X(n)$, $n = 0, 1, 2, \ldots$, be a regular Markov chain. Let $\pi = (\pi_1, \pi_2, \ldots, \pi_s)$ be the steady-state probability vector for the chain, and let T_j denote the time it takes to return to state j given that the chain is currently in state j, $j = 1, 2, \ldots, s$. Then

$$E(T_j) = \frac{1}{\pi_j}.$$

EXAMPLE **4.6-5** Suppose the chain in Example 4.6-1 begins in the busy state (1). Since the steady-state probability for the busy state is .2, the expected return time to the busy state is $1/.2 = 5$ transitions (5 minutes in this case). This is an intuitively reasonable result. Since the chain is in the busy state an average of $1/5$ of the time, we would expect, on average, a return to the busy state once every five transitions.

There is another interpretation of the steady-state distribution. Suppose that the initial state of a regular Markov chain is a random variable and that the initial probability vector is $\pi(0)$. From the discussion in Section 4.3, recall that the kth-step probability vector is

$$\pi(k) = \pi(0)P^k.$$

If the steady-state probability vector happens to be the same as the initial probability vector, then $\pi(0) = \pi$, and it follows from Theorem 4.6-3 that $\pi(1) = \pi P = \pi$, $\pi(2) = \pi P^2 = \pi P = \pi$, and $\pi(k) = \pi$, for all $k > 0$. Thus, if the initial distribution is the same as the steady-state distribution, the probability of being in any state does not change with time. For this reason the steady-state distribution is also called a *stationary distribution*.

Exercises 4.6

4.6-1 Suppose that the pattern of sunny days and cloudy days follows the Markov chain with the one-step transition matrix given in Exercise 4.3-2. In the long run, what fraction of days will be sunny and what fraction will be cloudy?

4.6-2 A machine can be in one of two states, up (1) or down (0). The machine changes state from one hour to the next according to the Markov chain given in Exercise 4.3-1.

- **a** If we observe the machine after it has been in operation a long time, what is the probability that the machine will be up? down?
- **b** If the machine begins up, what is the expected amount of time that it will take for the machine to be up again?

4.6-3 A taxicab moves between the airport, Hotel A, and Hotel B according to a Markov chain given in Exercise 4.3-4.

- **a** In the long run, what fraction of visits will the taxicab make to each of the three locations?
- **b** If the taxicab begins at the airport, what is the expected number of transitions that it will take for it to return to the airport?

4.6-4 Variable demand for a product causes a company's inventory to fluctuate. Suppose that the inventory is classified as high (1), normal (2), and low (3) and that it changes states from week to week according to the Markov chain with one-step transition matrix

$$P = \begin{array}{c} \\ 1 \\ 2 \\ 3 \end{array} \begin{array}{c} \begin{array}{ccc} 1 & 2 & 3 \end{array} \\ \begin{bmatrix} .4 & .5 & .1 \\ .2 & .5 & .3 \\ .3 & .4 & .3 \end{bmatrix} \end{array}.$$

In the long run, what fraction of the time will the inventory be in each of the states?

4.6-5 Suppose that a Markov chain has the following one-step transition matrix where $0 < a < 1$ and $0 < b < 1$:

$$P = \begin{array}{c} \\ 0 \\ 1 \end{array} \begin{array}{c} \begin{array}{cc} 0 & 1 \end{array} \\ \begin{bmatrix} 1-a & a \\ b & 1-b \end{bmatrix} \end{array}.$$

Find a formula for the steady-state probabilities in terms of a and b.

4.6-6 Find the steady-state probabilities for the simple queuing system in Section 4.4. Interpret the steady-state probabilities.

4.6-7 Show that the following chains are not regular.

a

$$P = \begin{array}{c} \\ 0 \\ 1 \end{array} \begin{array}{c} \begin{array}{cc} 0 & 1 \end{array} \\ \begin{bmatrix} 0 & 1 \\ 1 & 0 \end{bmatrix} \end{array}.$$

b

$$P = \begin{array}{c} \\ 1 \\ 2 \\ 3 \end{array} \begin{array}{c} \begin{array}{ccc} 1 & 2 & 3 \end{array} \\ \begin{bmatrix} .6 & .3 & .1 \\ .1 & .5 & .4 \\ 0 & 0 & 1 \end{bmatrix} \end{array}.$$

4.6-8 Consider a vending machine that can be in one of three possible states each day: 0 = working in good condition; 1 = working in need of minor repair; 2 = working in need of major repair; and 3 = out-of-order. The one-step transition matrix corresponding to this vending machine's movements from state to state is

$$P = \begin{array}{c} \\ 0 \\ 1 \\ 2 \\ 3 \end{array} \begin{array}{c} \begin{array}{cccc} 0 & 1 & 2 & 3 \end{array} \\ \begin{bmatrix} .80 & .14 & .04 & .02 \\ 0 & .60 & .30 & .10 \\ 0 & 0 & .65 & .35 \\ .90 & 0 & 0 & .10 \end{bmatrix} \end{array}.$$

a Verify that this is a regular Markov chain.

b Find the steady-state probability vector for this Markov chain.

c If the machine is currently in the out-of-order state, how many days, on average, will it take the machine to return to this state?

4.6-9 Find the steady-state probability vector for the Markov chain in Exercise 4.3-8.

4.6-10 A Markov chain is called doubly stochastic if the entries in each column add to 1. The term *doubly* comes from the fact that the entries in both the rows and columns add to 1. Exercise 4.3-8 is an example of a regular Markov chain that is doubly stochastic. If a regular, doubly stochastic Markov chain has k states, verify that the steady state probability vector is

$$\pi = (1/k, 1/k, \ldots, 1/k).$$

4.6-11 For the sequential file search in Example 4.2-10 and Exercise 4.2-6, the steady-state probability for the state ABC is

$$P(A)\frac{P(B)}{1-P(A)}\frac{P(C)}{1-P(A)-P(B)}.$$

Using this general pattern, obtain the steady-state probabilities for the other five states. Verify that these six probabilities constitute the steady-state probability vector.

4.6-12 A merchant begins with an inventory of n items. Assume that the inventory is reduced by at most one item a day according to the transition probabilities

$$P(i \to i-1) = p, \quad P(i \to i) = 1-p, \quad \text{for } i = 1, 2, \ldots, n.$$

After the inventory has been completely depleted, the merchant takes one day to replenish it. That is, $P(0 \to n) = 1$.

a Give the one-step transition matrix for this Markov chain.

b Find the steady-state probability vector.

c If the current state is 0, what is the expected number of days that it will take to return to this state?

4.6-13 If a Markov chain starts in state s_0, show

$$E(V(s,k)) = \frac{1}{k}\sum_{i=1}^{k} P^{(i)}(s_0 \to s).$$

(*Hint*: Let $I_s^{(i)} = 1$ if Markov chain is in state s after i transitions, and 0 otherwise. Note that $V(s,k) = (1/k)\sum_{i=1}^{k} I_s^{(i)}$.)

4.6-14 Use the result in Exercise 4.6-13 to prove Theorem 4.6-2.

SECTION

4.7 Absorbing States and First Passage Times

The regular Markov chains discussed in Section 4.6 have the property that a path can always be found that connects any state to any other state. This follows from the fact that the k-step transition matrix eventually will have all positive probabilities. Not all chains are like this. For example, the Markov chain in Example 4.5-2, a model for the deterioration of an automobile, cannot move from the poor state to the fair state or from the fair state to the good state. This simply reflects the fact that the automobile does not improve with age. Of special interest are states that cannot be exited when entered. For instance, when the automobile in Example 4.5-2 reaches the poor state, it stays there. Such states are called *absorbing states*.

Definition 4.7-1

A state j is said to be an absorbing state if $P(j \to j) = 1$; otherwise it is a nonabsorbing state.

Throughout this section we assume there is a path that leads from each nonabsorbing state to an absorbing state. We are interested in the number of transitions that it takes a chain to reach an absorbing state given that it starts in a nonabsorbing state. We dealt with this kind of problem by the simulation method in Example 4.5-2, where we considered how long it would take an automobile to go from the excellent state to the poor state. We now consider this problem and problems like it from a theoretical point.

Since a chain cannot leave an absorbing state once it enters, we can tell whether there has been a visit to an absorbing state in k or fewer transitions simply by observing the chain at the kth step and noting whether it is in an absorbing state. Thus,

$$\begin{aligned} &P(\text{chain reaches an absorbing state in } k \text{ or fewer transitions}) \\ &\quad = P(\text{chain is in an absorbing state at the } k\text{th step}). \end{aligned}$$

For instance, the probability that the automobile of Example 4.5-2 goes from the excellent state to the poor state in seven or fewer years is the probability that it is in the poor condition at the seventh year. This result is summarized in the following theorem.

THEOREM 4.7-1 Let A denote the set of absorbing states in a finite-state Markov chain. Assume that A has at least one state and that there is a path from every nonabsorbing state to at least one state in A. Let T_i denote the number of transitions it takes to go from the nonabsorbing state i to A. Then

$$P(T_i \le k) = \sum_{j \in A} P^{(k)}(i \to j).$$

The number of transitions that it takes a chain to reach an absorbing state is called the *time to absorption.* Theorem 4.7-1 expresses the cumulative distribution function of time to absorption in terms of the k-step transition probabilities.

EXAMPLE 4.7-1 Consider the automobile deterioration model in Example 4.5-2. The seventh-step transition matrix, with entries rounded to two decimals, is

$$P^7 = \begin{array}{c} \\ 1 \\ 2 \\ 3 \\ 4 \end{array} \begin{array}{c} \begin{array}{cccc} 1 & 2 & 3 & 4 \end{array} \\ \begin{bmatrix} .08 & .16 & .21 & .55 \\ .00 & .03 & .08 & .89 \\ .00 & .00 & .01 & .99 \\ .00 & .00 & .00 & 1.00 \end{bmatrix} \end{array}.$$

The probability that the automobile goes from the excellent state (1) to the poor state (4) in seven or fewer years is the (1,4) entry of P^7. Thus, $P(T_1 \le 7) = .55$. This is consistent with the results of the simulation in Example 4.5-2. Similarly, if the automobile begins in the good state (2), it has a probability of .89 of reaching the poor state in seven or fewer years, and if it begins in the fair state (3) it has a probability of .99 of reaching the poor state in seven or fewer years.

EXAMPLE 4.7-2 A gambler bets \$10 a game on a sequence of baseball games. He starts with \$20 and places bets one at a time until he either doubles his money or loses it all. Suppose he has a probability of .6 of losing and a probability .4 of winning each time. We are interested in the number of bets he makes until he quits. The amount of money he has after each game is a Markov chain with one-step transition matrix

$$P = \begin{array}{c} \\ 0 \\ 10 \\ 20 \\ 30 \\ 40 \end{array} \begin{array}{c} \begin{array}{ccccc} 0 & 10 & 20 & 30 & 40 \end{array} \\ \begin{bmatrix} 1 & 0 & 0 & 0 & 0 \\ .6 & 0 & .4 & 0 & 0 \\ 0 & .6 & 0 & .4 & 0 \\ 0 & 0 & .6 & 0 & .4 \\ 0 & 0 & 0 & 0 & 1 \end{bmatrix} \end{array}.$$

This chain has two absorbing states, 0 and 40. For illustration, let us consider the probability that the gambler quits in six or fewer bets. The sixth-step transition matrix, rounded to three decimals, is

$$P^6 = \begin{array}{c} \\ 0 \\ 10 \\ 20 \\ 30 \\ 40 \end{array} \begin{array}{c} \begin{array}{ccccc} 0 & 10 & 20 & 30 & 40 \end{array} \\ \left[\begin{array}{ccccc} 1.000 & 0.000 & 0.000 & 0.000 & 0.000 \\ 0.813 & 0.055 & 0.000 & 0.037 & 0.095 \\ 0.616 & 0.000 & 0.111 & 0.000 & 0.274 \\ 0.320 & 0.083 & 0.000 & 0.055 & 0.542 \\ 0.000 & 0.000 & 0.000 & 0.000 & 1.000 \end{array}\right] \end{array}.$$

Beginning with $20, the gambler has the following probability of being in an absorbing state at time 6:

$$P^{(6)}(20 \to 0) + P^{(6)}(20 \to 40) = .616 + .274 = .890.$$

Thus, the gambler has an 89% chance of placing six or fewer bets.

If there is a path from every nonabsorbing state to an absorbing state in a finite Markov chain, then the chain is certain to reach an absorbing state. We consider the *mean time to absorption*. The results of Theorem 4.7-1 provide us the cumulative distribution function of the time to absorption from which the probability mass function may be obtained. Given the probability mass function, we can compute the mean time to absorption.

EXAMPLE 4.7-3 The k-step transition matrices were obtained for the automobile deterioration model in Example 4.5-2. The cumulative distribution function for the time to absorption T, assuming that the automobile begins in the excellent state, is obtained by application of Theorem 4.7-1. The probability mass function is then obtained from the cumulative distribution function. Since the smallest value for the time to absorption is 3 years, and since there is very little chance that the time to absorption will be more than 18 years, the cumulative distribution function is nonzero for this range with probabilities

t	3	4	5	6	7	8	9	10	11	12	13	14	15	16	17	18
$F(t)$	.06	.17	.30	.43	.55	.65	.74	.80	.85	.89	.92	.94	.96	.97	.98	.99
$p(t)$	.06	.11	.13	.13	.12	.10	.09	.06	.05	.04	.03	.02	.02	.01	.01	.01

An approximation for the mean time to absorption can be found by computing $\Sigma tp(t)$ for the values of t in the table. Thus, $\mu \simeq 7.6$ years. That is, it takes an average of about 7.6 years for the automobile to go from the excellent state to the poor state. The true mean will be somewhat larger than this value due to the exclusion of values of t greater than 18 from the computation of μ.

There is an alternative method of finding the mean time to absorption. It involves solving a system of linear equations. This system of equations can be derived as follows. Let a chain consist of nonabsorbing states $1, 2, \ldots, r$ and let A denote the set of absorbing states. Let $\mu_1, \mu_2, \ldots, \mu_r$ denote the mean time to absorption from states $1, 2, \ldots, r$, respectively. Beginning in a nonabsorbing state i, a path may lead directly to an absorbing state in one step, and this occurs with probability $\sum_{j \in A} P(i \to j)$. On the other hand, a path may lead in the first step to a nonabsorbing state j. Given this happens, the mean number of steps to an absorbing state is $1 + \mu_j$. Thus,

$$\mu_i = \sum_{j \in A} P(i \to j) + \sum_{j=1}^{r} P(i \to j)(1 + \mu_j).$$

Since $\sum_{j \in A} P(i \to j) + \sum_{j=1}^{r} P(i \to j) = 1$, we have

$$\mu_i = 1 + \sum_{j=1}^{r} P(i \to j)\,\mu_j, \quad i = 1, 2, \ldots, r.$$

This system of equations can be conveniently expressed in matrix form. In doing this, we note that the transition probabilities in this system represent the transitions among the nonabsorbing states. The result is summarized by the following theorem.

THEOREM 4.7-2 Let Q denote the matrix consisting of transition probabilities among the nonabsorbing states. The mean times to absorption satisfy the system of equations

$$\begin{bmatrix} \mu_1 \\ \mu_2 \\ \vdots \\ \mu_r \end{bmatrix} = \begin{bmatrix} 1 \\ 1 \\ \vdots \\ 1 \end{bmatrix} + Q \begin{bmatrix} \mu_1 \\ \mu_2 \\ \vdots \\ \mu_r \end{bmatrix}.$$

The solution to the system of equations in Theorem 4.7-2 may be solved by the usual algebraic substitutions, or it may be solved in matrix form as

$$\begin{bmatrix} \mu_1 \\ \mu_2 \\ \vdots \\ \mu_r \end{bmatrix} = (I - Q)^{-1} \begin{bmatrix} 1 \\ 1 \\ \vdots \\ 1 \end{bmatrix},$$

where I is the $r \times r$ identity matrix and $(I - Q)^{-1}$ is the matrix inverse of $I - Q$.

EXAMPLE 4.7-4 For the automobile deterioration problem of Example 4.5-2, the matrix Q is

$$Q = \begin{array}{c} \\ 1 \\ 2 \\ 3 \end{array} \begin{array}{c} \begin{array}{ccc} 1 & 2 & 3 \end{array} \\ \begin{bmatrix} .7 & .3 & 0 \\ 0 & .6 & .4 \\ 0 & 0 & .5 \end{bmatrix} \end{array}.$$

It follows that the system of equations defined in Theorem 4.7-2 is

$$\begin{aligned} \mu_1 &= 1 + .7\mu_1 + .3\mu_2, \\ \mu_2 &= 1 + .6\mu_2 + .4\mu_3, \\ \mu_3 &= 1 + .5\mu_3. \end{aligned}$$

The solution to this system is $\mu_1 = 7.83$, $\mu_2 = 4.5$, $\mu_3 = 2$. Thus, the mean times for the automobile to reach the poor state from the excellent, good, and fair states are 7.83 years, 4.5 years, and 2 years, respectively.

EXAMPLE 4.7-5 Consider the gambler's problem in Example 4.7-2. The nonabsorbing states are 10, 20, and 30, and the matrix Q containing the transition probabilities among the non-absorbing states is

$$Q = \begin{array}{c} \\ 10 \\ 20 \\ 30 \end{array} \begin{array}{c} \begin{array}{ccc} 10 & 20 & 30 \end{array} \\ \begin{bmatrix} 0 & .4 & 0 \\ .6 & 0 & .4 \\ 0 & .6 & 0 \end{bmatrix} \end{array}.$$

Thus, the system of equations defined in Theorem 4.7-2 is

$$\begin{aligned}\mu_{10} &= 1 + .4\mu_{20},\\ \mu_{20} &= 1 + .6\mu_{10} + .4\mu_{30},\\ \mu_{30} &= 1 + .6\mu_{20}.\end{aligned}$$

The solution is

$$\begin{aligned}\begin{bmatrix}\mu_{10}\\ \mu_{20}\\ \mu_{30}\end{bmatrix} &= (I-Q)^{-1}\begin{bmatrix}1\\1\\1\end{bmatrix}\\ &= \begin{bmatrix}1 & -.4 & 0\\ -.6 & 1 & -.4\\ 0 & -.6 & 1\end{bmatrix}^{-1}\begin{bmatrix}1\\1\\1\end{bmatrix}\\ &= \begin{bmatrix}1.46 & 0.77 & 0.31\\ 1.15 & 1.92 & 0.77\\ 0.69 & 1.15 & 1.46\end{bmatrix}\begin{bmatrix}1\\1\\1\end{bmatrix}\\ &= \begin{bmatrix}2.54\\ 3.85\\ 3.31\end{bmatrix}.\end{aligned}$$

The gambler who begins with \$20 will bet an average of 3.85 times before either losing all his or her money or doubling it. A gambler who begins with \$10 will bet an average of 2.54 times before he or she quits, and a gambler who begins with \$30 will bet an average of 3.31 times before he or she quits.

In Theorem 4.6-4, we dealt with the problem of the expected return time for regular Markov chains. With the tools of this section, we can consider the amount of time that it takes a regular chain to go from state i to state j, where i is different from j. This time is called a *first passage time*. The trick is to define a new chain that has probabilities identical to the original chain except for state j. In the new chain, state j is redefined to be an absorbing state. The time that it takes the original chain to go from state i to state j is just the time to absorption in the new chain.

EXAMPLE **4.7-6** Consider the transitions among speakers described in Example 4.5-1. The quantity of interest here is the mean number of transitions that it takes for the conversation to go from speaker A to speaker D. If we make state D an absorbing state, the resulting Markov chain is

$$\begin{array}{c} \\ A \\ B \\ C \\ D \end{array} \begin{array}{c} \begin{array}{cccc} A & B & C & D \end{array} \\ \begin{bmatrix} 0 & 1/3 & 1/3 & 1/3 \\ 1/3 & 0 & 1/3 & 1/3 \\ 1/3 & 1/3 & 0 & 1/3 \\ 0 & 0 & 0 & 1 \end{bmatrix} \end{array}.$$

If μ_A, μ_B, and μ_C denote the mean number of transitions from states A, B, and C, respectively, to state D, the system of equations defined in Theorem 4.7-2 is

$$\begin{aligned} \mu_A &= 1 + (1/3)\mu_B + (1/3)\mu_C, \\ \mu_B &= 1 + (1/3)\mu_A + (1/3)\mu_C, \\ \mu_C &= 1 + (1/3)\mu_A + (1/3)\mu_B. \end{aligned}$$

By inspection, we see $\mu_A = 3$, $\mu_B = 3$, and $\mu_C = 3$. Thus, it will take an average of three transitions, or changes of speaker, for D to become the speaker regardless of who among A, B, or C begins the conversation.

It is interesting to consider the limit of the kth-step transition matrix in a chain that has absorbing states. Let $f(i \to j)$ denote the probability that the chain is eventually absorbed into state j given that it starts in state i. Again, under the assumption that there is a path from every nonabsorbing state to the set of absorbing states, we have

$$\lim_{k \to \infty} P^{(k)}(i \to j) = f(i \to j).$$

The limiting value will be zero if j is a nonabsorbing state. If there is only one absorbing state j_0 in the chain, then $f(i \to j_0) = 1$. The interesting application of this equation comes in chains with more than one absorbing state.

EXAMPLE **4.7-7** In the gambler's problem in Example 4.7-2, we will find that by about the 20th step, the kth-step transition matrix has essentially reached its limiting value. Thus,

$$\lim_{k\to\infty} P^k = \begin{array}{c} 0 \\ 10 \\ 20 \\ 30 \\ 40 \end{array} \overset{\begin{array}{ccccc} 0 & 10 & 20 & 30 & 40 \end{array}}{\begin{bmatrix} 1.00 & 0.00 & 0.00 & 0.00 & 0.00 \\ 0.88 & 0.00 & 0.00 & 0.00 & 0.12 \\ 0.69 & 0.00 & 0.00 & 0.00 & 0.31 \\ 0.42 & 0.00 & 0.00 & 0.00 & 0.58 \\ 0.00 & 0.00 & 0.00 & 0.00 & 1.00 \end{bmatrix}}.$$

For instance, beginning with \$30 the gambler has a 42% chance of going broke and a 58% chance of doubling his or her money, but beginning with \$10 he or she has only a 12% chance of doubling his or her money.

There is a convenient matrix formula for computing the probability of eventual absorption into any of the absorbing states. First we rearrange the states of the matrix to group all of the absorbing states. That is, we put the one-step transition matrix in the form

$$P = \begin{array}{r} \text{Absorbing states} \\ \text{Nonabsorbing states} \end{array} \left[\begin{array}{c|c} I & 0 \\ \hline R & Q \end{array}\right]. \tag{4.7-1}$$

The matrix I is an identity matrix, 0 is a matrix of zeros, R is a matrix of transition probabilities from nonabsorbing states to absorbing states, and Q is the matrix of transition probabilities among nonabsorbing states. For instance, the transition matrix for Example 4.7-2 is

$$P = \begin{array}{c} 0 \\ 40 \\ \\ 10 \\ 20 \\ 30 \end{array} \overset{\begin{array}{cc|ccc} 0 & 40 & 10 & 20 & 30 \end{array}}{\left[\begin{array}{cc|ccc} 1 & 0 & 0 & 0 & 0 \\ 0 & 1 & 0 & 0 & 0 \\ \hline .6 & 0 & 0 & .4 & 0 \\ 0 & 0 & .6 & 0 & .4 \\ 0 & .4 & 0 & .6 & 0 \end{array}\right]}$$

THEOREM 4.7-3 Let the one-step transition matrix of a finite Markov chain have the form of Equation (4.7-1). Let F denote the matrix whose (i, j)th element is $f(i \to j)$. Then

$$F = (I - Q)^{-1}R.$$

A proof of this result is outlined in Exercise 4.7-7.

EXAMPLE 4.7-8 For the gambler's chain in Example 4.7-2, we have

$$I-Q = \begin{bmatrix} 1 & -.4 & 0 \\ -.6 & 1 & -.4 \\ 0 & -.6 & 1 \end{bmatrix},$$

$$\begin{aligned} F &= (I-Q)^{-1}R \\ &= \begin{bmatrix} 1.46 & 0.77 & 0.31 \\ 1.15 & 1.92 & 0.77 \\ 0.69 & 1.15 & 1.46 \end{bmatrix} \begin{bmatrix} .6 & 0 \\ 0 & 0 \\ 0 & .4 \end{bmatrix} \\ &= \begin{bmatrix} 0.88 & 0.12 \\ 0.69 & 0.31 \\ 0.42 & 0.58 \end{bmatrix}. \end{aligned}$$

Of course, these results are consistent with those found in Example 4.7-7.

As a final note, we state a theorem that shows that the matrix $(I-Q)^{-1}$ can be interpreted as an expected value.

THEOREM 4.7-4 Let i and j be nonabsorbing states. Let μ_{ij} denote the expected number of visits to state j beginning in state i before the chain reaches an absorbing state. (If we are interested in returns to state i itself, the initial state i is counted as one visit.) Let U be the matrix whose ijth element is μ_{ij}. Then $U = (I-Q)^{-1}$.

The proof of this result is outlined in Exercise 4.7-9.

EXAMPLE 4.7-9 Using the computation of $(I - Q)^{-1}$ in Example 4.7-5, we see that the gambler who begins with \$20 will visit state \$10 an average of 1.15 times before reaching an absorbing state.

Exercises 4.7

4.7-1 To take out the family car you must request permission from your parents. Suppose there is an equal chance of Mom or Dad being asked first. When Dad is asked, there is a .7 probability you will be told to ask Mom and a .3 probability of getting permission to take out the car. When Mom is asked, there is a .4 probability you will be told to ask Dad and .6 probability of getting permission to take out the car. Suppose that you continue making requests until permission is given and that this process can be modeled as a Markov chain. The transition matrix is

$$P = \begin{array}{ll} & \begin{array}{cccc} 0 & 1 & 2 & 3 \end{array} \\ \begin{array}{lc} \text{Start process} & 0 \\ \text{Ask Dad} & 1 \\ \text{Ask Mom} & 2 \\ \text{Permission given} & 3 \end{array} & \begin{bmatrix} 0.0 & 0.5 & 0.5 & 0.0 \\ 0.0 & 0.0 & 0.7 & 0.3 \\ 0.0 & 0.4 & 0.0 & 0.6 \\ 0.0 & 0.0 & 0.0 & 1.0 \end{bmatrix} \end{array}$$

a What is the probability of getting permission in three or fewer requests?

b Find the probability of getting permission in exactly two requests.

c Find the mean number of requests it takes to get permission to take out the car.

4.7-2 A coin is tossed until five consecutive heads appear. Model this process as a Markov chain where the states are the numbers of consecutive heads $(0, 1, \ldots, 5)$.

a Find the probability that it takes 10 or fewer tosses to observe five consecutive heads.

b Find the mean number of tosses it takes to obtain five consecutive heads.

4.7-3 As a computer program is executed, it may access modules 1, 2, 3, 4, or it may terminate in state 5. Assume that the transitions among these modules behave as a Markov chain with the transition matrix

$$P = \begin{array}{cc} & \begin{array}{ccccc} 1 & 2 & 3 & 4 & 5 \end{array} \\ \begin{array}{c} 1 \\ 2 \\ 3 \\ 4 \\ 5 \end{array} & \begin{bmatrix} .1 & .2 & .2 & .1 & .4 \\ .1 & .1 & .2 & .2 & .4 \\ 0 & .1 & .2 & .2 & .5 \\ 0 & .1 & .1 & .2 & .6 \\ 0 & 0 & 0 & 0 & 1 \end{bmatrix} \end{array},$$

and assume that the program begins in module 1.

a Find the cumulative distribution function for the number of transitions that it takes for the program to terminate.

b Find the mean number of transitions that it takes for the program to terminate.

4.7-4 Refer to the rat in the maze in Example 4.3-1. Find the mean number of transitions it takes for the rat to go from state 1 to state 4.

4.7-5 A commuter coming from the university has three parking lots, A, B, and C, to choose from. She enters one lot, then another, and another until she finds a place to park. Her pattern of movement is always cyclical from A to B to C back to A, and so forth. Each time she enters lots A or B, she has a probability of .6 of finding a parking place and a probability of .4 of moving on to the next lot. In lot C, she has a probability of .3 of finding a place and a probability of .7 of moving along. Denote the states of this system as home (H), enter lot A and park (AP), enter lot A and not park (ANP), (BP), (BNP), (CP), (CNP).

a Give the transition matrix for this chain.

b Compute the average number of transitions the commuter must make until she finds a parking place. Assume that the commuter begins at home.

4.7-6 Suppose that the instructions received by a CPU of a computer can be of three types: read, process, and write. Assume that the instructions arrive at the CPU one after the other according to a Markov chain with transition matrix

$$P = \begin{array}{ll} & \begin{array}{ccc} 1 & 2 & 3 \end{array} \\ \begin{array}{lc} \text{Read} & 1 \\ \text{Process} & 2 \\ \text{Write} & 3 \end{array} & \begin{bmatrix} .3 & .6 & .1 \\ .4 & .3 & .3 \\ .7 & .2 & .1 \end{bmatrix} \end{array}.$$

a Find the expected number of instructions it takes to go from read to write.

b Find the expected number of instructions it takes to go from process to write.

4.7-7 Derive Theorem 4.7-3 by using the following steps. Let NA denote the nonabsorbing states. Beginning in a nonabsorbing state i, the first step can be directly to absorbing state j, and this occurs with probability $P(i \to j)$; otherwise, the first step is to nonabsorbing state s with probability $P(i \to s)$ from which the probability of eventual absorption into state j is $f(s \to j)$. Thus,

$$f(i \to j) = P(i \to j) + \sum_{s \in \text{NA}} P(i \to s) f(s \to j).$$

In matrix form, $F = R + Q \cdot F$, from which Theorem 4.7-3 follows.

4.7-8 A gambler begins a game with $\$A$ and plays an opponent with $\$B$. Each time a bet is made, the gambler has a probability of .5 of winning \$1 and .5 of losing \$1. The game is continued until the gambler or his opponent is out of money.

a Find the probability that the gambler loses his money.

b Show that if B is much larger than A, then the gambler has a very large probability of losing all his money. Thus, a gambler who plays a fair game against an opponent who has infinite resources is certain to go broke.

4.7-9 Use the following steps to prove Theorem 4.7-4. Let A denote the absorbing states and NA the nonabsorbing states. Consider μ_{ii}. The first step can be to an absorbing state, in which case the expected number of visits to state i is 1, or the first step can be to a nonabsorbing state s, in which case the expected number of visits to state i before absorption is $1 + \mu_{si}$. Thus,

$$\mu_{ii} = \sum_{j \in A} P(i \to j) + \sum_{s \in NA} P(i \to s)\,(1 + \mu_{si})$$

$$= 1 + \sum_{s \in NA} P(i \to s)\,\mu_{si}.$$

For $i \neq j$, if the first step is to an absorbing state, then the expected number of visits to the nonabsorbing state j is 0; otherwise, if the first step is to a nonabsorbing state s, then the expected number of visits to state j before absorption is μ_{sj}. Thus if $i \neq j$,

$$\mu_{ij} = \sum_{s \in NA} P(i \to s)\mu_{sj}.$$

Combining these two results in matrix form, we have $U = I + Q \cdot U$, from which the result follows.

4.7-10 Compute the matrix U for chain in Exercise 4.7-3 and interpret the results.

5

Continuous Random Variables

SECTION

5.1 Probability Density Functions

We now consider random variables that can be measured on a continuous scale such as time, distance, height, and weight. Unlike discrete random variables, the possible values of random variables measured on a continuous scale are uncountable. It is not possible to assign a positive probability mass to all the outcomes. Instead, we develop the idea of a *probability density function.*

To illustrate the concept of a probability density function, consider the following example. Let T denote the amount of time it takes a driver to go from home to work. Although the trip takes around 30 minutes on average, the actual time is a random variable because of unpredictable conditions along the way. Having driven the route many times, the driver was able to determine the probabilities empirically for the various travel times. Results to the nearest 2 minutes are given in Table 5.1-1. For instance, the probability that the travel time is between 25 and 27 minutes is .10.

TABLE **5.1-1**
Probability Distribution of Travel Time to Nearest Two Minutes

Travel Time in Minutes	Probability
$25 < T \leq 27$	.10
$27 < T \leq 29$	.17
$29 < T \leq 31$	.44
$31 < T \leq 33$	.20
$33 < T \leq 35$	.09

Figure 5.1-1 is a *probability histogram* of the data. In a probability histogram the vertical scale is constructed so that the area of each rectangle equals the probability of the corresponding interval and the sum of the rectangular areas is 1. A probability histogram equates areas with probability. If we want to find the probability that the travel time is between, say, 27 and 33 minutes, we simply add the areas of the corresponding rectangles, which in this case gives a value of 2(.085) + 2(.22) + 2(.10) = .81.

FIGURE 5.1-1
Histogram of Travel Times to the Nearest Two Minutes

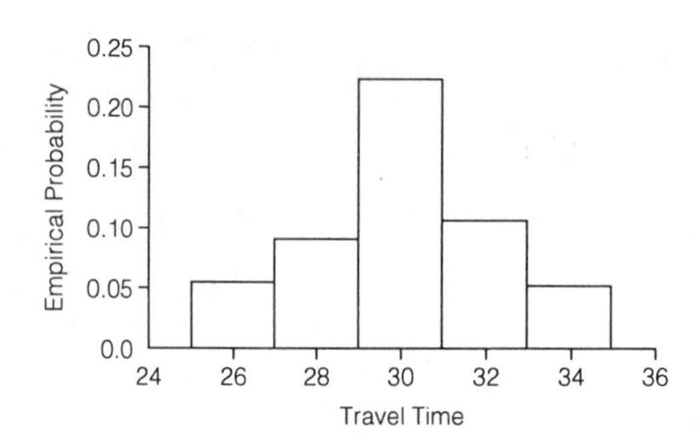

Continuing in this fashion, suppose instead of recording time to the nearest two minutes, the driver recorded time to the nearest minute. These empirical probabilities are given in Table 5.1-2.

TABLE 5.1-2
Probability Distribution of Travel Times to the Nearest Minute

Travel Time in Minutes	Probability
$25 < T \leq 26$	.04
$26 < T \leq 27$	.06
$27 < T \leq 28$	.07
$28 < T \leq 29$	.10
$29 < T \leq 30$	.20
$30 < T \leq 31$	.24
$31 < T \leq 32$	.12
$32 < T \leq 33$	.08
$33 < T \leq 34$	.06
$34 < T \leq 35$	.03

The probability histogram corresponding to the probabilities in Table 5.1-2 is given in Figure 5.1-2. Again the areas of the rectangles correspond to probabilities. For instance, the probability of a travel time between 28 and 32 minutes is the sum of the areas of the rectangles between 28 and 32, which is .66.

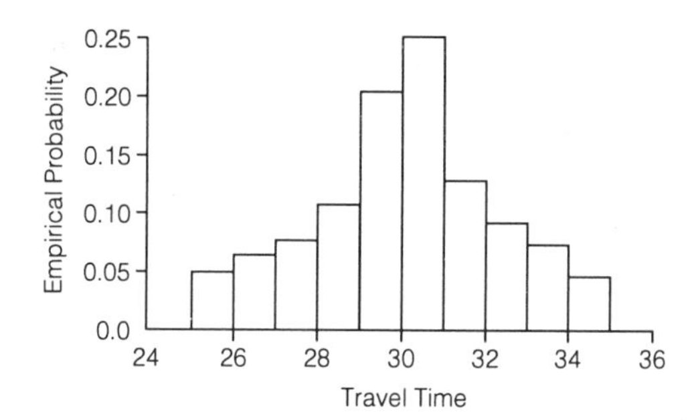

FIGURE 5.1-2
Histogram of Travel Times to the Nearest Minute

It is not hard to imagine that if we kept recording time in smaller and smaller increments the histogram would take on the appearance of the curve given in Figure 5.1-3. This curve is called the probability density function of travel time. The area under the curve between any two numbers is the probability that the travel time falls between those two numbers.

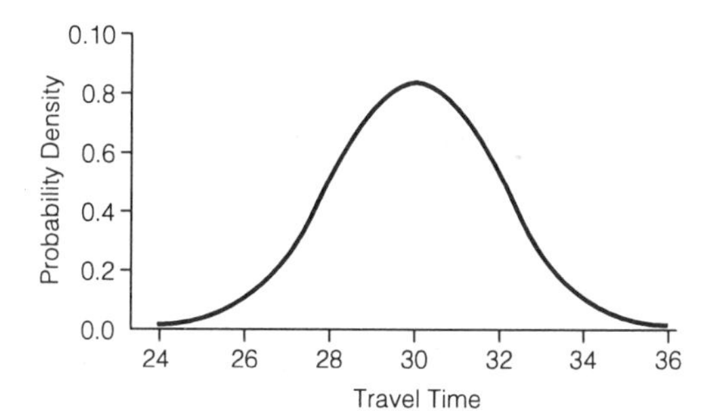

FIGURE 5.1-3
Probability Density Function of Travel Time

Definition 5.1-1

A random variable X is said to have a *probability density function* $f(x)$ if for all x

$$f(x) \geq 0,$$

and if for all values a and b

$$P(a \leq X \leq b) = \int_a^b f(x)\,dx.$$

Random variables having such density functions are called *continuous random variables*.

There is a sometimes confusing technical point that needs to be addressed. Consider the probability that the random variable X equals the value a. According to Definition 5.1-1,

$$\begin{aligned} P(X = a) &= P(a \leq X \leq a) \\ &= \int_a^a f(x)\,dx \\ &= 0. \end{aligned}$$

At first, it might seem contrary to intuition that the probability a random variable takes on any specific value is zero. However, suppose we ask in the previous example, "What is the probability that the travel time is exactly 29.65495216 minutes?" Surely, we would say that this probability is extremely small. Likewise, it would be equally true that the probability of observing any other time would also be extremely small provided we could record the time as precisely as we wish. As a mathematical idealization, we therefore say that the probability of observing any specific value of a continuous random variable is zero. Unlike the discrete case, zero probability in the continuous case is not the same as an impossibility. In practice, we use the probability density function to assign probabilities to intervals of numbers. The probability of observing the random variable in an interval is the area under the probability density function corresponding to that interval.

EXAMPLE 5.1-1 Let random variable X have probability density function

$$f(x) = \begin{cases} 2x, & 0 \leq x \leq 1, \\ 0, & \text{otherwise.} \end{cases}$$

This density function is shown in Figure 5.1-4.

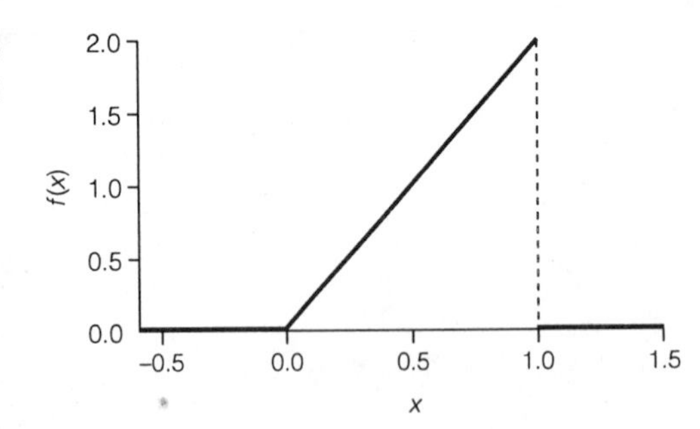

FIGURE 5.1-4 Probability Density Function for Example 5.1-1

To find $P(.25 \le X \le .75)$, we determine the area under the curve from .25 to .75:

$$\int_{.25}^{.75} 2x\,dx = x^2 \Big|_{.25}^{.75}$$
$$= .75^2 - .25^2$$
$$= .50.$$

To find $P(X > .50)$, note that the probability density function is positive only for values less than 1. That is, values of x bigger than 1 do not contribute to the probability. Therefore, this probability is given by the area under the curve from .50 to 1. This area is

$$\int_{.5}^{1} 2x\,dx = x^2 \Big|_{.5}^{1}$$
$$= 1 - .5^2$$
$$= .75.$$

EXAMPLE 5.1-2 Let the random variable X have a probability density function

$$f(x) = \begin{cases} cx(2-x), & 0 \le x \le 2, \\ 0, & \text{otherwise.} \end{cases}$$

To find the constant c, note that the area under the curve must be 1. This area is

$$\int_0^2 cx(2-x)\,dx = c\left(x^2 - \frac{x^3}{3}\right)\Bigg|_0^2$$
$$= c\left(\frac{4}{3}\right).$$

Therefore, $c = 3/4$. The density function is graphed in Figure 5.1-5.

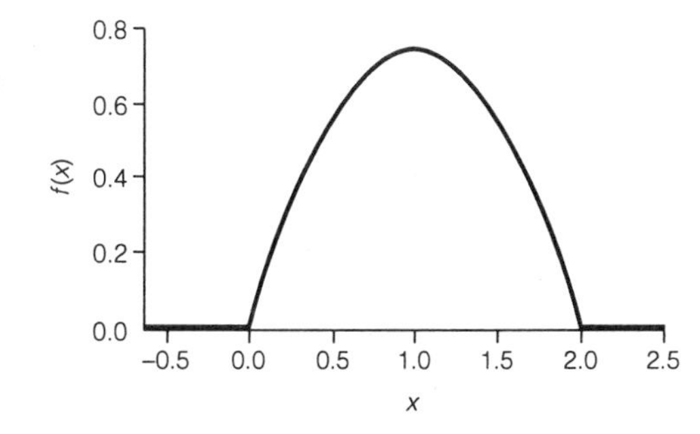

FIGURE 5.1-5 Probability Density Function for Example 5.1-2

Definition 5.1-2

A random variable X is said to have the *uniform distribution* on the interval $c \le x \le d$ if

$$f(x) = \begin{cases} \dfrac{1}{d-c}, & c \le x \le d, \\ 0, & \text{otherwise.} \end{cases}$$

The random variable X will be referred to as a uniform $[c, d]$ random variable.

The shape of the uniform probability density function, which is rectangular, is shown in Figure 5.1-6.

FIGURE **5.1-6** Uniform $[c, d]$ Probability Density Function

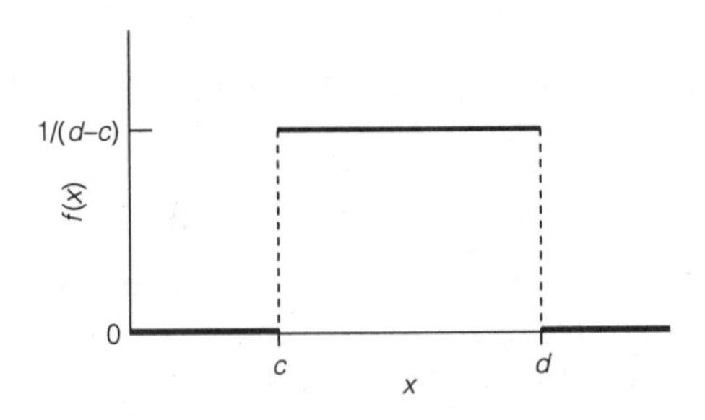

The uniform $[c, d]$ random variable represents a number selected at random from the interval $c \le x \le d$. If $[a, b]$ is any subinterval of $[c, d]$, then

$$\begin{aligned} P(a \le X \le b) &= \int_a^b \frac{1}{d-c}\, dx \\ &= \frac{x}{d-c} \Big|_a^b \\ &= \frac{b-a}{d-c}. \end{aligned}$$

Thus, the probability that a uniform $[c, d]$ random variable is observed to fall in the subinterval $[a, b]$ is the length of the subinterval divided by the length of the interval $[c, d]$. The unit random numbers that we have been using in our computer simulations can be thought of as observed values of a uniform $[0, 1]$ random variable.

In the previous examples, the sets for which the density functions were positive were bounded intervals. In the following example, this set is unbounded.

EXAMPLE 5.1-3 Let X be the number of minutes that it takes a randomly selected person to solve a word puzzle. The probability density function of X is

$$f(x) = \begin{cases} \dfrac{3}{x^4}, & x \geq 1, \\ 0, & x < 1. \end{cases}$$

The probability that it takes a person more than two minutes to solve the puzzle is

$$\begin{aligned} P(X > 2) &= \int_2^\infty \frac{3}{x^4}\,dx \\ &= \frac{-1}{x^3}\bigg|_2^\infty \\ &= \frac{1}{8}. \end{aligned}$$

Figure 5.1-7 is a graph of the density function. The shaded area in Figure 5.1-7 is for $P(X > 2)$.

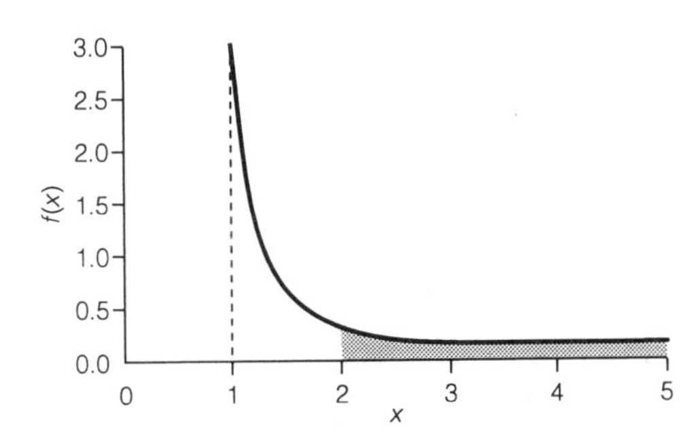

FIGURE 5.1-7 Probability Density Function for Example 5.1-3

The definition of the cumulative distribution function of a continuous random variable is analogous to that of the discrete case.

Definition 5.1-3

The cumulative distribution function of a continuous random variable X with probability density function $f(x)$ is

$$\begin{aligned} F(x) &= P(X \le x) \\ &= \int_{-\infty}^{x} f(v)\, dv. \end{aligned}$$

The cumulative distribution function is just the area under the density function falling less than or equal to x. Note that in the computation of $F(x)$ the actual limits of integration will depend on the domain over which $f(x) > 0$. This is illustrated in the following example.

EXAMPLE **5.1-4** Let the random variable X be defined as in Example 5.1-1. Since the density function is positive only for $0 \le x \le 1$, the computation of $F(x)$ will depend on whether x is in this region. The shaded area in Figure 5.1-8 represents $F(x)$ when $0 \le x \le 1$.

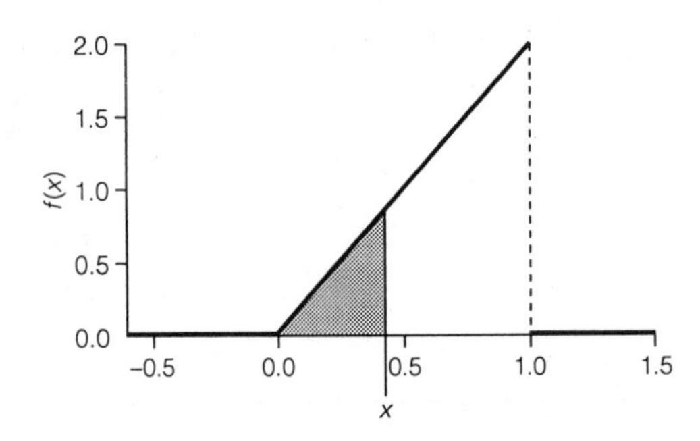

FIGURE **5.1-8** Area Representing Cumulative Probability for $0 \le x \le 1$

The computation of this probability is

$$\begin{aligned} F(x) &= \int_{-\infty}^{x} f(v)\, dv \\ &= \int_{0}^{x} 2v\, dv \\ &= x^2, \quad 0 \le x \le 1. \end{aligned}$$

For x less than 0, the area under the curve is 0. That is,

$$F(x) = 0, \quad x < 0.$$

For x greater than 1, the area under the curve is 1, so

$$F(x) = 1, \quad x > 1.$$

Therefore,

$$F(x) = \begin{cases} 0, & x < 0, \\ x^2, & 0 \le x \le 1, \\ 1, & x > 1. \end{cases}$$

The graph of the cumulative distribution function is given in Figure 5.1-9.

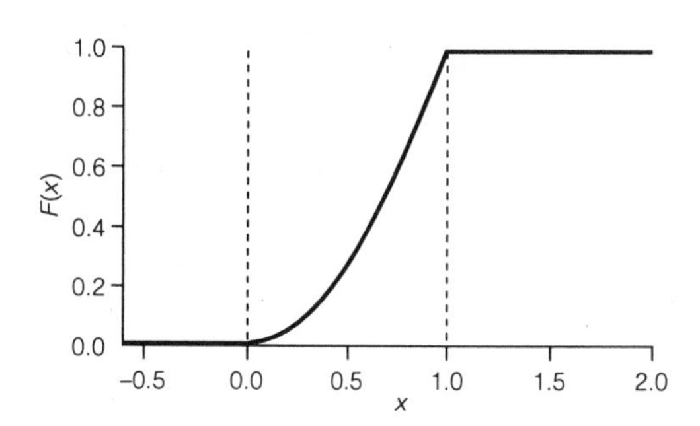

FIGURE **5.1-9** Cumulative Distribution Function for Example 5.1-4

The cumulative distribution function can be used directly to compute probabilities of interest. For instance,

$$P(X \le .2) = F(.2) = .2^2 = .04,$$

$$P(X > .7) = 1 - P(X \le .7) = 1 - F(.7) = 1 - .7^2 = .51,$$

$$P(.3 < X \le .8) = P(X \le .8) - P(X \le .3) = F(.8) - F(.3) = .8^2 - .3^2 = .55.$$

EXAMPLE **5.1-5** Consider the uniform $[c, d]$ random variable. Since the density function has the value $1/(d-c)$ over the interval $c \le x \le d$, we have

$$F(x) = \int_c^x \frac{1}{d-c}\, dv$$

$$= \frac{x-c}{d-c}, \quad c \le x \le d.$$

Moreover, $F(x) = 0$ for $x < c$ and $F(x) = 1$ for $x > d$. The graph of $F(x)$ is shown in Figure 5.1-10.

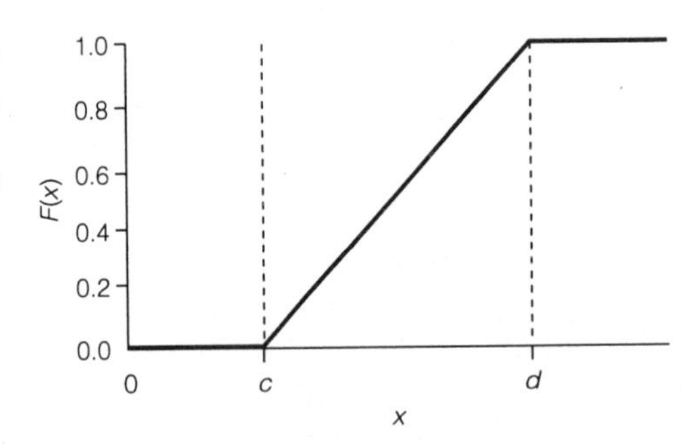

FIGURE **5.1-10** Cumulative Distribution Function for Uniform $[c, d]$ Random Variable

EXAMPLE **5.1-6** Let X denote the time it takes a person to solve a word puzzle as defined in Example 5.1-3. The cumulative distribution function is computed as

$$F(x) = \int_1^x \frac{3}{v^4}\,dv$$

$$= 1 - \frac{1}{x^3}, \quad x \geq 1,$$

and $F(x) = 0, x < 1$. The probability that it takes a person 1.5 minutes or less to solve the puzzle is $F(1.5) = 1 - (1.5)^{-3} = .70$. The graph of the cumulative distribution function is shown in Figure 5.1-11.

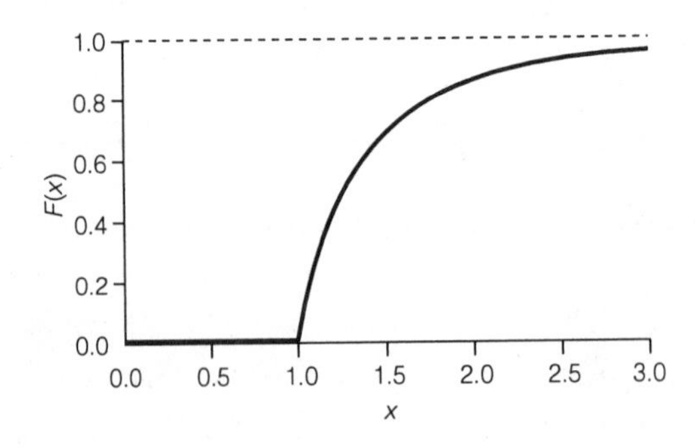

FIGURE **5.1-11** Cumulative Distribution Function for Example 5.1-6

Since the cumulative distribution function is the integral of the probability density function, it follows that the probability density function of a continuous random variable is the derivative of the cumulative distribution function. That is,

$$f(x) = \frac{d}{dx} F(x) = F'(x).$$

The following example illustrates the use of this fact in deriving the probability density function of a random variable.

EXAMPLE 5.1-7 A dart is thrown at a circular dart board of radius 5 units. The probability that the dart lands in any given region of the board is assumed to be equal to the area of the region divided by the area of the dart board. In other words, the dart striking the board is like selecting a point at random on the board. Let X denote the distance that the dart lands from the center. We can find the cumulative distribution function of X directly. We see that for any value of x between 0 and 5

$$\begin{aligned} F(x) &= P(X \leq x) \\ &= P(\text{dart lands within a distance } x \text{ from the center}) \\ &= \frac{\text{Area of circle of radius } x}{\text{Area of circle of radius } 5} \\ &= \frac{\pi x^2}{\pi 5^2} \\ &= \frac{x^2}{25}, \quad 0 \leq x \leq 5. \end{aligned}$$

Since the dart is assumed to strike the board, the probability density function is positive only for $0 \leq x \leq 5$, and its value can be computed by differentiating $F(x)$. Thus,

$$f(x) = \begin{cases} \dfrac{2x}{25}, & 0 \leq x \leq 5, \\ 0, & \text{otherwise.} \end{cases}$$

Exercises 5.1

5.1-1 Let X have a probability density function defined by

$$f(x) = \begin{cases} cx, & 0 \le x \le 4, \\ 0, & \text{otherwise.} \end{cases}$$

a Find the constant c.

b Find $P(1 \le X \le 2)$.

5.1-2 The number of seconds it takes a certain runner to complete 400 meters is a random variable with a uniform distribution on the interval $58 \le x \le 62$.

a Find the probability that the distance is covered in less than 60 seconds.

b Find the cumulative distribution function of the time that it takes the runner to run 400 meters.

5.1-3 The time, in seconds, to compile computer programs written by students in a certain class is a random variable X with probability density function

$$f(x) = \begin{cases} \dfrac{3}{(x+1)^4}, & x \ge 0, \\ 0, & \text{otherwise.} \end{cases}$$

a Find the probability that it takes more than 1.5 seconds to compile a program.

b Find the time m at which $P(X \le m) = .5$. This value is called the *median* of the random variable X.

5.1-4 A random variable X has the probability density function

$$f(x) = \begin{cases} .75x(2-x), & 0 \le x \le 2, \\ 0, & \text{otherwise.} \end{cases}$$

Find and sketch the cumulative distribution function of X.

5.1-5 The number of minutes between successive arrivals of airplanes at an airport is a random variable X with cumulative distribution function

$$F(x) = \begin{cases} 1 - e^{-.2x}, & x > 0, \\ 0, & \text{otherwise.} \end{cases}$$

a Find $P(5 \le X \le 10)$.

b Find the probability that there are more than 10 minutes between successive arrivals.

c Find the probability density function of X.

5.1-6 A bottling machine is supposed to fill containers with 16 ounces of liquid. The number of ounces X by which the machine misses the target value of 16 is a random variable with probability density function

$$f(x) = \begin{cases} 1+x, & -1 \le x \le 0, \\ 1-x, & 0 \le x \le 1, \\ 0, & \text{otherwise} \end{cases}$$

a Sketch the density function.

b Find the probability that the machine misses the target value by a magnitude of less than .5 ounce.

c Find the probability that the machine puts in less than 15.75 ounces of liquid in a bottle.

SECTION

5.2 Expected Value and Distribution of a Function of a Random Variable

The definitions of expected value, variance, standard deviation, and the various theorems pertaining to these concepts carry over from the discrete case to the continuous case. The transition is simply a matter of substituting $f(x)\,dx$ for the probability mass function $p(x)$ and substituting $\int$ for Σ. For example, if X is a continuous random variable with probability density function $f(x)$, then the expected value is

$$E(X) = \int_{-\infty}^{\infty} x f(x)\,dx.$$

For a function $Y = \Phi(X)$ of a continuous random variable,

$$E(Y) = E(\Phi(X)) = \int_{-\infty}^{\infty} \Phi(x) f(x)\,dx.$$

We will use the theorems about expectations from Chapter 2 as we have need for them without first restating them for the continuous case.

EXAMPLE 5.2-1 Let X have the continuous probability density function defined by

$$f(x) = \begin{cases} 2x, & 0 \le x \le 1, \\ 0, & \text{otherwise.} \end{cases}$$

In computing the expected value we need only consider the interval $[0, 1]$ in evalu-

ating the integral because the density is zero outside this range. Thus,

$$\begin{aligned} E(X) &= \int_0^1 x(2x)\,dx \\ &= \left.\frac{2x^3}{3}\right|_0^1 \\ &= \frac{2}{3}. \end{aligned}$$

Moreover,

$$\begin{aligned} E(X^2) &= \int_0^1 x^2(2x)\,dx \\ &= \left.\frac{2x^4}{4}\right|_0^1 \\ &= \frac{1}{2}. \end{aligned}$$

Therefore,

$$\begin{aligned} \mathrm{VAR}(X) &= E(X^2) - [E(X)]^2 \\ &= \frac{1}{2} - \left(\frac{2}{3}\right)^2 \\ &= \frac{1}{18} \end{aligned}$$

and

$$\mathrm{STD}(X) = \sqrt{\frac{1}{18}} = .24.$$

The two-standard-deviation interval about the mean is $.67 \pm .48$, which is the interval .19 to 1.15. This interval is shown in Figure 5.2-1. The probability that X falls in this interval is

$$\begin{aligned} \int_{.19}^1 2x\,dx &= 1 - (.19)^2 \\ &= .96. \end{aligned}$$

Again note that the upper limit of the integral is 1, not 1.15, since the probability density function is 0 for values of $x > 1$ As expected, the two-standard-deviation interval about the mean contains most of the probability.

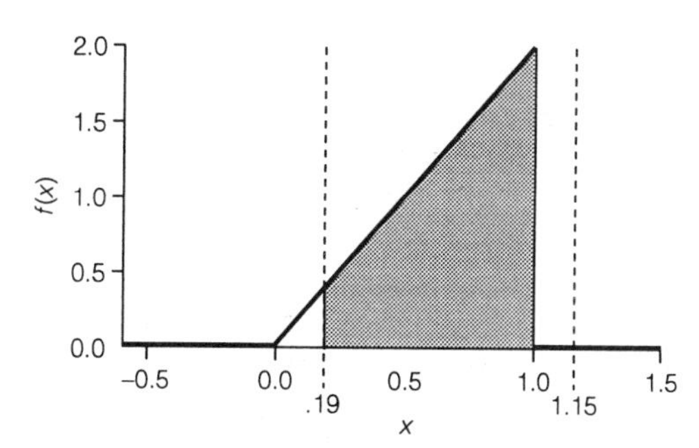

FIGURE **5.2-1** Two Standard Deviation Interval for Example 5.2-1

EXAMPLE **5.2-2** Suppose the amount of time, in hours, it takes a novice golfer to finish nine holes of golf is a random variable X with probability density function

$$f(x) = \begin{cases} \dfrac{8}{x^3}, & 2 \le x, \\ 0, & \text{otherwise.} \end{cases}$$

The expected time to finish is

$$\begin{aligned} E(X) &= \int_2^\infty x\left(\frac{8}{x^3}\right) dx \\ &= -\frac{8}{x}\bigg|_2^\infty \\ &= 4 \text{ hours.} \end{aligned}$$

Note that

$$\begin{aligned} E(X^2) &= \int_2^\infty x^2\left(\frac{8}{x^3}\right) dx \\ &= \int_2^\infty \left(\frac{8}{x}\right) dx \\ &= 8\ln(x)\bigg|_2^\infty \\ &= \infty. \end{aligned}$$

Thus, the variance of X is infinite. If this distribution is truly representative of the amount of time it takes novice golfers to finish a round, then the infinite variance tells us that there must be some poor golfers who get caught in sand traps for very long times.

EXAMPLE 5.2-3 The mean and variance of a uniform $[c, d]$ random variable are derived as follows.

$$
\begin{aligned}
E(X) &= \int_c^d \frac{x}{d-c}\,dx \\
&= \frac{d^2 - c^2}{2(d-c)} \\
&= \frac{1}{2}(d+c),
\end{aligned}
$$

$$
\begin{aligned}
E(X^2) &= \int_c^d \frac{x^2}{d-c}\,dx \\
&= \frac{d^3 - c^3}{3(d-c)} \\
&= \frac{1}{3}(d^2 + cd + c^2).
\end{aligned}
$$

Thus,

$$
\begin{aligned}
\mathrm{VAR}(X) &= \frac{1}{3}(d^2 + cd + c^2) - \left[\frac{1}{2}(d+c)\right]^2 \\
&= \frac{1}{12}(d-c)^2.
\end{aligned}
$$

In particular, a uniform $[0, 1]$ random variable has a mean of $1/2$ and a variance of $1/12$.

Given a random variable X and its probability distribution, we may be interested in finding the distribution of a new random variable $Y = \Phi(X)$. Since Y is a function of X, we call this a "change of variables" problem. To solve the problem, we will first find the cumulative distribution function $G(y)$ in terms of $F(x)$, using probability arguments. The density $g(y)$ can then be found by differentiating $G(y)$ with respect to y. Following this discussion we present a more direct transformation of variables method to find the distribution of Y.

EXAMPLE 5.2-4 Let X have the probability density function defined in Example 5.2-1, and let $Y = 5X + 1$. Note that

cumulative distribution

$$P(Y \le y) = P(5X + 1 \le y)$$
$$= P\left(X \le \frac{y-1}{5}\right)$$
$$= F\left(\frac{y-1}{5}\right).$$

$\int f(x)$

Since $F(x) = 0$, for $x < 0$, we have $G(y) = 0$ for $(y-1)/5 < 0$ or for $y < 1$. Also, since $F(x) = 1$ for $x > 1$, we have $G(y) = 1$ for $(y-1)/5 > 1$ or for $y > 6$. Finally $F(x) = x^2$ for $0 \le x \le 1$; therefore,

$$G(y) = \left[\frac{y-1}{5}\right]^2$$

for $0 \le (y-1)/5 \le 1$ or for $1 \le y \le 6$. Differentiating $G(y)$, with respect to y we find

$$g(y) = \frac{2(y-1)}{25}, \quad 1 \le y \le 6,$$

and $g(y) = 0$ for $y < 1$ or $y > 6$. The density function of Y is graphed in Figure 5.2-2.

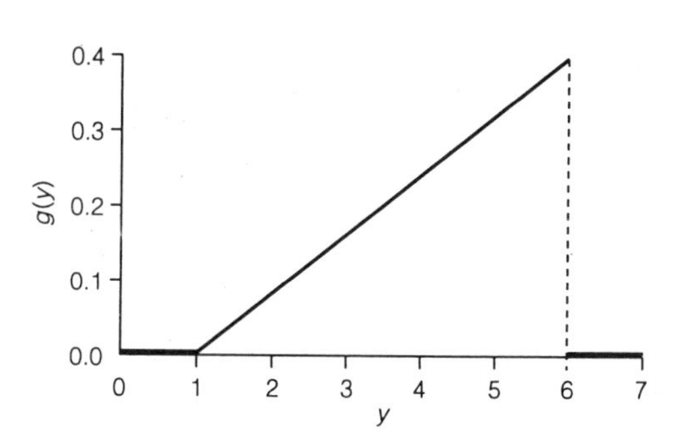

FIGURE **5.2-2** Probability Density Function for Y in Example 5.2-4

EXAMPLE **5.2-5** Let X be a uniform $[-1, 1]$ random variable. From Example 5.1-5,

$$F(x) = \begin{cases} 0, & x < 1, \\ \dfrac{x+1}{2}, & -1 \le x \le 1, \\ 1, & x > 1. \end{cases}$$

Let $Y = X^2$. As x transverses the interval $[-1, 1]$, $y = x^2$ traverses the interval $[0, 1]$. Thus, for $0 \le y \le 1$,

$$\begin{aligned} G(y) &= P(Y \le y) \\ &= P(X^2 \le y) \\ &= P(-\sqrt{y} \le X \le \sqrt{y}) \\ &= F(\sqrt{y}) - F(-\sqrt{y}) \\ &= \frac{\sqrt{y}+1}{2} - \frac{-\sqrt{y}+1}{2} \\ &= \sqrt{y}. \end{aligned}$$

Therefore,

$$g(y) = \begin{cases} \dfrac{1}{2\sqrt{y}}, & 0 \le y \le 1, \\ 0, & \text{otherwise.} \end{cases}$$

If $Y = \Phi(X)$ is an increasing or decreasing function, the probability density function can be written directly without the use of probability arguments. The function $\Phi(X)$ is increasing (decreasing) if for $x_1 \le x_2$, $\Phi(x_1) \le \Phi(x_2)$ $(\Phi(x_1) \ge \Phi(x_2))$.

Suppose $Y = \Phi(X)$ is an increasing function. Then

$$\begin{aligned} G(y) &= P(Y \le y) \\ &= P(\Phi(x) \le \Phi(x)) \\ &= P(X \le x) \\ &= F(x). \end{aligned}$$

Therefore,

$$\begin{aligned} g(y) &= \frac{d}{dy} G(y) \\ &= \frac{d}{dy} F(x) \\ &= F'(x) \frac{dx}{dy} \\ &= f(x) \frac{dx}{dy}, \end{aligned}$$

where x is expressed in terms of y by solving $y = \Phi(x)$ for x.

If $Y = \Phi(X)$ is a decreasing function, then

$$\begin{aligned} G(y) &= P(Y \le y) \\ &= P(\Phi(X) \ge \Phi(x)) \\ &= P(X \ge x) \\ &= 1 - F(x) \end{aligned}$$

and

$$\begin{aligned} g(y) &= \frac{d}{dy} G(y) \\ &= \frac{d}{dy}(1 - F(x)) \\ &= -f(x)\frac{dx}{dy}. \end{aligned}$$

Because dx/dy is negative when $\Phi(X)$ is a decreasing function and dx/dy is positive when $\Phi(X)$ is an increasing function, the following expression can be used for either case.

Transformation of Variables

Let the random variable X have probability density function $f(x)$. For $Y = \Phi(X)$, either an increasing or decreasing function of X, the probability density function for Y is

$$g(y) = f(x)\left|\frac{dx}{dy}\right|,$$

where x is expressed in terms of y by solving $y = \Phi(x)$ for x.

EXAMPLE 5.2-6 Let X and Y be defined as in Example 5.2-4. That is,

$$f(x) = \begin{cases} 2x, & 0 \le x \le 1, \\ 0, & \text{otherwise.} \end{cases}$$

and $Y = 5X + 1$. Then $x = (y-1)/5$, $dx/dy = 1/5$, and

$$g(y) = \frac{2(y-1)}{5}\left|\frac{1}{5}\right|$$
$$= \frac{2}{25}(y-1)$$

for $1 \le y \le 6$. Note that $g(y) = 0$ for $y < 1$ or $y > 6$.

EXAMPLE 5.2-7 Let X be the number of minutes it takes a person to solve a word puzzle as in Example 5.1-3. The probability density function is

$$f(x) = \begin{cases} \frac{3}{x^4}, & 1 \le x, \\ 0, & x < 1. \end{cases}$$

Let $Y = 1/X$. First note that as x traverses the interval $1 \le x < \infty$, then $y = 1/x$ traverses the interval $0 < y \le 1$. Now, $x = 1/y$, $dx/dy = -1/y^2$, and

$$g(y) = \frac{3}{(1/y)^4}\left|\frac{-1}{y^2}\right|$$
$$= 3y^2,$$

for $0 < y \le 1$ and $g(y) = 0$ for $y \le 0$ or $y > 1$.

As in the discrete case,

$$E(a+bX) = a + bE(X) \text{ and } \mathrm{STD}(a+bX) = |b|\mathrm{STD}(X)$$

for a continuous random variable X.

EXAMPLE 5.2-7 Consider the random variable X from Examples 5.1-3 and 5.2-6. Now,

$$E(X) = \int_1^\infty x\frac{3}{x^4}\,dx = 1.5 \text{ minutes},$$

$$E(X^2) = \int_1^\infty x^2\frac{3}{x^4}\,dx = 3,$$

$$\text{VAR}(X) = 3 - (1.5)^2 = .75,$$

$$\text{STD}(X) = \sqrt{.75} = .866 \text{ minute.}$$

Suppose a new puzzle is introduced that takes three times as long as the current puzzle. Let the random variable $Y = 3X$ denote the time that it takes to solve the new puzzle. A two-standard-deviation interval for Y is computed as follows. First, $E(Y) = 3E(X) = 4.5$ minutes and $\text{STD}(Y) = 3\text{STD}(X) = 2.598$ minutes. Therefore, $\mu_Y \pm 2\sigma = 4.5 \pm 5.196$, which is the interval from $-.696$ to 9.696 minutes.

Exercises 5.2

5.2-1 A random variable X has probability density function

$$f(x) = \begin{cases} 1 - \dfrac{x}{2}, & 0 \le x \le 2, \\ 0, & \text{otherwise.} \end{cases}$$

a Find $E(X)$ and $\text{STD}(X)$.

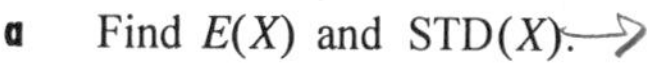

b Compute the probability that X falls within two standard deviations of $E(X)$.

5.2-2 Let X have the uniform distribution on the interval $[0, 1]$. Find $E(X^n)$ for any integer $n > 0$.

5.2-3 The number of minutes that it takes to load an airplane with supplies is a random variable X with probability density function

$$f(x) = \begin{cases} \dfrac{x - 10}{200}, & 10 \le x \le 30, \\ 0, & \text{otherwise.} \end{cases}$$

Find $E(X)$ and $\text{STD}(X)$.

5.2-4 The temperature X (Celsius) at a randomly selected point in a commercial refrigerator is a random variable with probability density function

$$f(x) = \begin{cases} \dfrac{x^2}{9}, & 0 \le x \le 3, \\ 0, & \text{otherwise.} \end{cases}$$

a Find $E(X)$ and $\text{STD}(X)$.

b Let $Y = 9X/5 + 32$, the Fahrenheit temperature. Find $E(Y)$ and $\text{STD}(Y)$.

5.2-5 Let X have the probability density function defined in Exercise 5.2-1, and let $Y = 2X + 3$. Find the cumulative distribution function and the probability density function of Y.

5.2-6 Let X have probability density function $f(x)$ and cumulative distribution function $F(x)$. Let $Y = aX + b$, where $a > 0$.

a Show that the cumulative distribution function of Y is

$$G(y) = F\left(\frac{y-b}{a}\right).$$

b Differentiate $G(y)$ to show that the probability density function of Y is

$$g(y) = \frac{1}{a} f\left(\frac{y-b}{a}\right).$$

5.2-7 Let X be a $[0, 1]$ uniform random variable. Let $Y = -\ln(X)$. Find the probability density function of Y.

5.2-8 Let X have probability density function

$$f(x) = \begin{cases} 4x^3, & 0 \le x \le 1, \\ 0, & \text{otherwise.} \end{cases}$$

Let $Y = 1/\sqrt{X}$. Find the probability density function of Y.

5.2-9 Let X have uniform distribution on the interval $[-2, 2]$. Let $Y = 4 - X^2$.

a Find the cumulative distribution function of Y.

b Find the probability density function of Y.

5.2-10 The radius of a circle X is a random variable with probability density function

$$f(x) = \begin{cases} 2x, & 0 \le x \le 1, \\ 0, & \text{otherwise.} \end{cases}$$

Let Y denote the area of the circle with radius X. Find the probability density function of Y.

5.2-11 The probability density function for X is

$$f(x) = \begin{cases} e^{-x}, & x > 0, \\ 0, & \text{otherwise.} \end{cases}$$

Let $Y = X^2$. Find the probability density function of Y.

5.2-12 Let the random variable X denote the length of a piece of software in terms of number of tokens. Assume for simplicity that we can approximate the mass function of X with the continuous probability density function

$$f(x) = \begin{cases} \frac{1}{100} e^{-x/100}, & x > 0, \\ 0, & \text{otherwise.} \end{cases}$$

The time in hours it takes to code the software is the random variable $Y = 8 + X/50$. Find the probability density function of Y. Find the probability that it will take at most 16 hours to code the software.

SECTION

5.3 Simulating Continuous Random Variables

In this section we consider the problem of simulating values of continuous random variables. The method for discrete random variables does not apply directly to the continuous case, but there are similarities. We use the discrete case to motivate a method for continuous random variable simulation.

Suppose a discrete random variable X with cumulative distribution function $F(x)$ has possible values $a_1 < a_2 < \cdots < a_n$, and suppose we wish to simulate a value of X. If a unit random number u is selected such that $F(a_{i-1}) \leq u < F(a_i)$, for $i = 2, 3, \ldots, n$, the value assigned to X is $X = a_i$. (If $u < F(a_1)$, then X is assigned a_1.) Now consider what would happen if X has many possible values that are close together, each having small probability. An example of this would occur if the height of a randomly selected individual were rounded to the nearest tenth of an inch. Such a random variable could, for practical purposes, be treated as a continuous random variable. The cumulative probabilities $F(a_{i-1})$ and $F(a_i)$ for two adjacent values of X would have almost the same values, and the interval $F(a_{i-1}) \leq u < F(a_i)$ would be very small. Thus, the simulated value of X would satisfy the relationship $u \approx F(a_i)$. This suggests the following method for simulating a value of a continuous random variable.

Simulation of Continuous Random Variables

Let X be a continuous random variable with cumulative distribution function $F(x)$. To simulate a value x for X,

1 Obtain a unit random number u.

2 Set $u = F(x)$.

3 Solve for x.

The preceding method is frequently called the *cumulative distribution function method* and is justified by the following theorem.

THEOREM 5.3-1 Let X be a continuous random variable with cumulative distribution function $F(x)$. Then the random variable $U = F(X)$ has the uniform $[0, 1]$ distribution.

Verification of Theorem 5.3-1 is outlined in Exercise 5.3-10. Theorem 5.3-1 tells us that we can relate a random variable X to a uniform $[0, 1]$ random variable U by the equation $U = F(X)$. Thus, given an observed value of U we can obtain an observed value of X by solving $u = F(x)$. If there is more than one x that satisfies this equation, choose the smallest.

EXAMPLE 5.3-1 Suppose we want to generate a random variable X that has probability density function

$$f(x) = \begin{cases} \dfrac{x}{2}, & 0 \le x \le 2, \\ 0, & \text{otherwise.} \end{cases}$$

The cumulative distribution function is

$$F(x) = \begin{cases} 0, & x < 0, \\ \dfrac{x^2}{4}, & 0 \le x \le 2, \\ 1, & 2 < x. \end{cases}$$

To generate a value of X, we need only consider a solution for $u = F(x)$ for x in the interval $[0, 2]$, because $f(x)$ is zero outside the interval $[0, 2]$. Since x is always nonnegative, the solution to $u = x^2/4$ is the positive square root of $4u$, or $x = \sqrt{4u}$. For instance if $u = .51679$, then $x = \sqrt{4(.51679)} = 1.44$. That is, 1.44 would be a simulated value of the random variable X.

EXAMPLE 5.3-2 Suppose we want to generate a random variable X that has the uniform distribution on the interval $[c, d]$. According to Example 5.1-5, the cumulative distribution function is

$$F(x) = \begin{cases} 0, & x < c, \\ \dfrac{x-c}{d-c}, & c \le x \le d, \\ 1, & x > d. \end{cases}$$

If u is a unit random number, then we solve for x in the equation

$$u = \frac{x-c}{d-c}.$$

Thus,

$$x = c + u(d - c).$$

This equation represents a rescaling of the unit number u. The unit random number is multiplied by the length of the interval $[c, d]$ and then shifted in location by adding a constant c. For instance, if $u = .74431$, then $x = 1 + 3(.74431) = 3.23293$ is a simulated value of a uniform $[1, 4]$ random variable.

EXAMPLE **5.3-3** Let X have probability density function

$$f(x) = \begin{cases} \dfrac{3}{x^4}, & x \geq 1, \\ 0, & x < 1. \end{cases}$$

As shown in Example 5.1-6, the cumulative distribution function of X is

$$F(x) = \begin{cases} 0, & x < 1, \\ 1 - \dfrac{1}{x^3}, & x \geq 1. \end{cases}$$

If we set $u = 1 - x^{-3}$ and solve for x, then $x = (1 - u)^{-1/3}$. For instance, if $u =$ .91213, then $x = (1 - .91213)^{-1/3} = 2.25$ would be the simulated value of the random variable X.

The following example illustrates practical limitations of the cumulative distribution function method of generating continuous random variables.

EXAMPLE **5.3-4** Let $f(x)$ have probability density function

$$f(x) = \begin{cases} c(16x^3 - x^4 - x^5), & 0 \leq x \leq 3, \\ 0, & \text{otherwise}, \end{cases}$$

where c is a constant chosen so that the area under $f(x)$ is 1. The cumulative distribution function has the form

$$F(x) = \begin{cases} 0, & x < 0, \\ c\left(4x^4 - \dfrac{x^5}{5} - \dfrac{x^6}{6}\right), & 0 \le x \le 3, \\ 1, & x > 3. \end{cases}$$

Thus, the equation $u = F(x)$ requires finding the roots of a sixth-degree polynomial. Numerical techniques are available for solving such equations, but these are beyond the scope of this book. The point is that there may be no simple solution to the equation $u = F(x)$.

The practical limitations illustrated in Example 5.3-4 have led to a variety of techniques other than the cumulative distribution function method for simulating continuous random variables. The discussion of the techniques is beyond the scope and purpose of this text. The reader is referred to sources such as Kennedy and Gentle (1980) and Knuth (1973).

Let $X_1, X_2, \ldots, X_n$ be a random sample of the continuous random variable X. Just as in the case of discrete random variables, the sample mean, sample variance, and sample standard deviation are statistical estimates for $E(X)$, $\mathrm{VAR}(X)$, and $\mathrm{STD}(X)$. The idea is illustrated in the following example.

EXAMPLE 5.3-5 Let X have the probability density function defined in Example 5.3-1. For this random variable, $E(X) = 4/3 = 1.33$, $\mathrm{VAR}(X) = 2/9 = 0.22$, and $\mathrm{STD}(X) = 0.47$. Now consider the following five simulated values of X in Table 5.3-1.

TABLE 3.3-1
Simulated Values for X with $f(x) = \frac{x}{2}$, $0 \le x \le 2$

u	$x = \sqrt{4u}$
.74211	1.7229
.10119	0.6362
.95452	1.9540
.14267	0.7554
.41744	1.2922

The sample mean is $\bar{X} = 1.2721$, the sample variance is $S^2 = .2680$, and the sample standard deviation is $S = .5177$. The values compare favorably to $E(X)$, $\mathrm{VAR}(X)$, and $\mathrm{STD}(X)$.

In order to display data from a random sample so that the picture reflects the shape of the probability density function $f(x)$, we construct an empirical probability histogram of the data. We do this by dividing the range of the data into equal-length subintervals and making a bar graph of the percentage of observations falling in each subinterval. The vertical axis is scaled so that the areas of the rectangles correspond to empirical probabilities in each interval and so that the total area is 1. A rule of thumb for the number of subintervals to use for the histogram to have a reasonable appearance is

$$\text{Number of subintervals} = \log_2(\text{sample size}).$$

EXAMPLE **5.3-6** Let X have the probability density function defined in Example 5.3-1. We obtained 1000 simulated values of the random variable. Our rule of thumb indicates that we should classify the data into approximately $\log_2(1000) = 10$ intervals. The data are classified into the 10 subintervals defined by $0 \le x < .2$, $.2 \le x < .4$, $\ldots$, $1.8 \le x < 2$. The empirical probability histogram is shown in Figure 5.3-1. The height of each rectangle is the empirical probability of the interval divided by .2, the length of the interval. The probability density function is superimposed on this histogram for purposes of comparison. We see close agreement between the empirical probabilities as represented by areas of rectangles and the true probabilities as represented by areas under the density $f(x)$. The sample mean and sample standard deviation of the 1000 numbers are 1.34 and 0.49, respectively. These values are in close agreement with $E(X) = 1.33$ and $\text{STD}(X) = .47$.

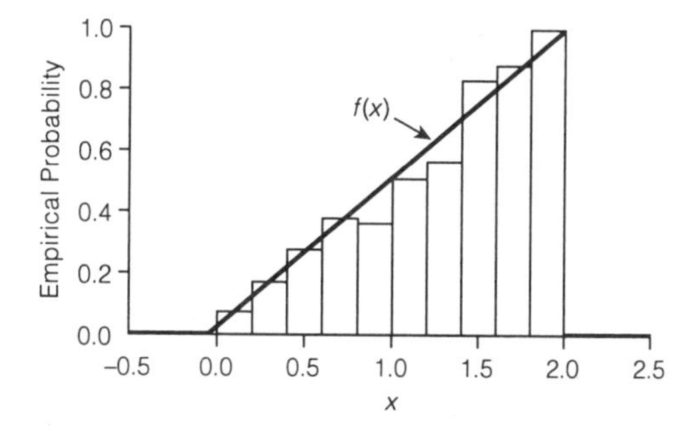

FIGURE **5.3-1** Empirical Probability Histogram and Probability Density Function for Example 5.3-6

EXAMPLE **5.3-7** A lending company uses three appraisers to determine the value of a house. Each appraiser gives a value, and the lending company chooses the middle value as its appraised value. For instance, if the appraisals were ($110,000, $95,000, $100,000), the lending company would use $100,000 as the appraised value of the house. The first appraiser's value is a random variable X with a uniform ($90,000, $110,000) distribution. The second appraiser's value is a random variable Y with a uniform (\$95,000, \$120,000) distribution. The third appraiser's value is a random variable W with the following triangular probability density function.

$$f(w) = \begin{cases} \dfrac{2}{(20{,}000)^2}(w - 90{,}000), & \$90{,}000 \le w \le \$110{,}000, \\ 0, & \text{otherwise.} \end{cases}$$

To simulate an appraised value for the house, one simulates values for X, Y, and W and the middle value is determined. The simulated values of X and Y can be determined from Example 5.3-2. Let u_1 and u_2 be unit random numbers. Then $x = 90{,}000 + 20{,}000u_1$ and $y = 95{,}000 + 25{,}000u_2$. To simulate values of W, we must first find the cumulative distribution function for W in the range \$90,000 to \$110,000. This is

$$F(w) = \frac{1}{(20{,}000)^2}(w - 90{,}000)^2, \quad \$90{,}000 \le w \le \$110{,}000.$$

Given a unit random number u_3, a simulated value of W is obtained by solving the equation $u_3 = F(w)$, which gives $w = 90{,}000 + 20{,}000\sqrt{u_3}$. Table 5.3-2 contains the empirical probabilities for the lending company's appraised value of the house based on 1000 simulations and broken into 10 intervals. Figure 5.3-2 shows a histogram of the number of simulated values falling in each interval along with a *jittered one-dimensional scattergram*. The jittered one-dimensional scattergram plots each simulated price on the horizontal axis with a randomly selected vertical value. In this example the horizontal values correspond to the simulated appraised values of the house. This display, along with the histogram, gives a visual sense of the data density in various intervals.

TABLE **5.3-2** Empirical Probabilities for Example 5.3-7

Appraised Value	Empirical Probability
\$ 92,000 to 93,799	.009
93,800 to 95,599	.020
95,600 to 97,399	.079
97,400 to 99,199	.077
99,200 to 100,999	.115
101,000 to 102,799	.149
102,800 to 104,599	.142
104,600 to 106,399	.163
106,400 to 108,199	.124
108,200 to 110,000	.122

FIGURE 5.3-2
Histogram and One-Dimensional Jittered Scattergram of Data in Table 5.3-1

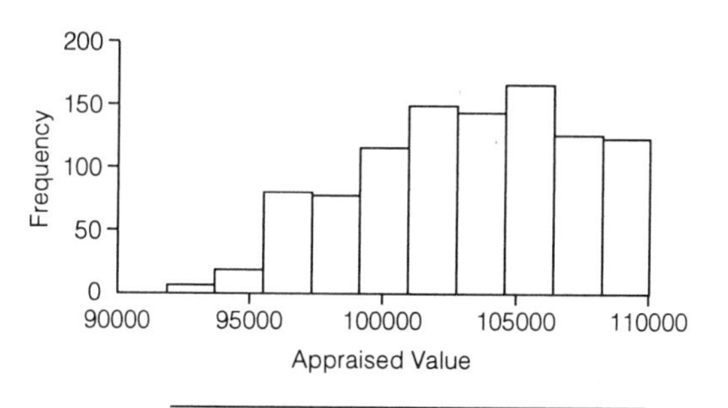

Average Appraisal = \$103,167.10
Standard Deviation = \$ 4,076.39
Minimum Appraisal = \$ 92,786.47
Maximum Appraisal = \$109,965.50

Exercises 5.3

5.3-1 The random variable X has probability density function

$$f(x) = \begin{cases} 2x, & 0 \le x \le 1, \\ 0, & \text{otherwise.} \end{cases}$$

- **a** Graph the probability density function of X.
- **b** Find and graph the cumulative distribution function of X.
- **c** Generate 1000 values for X and make an empirical probability histogram of the data.

5.3-2 The random variable X has probability density function

$$f(x) = \begin{cases} 2(1-x), & 0 \le x \le 1, \\ 0, & \text{otherwise.} \end{cases}$$

- **a** Graph the probability density function of X.
- **b** Find and graph the cumulative distribution function of X.
- **c** Generate 1000 values for X and make an empirical probability histogram of the data.

5.3-3 The random variable Y has probability density function

$$f(y) = \begin{cases} \dfrac{2(y-a)}{(b-a)^2}, & a \le y \le b, \\ 0, & \text{otherwise.} \end{cases}$$

- **a** Graph the probability density function of Y.
- **b** Find and graph the cumulative distribution function of Y.

c Show that Y can be simulated as $a + (b-a)x$, where x is the random variable in Exercise 5.3-1.

d Generate 1000 values for Y and make a histogram of the data.

5.3-4. The random variable Y has probability density function

$$f(y) = \begin{cases} \dfrac{2(b-y)}{(b-a)^2}, & a \le y \le b, \\ 0, & \text{otherwise.} \end{cases}$$

a Graph the probability density function of Y.

b Find and graph the cumulative distribution function of Y.

c Show that Y can be simulated as $a + (b-a)x$, where x is the random variable in Exercise 5.3-2.

d Generate 1000 values for Y and make an empirical probability histogram of the data.

5.3-5 The cumulative distribution function for the random variable X is

$$F(x) = \begin{cases} 0, & x \le 0, \\ \dfrac{x}{8}, & 0 < x < 2, \\ \dfrac{x^2}{16}, & 2 \le x \le 4, \\ 1, & 4 < x. \end{cases}$$

a Find and graph $f(x)$.

b Find $E(X)$ and $\text{STD}(X)$.

c Simulate 1000 values for Y and find the sample average and sample standard deviation. Compare these values to part (b).

5.3-6. In Exercise 5.2-4, the temperature X (Celsius) at a randomly selected point in a commercial refrigerator was a random variable with probability density function

$$f(x) = \begin{cases} \dfrac{x^2}{9}, & 0 \le x \le 3, \\ 0, & \text{otherwise.} \end{cases}$$

a Simulate 1000 values for X. Compute the sample mean and sample standard deviation and compare these to $E(X)$ and $\text{STD}(X)$ in Exercise 5.2-4.

b The Fahrenheit temperature is $Y = 9X/5 + 32$. Simulate 1000 values for Y. Compute the sample mean and sample standard deviation and compare these to $E(Y)$ and $\text{STD}(Y)$ in Exercise 5.2-4.

5.3-7. Consider Example 5.3-7. Repeat the simulation using the average of the appraisals rather than the middle value as the appraised value. Construct a histogram and compare your results to those in Example 5.3-7.

5.3-8. The total time it takes for the execution of a piece of software on a parallel processor is the maximum amount of time it takes any processor to complete its portion of the computation. Assume the parallel processor has eight processors. Let the random variable X denote the amount of time, in milliseconds, for one processor to complete its portion of the computation. Let the probability density function of X be

$$f(x) = \begin{cases} \dfrac{5}{4x^2}, & 1 < x < 5, \\ 0, & \text{otherwise.} \end{cases}$$

Assume that the processors function independently. Simulate 1000 execution times for this software. To do this simulate eight values for X and find the maximum. Compute the sample mean and sample standard deviation. Make a histogram of the data.

5.3-9. Suppose two of the processors in Exercise 5.3-8 are slower than the other six. Let the time it takes these processors to complete their portion of the computation have uniform distributions on the interval $[1, 5]$. Simulate 1000 execution times for the software under these new conditions. Compute the sample mean and sample standard deviation. Make a histogram of the data. Compare these results to those obtained in Exercise 5.3-8.

5.3-10 Prove Theorem 5.3-1. To do this find the cumulative distribution function of $U = F(X)$, noting that

$$P(F(X) \le x) = P(X \le F^{-1}(x)) = F[F^{-1}(x)] = x,$$

which shows that $F(X)$ has the cumulative distribution of a uniform $[0, 1]$ random variable.

SECTION

5.4 Joint Probability Distributions

For two continuous random variables X and Y, joint probabilities are computed as the volume under a surface $z = f(x, y)$ as shown in Figure 5.4-1. The function $f(x, y)$ is called a *joint probability density function.*

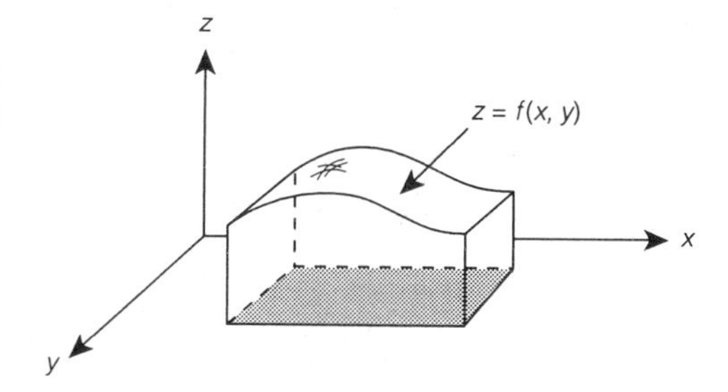

FIGURE **5.4-1** Probability Shown as Volume Under a Surface

Definition 5.4-1

A function $f(x, y)$ is said to be a *joint probability density function* for continuous random variables X and Y if

i $f(x, y) \geq 0$,

ii $\int_{-\infty}^{\infty} \int_{-\infty}^{\infty} f(x, y)\ dx\, dy = 1$,

iii For any region A,

$$P((X, Y) \in A) = \iint_{\{x, y:\ (x, y) \in A\}} f(x, y)\ dx\, dy.$$

In computing probabilities and expected values, the transition from the discrete case to the continuous case is just a matter of replacing the discrete probability mass function $p(x, y)$ with $f(x, y)\ dx\, dy$ and replacing double summations $\Sigma\Sigma$ with double integrals $\int\int$. The following examples illustrate this transition.

EXAMPLE 5.4-1 Let X and Y be continuous random variables with joint probability density function

$$f(x, y) = \begin{cases} 1 - \frac{x}{3} - \frac{y}{3}, & 0 \leq x \leq 2,\ \ 0 \leq y \leq 1, \\ 0, & \text{otherwise.} \end{cases}$$

First note that $f(x, y)$ satisfies Properties i and ii of Definition 5.4-1. Clearly $f(x, y) \geq 0$. The shaded area in Figure 5.4-2 represents the joint domain for $f(x, y)$, where $f(x, y) > 0$. From Figure 5.4-2, we see that the limits of integration for the integral in Property ii are $0 \leq x \leq 2$, $0 \leq y \leq 1$, since $f(x, y) = 0$ outside this region. Thus,

$$\begin{aligned} \int_{-\infty}^{\infty} \int_{-\infty}^{\infty} f(x, y)\ dx\, dy &= \int_{y=0}^{1} \int_{x=0}^{2} \left(1 - \frac{x}{3} - \frac{y}{3}\right) dx\, dy \\ &= \int_{y=0}^{1} \left(x - \frac{x^2}{6} - \frac{xy}{3}\right)\Bigg|_{x=0}^{2} dy \\ &= \int_{y=0}^{1} \left(\frac{4}{3} - \frac{2y}{3}\right) dy \\ &= \frac{4y}{3} - \frac{y^2}{3}\ \Bigg|_{y=0}^{1} = 1. \end{aligned}$$

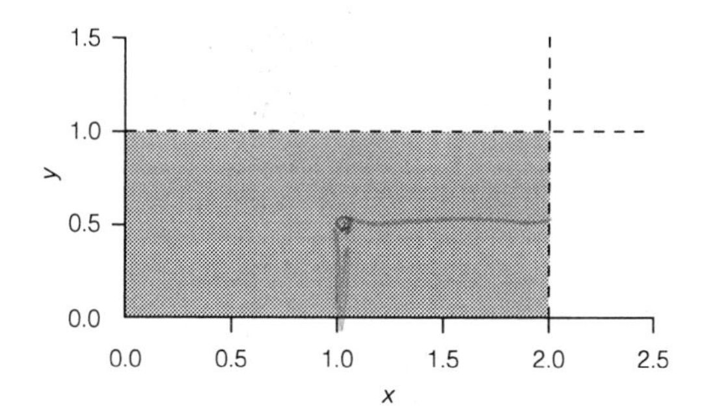

FIGURE **5.4-2** Region of Integration where $f(x, y) > 0$

The shaded area in Figure 5.4-3 represents the region of integration for $P(X \geq 1, Y \leq .5)$. The probability is

$$\begin{aligned} P(X \geq 1, Y \leq .5) &= \int_{x=1}^{2} \int_{y=0}^{.5} \left(1 - \frac{x}{3} - \frac{y}{3}\right) dy\, dx \\ &= \int_{x=1}^{2} \left(y - \frac{xy}{3} - \frac{y^2}{6}\right)\Bigg|_{y=0}^{.5} dx \\ &= \int_{x=1}^{2} \left(\frac{1}{2} - \frac{x}{6} - \frac{1}{24}\right) dx \\ &= \left(\frac{11x}{24} - \frac{x^2}{12}\right)\Bigg|_{1}^{2} \\ &= \left[\frac{22}{24} - \frac{4}{12}\right] - \left[\frac{11}{24} - \frac{1}{12}\right] \\ &= \frac{5}{24}. \end{aligned}$$

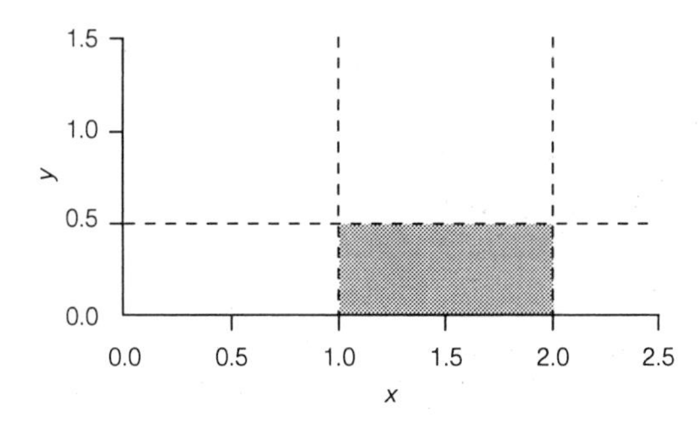

FIGURE **5.4-3** Region of Integration for $P(X \geq 1, Y \leq .5)$

EXAMPLE 5.4-2 Let X and Y have the joint probability density function defined in Example 5.4-1. We will find $P(X < Y)$. The required region of integration is the triangular region bounded by the lines $y = x$, $x = 0$, and $y = 1$. This region is shown in Figure 5.4-4. Therefore,

$$\begin{aligned} P(X<Y) &= \int_{y=0}^{1}\int_{x=0}^{y}\left(1-\frac{x}{3}-\frac{y}{3}\right)dx\,dy \\ &= \int_{y=0}^{1}\left(x-\frac{x^2}{6}-\frac{xy}{3}\right)\Bigg|_{x=0}^{y} dy \\ &= \int_{y=0}^{1}\left(y-\frac{y^2}{6}-\frac{y^2}{3}\right)dy \\ &= \left(\frac{y^2}{2}-\frac{y^3}{6}\right)\Bigg|_{0}^{1} \\ &= \frac{1}{3}. \end{aligned}$$

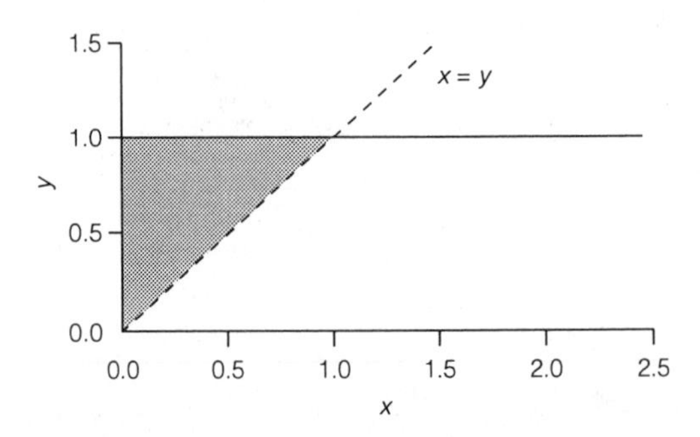

FIGURE 5.4-4 Region of Integration for $P(X < Y)$

EXAMPLE 5.4-3 Let X and Y have the probability density function defined in Example 5.4-1. The expected value of $T = XY$ can be found by a continuous version of Theorem 2.7-1. We have

$$\begin{aligned} E(XY) &= \int_{-\infty}^{\infty}\int_{-\infty}^{\infty} xyf(x, y)\,dx\,dy \\ &= \int_{y=0}^{2}\int_{x=0}^{2} xy\left(1-\frac{x}{3}-\frac{y}{3}\right)dx\,dy \end{aligned}$$

$$= \int_{y=0}^{1} \left(\frac{x^2 y}{2} - \frac{x^3 y}{9} - \frac{x^2 y^2}{6} \right) \Bigg|_{x=0}^{2} dy$$

$$= \int_{y=0}^{1} \left(\frac{10y}{9} - \frac{2y^2}{3} \right) dy$$

$$= \frac{5y^2}{9} - \frac{2y^3}{9} \Bigg|_0^1$$

$$= \frac{1}{3}.$$

The probability density functions of either X or Y may be found from their joint probability density function by "integrating out" the other variable. This is analogous to summing the entries over the rows and columns in a joint probability mass function table to find the marginal probability mass functions.

Definition 5.4-2

Let X and Y be continuous random variables with joint probability density function $f(x, y)$. Then the *marginal probability density functions* of X and Y are

$$f_X(x) = \int_{-\infty}^{\infty} f(x, y)\, dy, \qquad f_Y(y) = \int_{-\infty}^{\infty} f(x, y)\, dx.$$

EXAMPLE 5.4-4 Let X and Y have the joint probability density function defined in Example 5.4-1. The probability density function of X is

$$f_X(x) = \int_{-\infty}^{\infty} f(x, y)\, dy$$

$$= \int_{y=0}^{1} \left(1 - \frac{x}{3} - \frac{y}{3} \right) dy$$

$$= \left(y - \frac{xy}{3} - \frac{y^2}{6} \right) \Bigg|_{y=0}^{1}$$

$$= \frac{5}{6} - \frac{x}{3}, \qquad 0 \le x \le 2.$$

Also, $f_X(x) = 0$ for $x < 0$ and for $x > 2$ since $f(x, y) = 0$ for these values of x. A similar computation shows

$$f_Y(y) = \begin{cases} \frac{4}{3} - \frac{2y}{3}, & 0 \le y \le 1, \\ 0, & \text{otherwise.} \end{cases}$$

Conditional probability density functions are defined similarly to conditional probability mass functions.

Definition 5.4-3

Let X and Y be continuous random variables with joint probability density function $f(x, y)$. The *conditional probability density functions* of X and Y are defined for values of x and y such that $f_X(x) > 0$ and $f_Y(y) > 0$ as

$$f_{X|Y}(x \mid y) = \frac{f(x, y)}{f_Y(y)}$$

and

$$f_{Y|X}(y \mid x) = \frac{f(x, y)}{f_X(x)}.$$

EXAMPLE 5.4-5 Let X and Y have the joint probability density function defined in Example 5.4-1. The conditional probability density function of Y given $X = .5$ is

$$f_{Y|X}(y \mid .5) = \frac{f(.5, y)}{f_X(.5)}$$

$$= \frac{1 - .5/3 - y/3}{5/6 - .5/3},$$

$$= \frac{5}{4} - \frac{y}{2}, \qquad 0 \le y \le 1.$$

This density function can be used to find conditional probabilities for the random variable Y given $X = .5$. For instance,

$$\begin{aligned} P(Y \le .2 \mid X = .5) &= \int_{y \le .2} f_{Y|X}(y|.5)\, dy \\ &= \int_{y=0}^{.2} \left(\frac{5}{4} - \frac{y}{2}\right) dy \\ &= \frac{6}{25}. \end{aligned}$$

Conditional expectation can be found by taking the expected value of the appropriate conditional probability density function. For instance,

$$\begin{aligned} E(Y|X = .5) &= \int_{y=-\infty}^{\infty} y f_{Y|X}(y|.5)\, dy \\ &= \int_{y=0}^{1} y\left(\frac{5}{4} - \frac{y}{2}\right) dy \\ &= \frac{11}{24}. \end{aligned}$$

The notion of a uniform random variable extends to jointly distributed uniform random variables.

Definition 5.4-4

Let $|R|$ denote the area of the region R in the (x, y) plane. Random variables X and Y have a joint uniform distribution on R if

$$f(x, y) = \begin{cases} \frac{1}{|R|}, & (x, y) \in R, \\ 0, & \text{otherwise.} \end{cases}$$

Note that if X and Y have a joint uniform distribution on R, then for any region $A \subset R$,

$$P((X, Y) \in A) = \frac{\text{Area of } A}{\text{Area of } R}.$$

EXAMPLE 5.4-6 Two friends arrive at a particular spot between 10 A.M. and 11 A.M. Suppose that each arrives randomly during this hour, and each waits 10 minutes. What is the probability that they meet? Let X and Y denote the number of minutes after 10 A.M. that each arrives. Since each arrives randomly, we may assume that the joint arrival time (X, Y) has joint uniform distribution on the square $0 \le x \le 60$ minutes, $0 \le y \le 60$ minutes. Thus,

$$f(x, y) = \begin{cases} \dfrac{1}{3600}, & 0 \le x \le 60, \quad 0 \le y \le 60, \\ 0, & \text{otherwise.} \end{cases}$$

They will meet at the spot if and only if $X - 10 \le Y \le X + 10$. The desired probability may be found by computing the area of the shaded region in Figure 5.4-5 and dividing this by 3600. The probability that they meet is $11/36$.

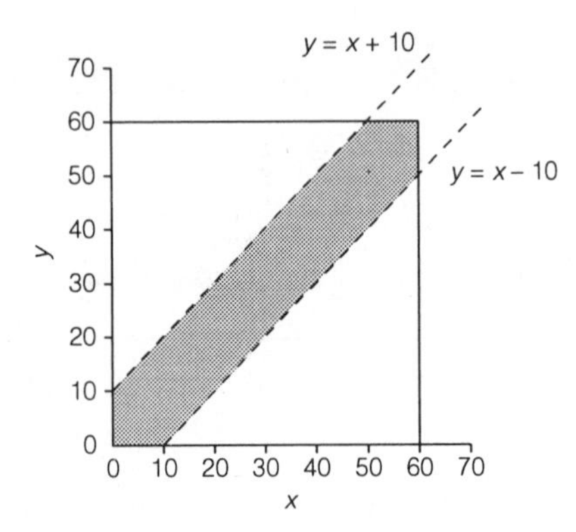

FIGURE 5.4-5 Meeting Times of the Two Friends in Example 5.4-6

Next we consider the definition of independent random variables in the continuous case.

Definition 5.4-5

Continuous random variables X and Y with joint probability density function $f(x, y)$ are *independent* if

$$f(x, y) = f_X(x) f_Y(y).$$

As in the discrete case, random variables intuitively are regarded as independent if the outcome of one random variable does not affect the outcome of the other.

EXAMPLE 5.4-7 The number of minutes it takes an average student to solve a certain arithmetic problem is a random variable W with probability density function

$$f(w) = \begin{cases} 3w^2, & 0 \le w \le 1, \\ 0, & \text{otherwise.} \end{cases}$$

Suppose that two students work independently of each other on the same problem. Let X denote the time it takes the first student to solve the problem and Y the time it takes for the second student to solve the problem. We will find the joint probability density function of X and Y. If the students are of the same ability, then it is reasonable to assume that the probability density functions of X and Y are the same as W. Thus,

$$f_X(x) = \begin{cases} 3x^2, & 0 \le x \le 1, \\ 0, & \text{otherwise.} \end{cases}$$

and

$$f_Y(y) = \begin{cases} 3y^2, & 0 \le y \le 1, \\ 0, & \text{otherwise.} \end{cases}$$

Since the students work independently of each other, the random variables may be assumed to be independent. Therefore, we may apply Definition 5.4-5 to find the joint probability density function of X and Y. We have

$$f(x, y) = \begin{cases} 9x^2y^2, & 0 \le x \le 1,\ 0 \le y \le 1, \\ 0, & \text{otherwise.} \end{cases}$$

This density function may then be used to find probabilities of interest. For example, the probability that the first student finishes at least 1/2 minute before the second student is the integral of the joint probability density function over the triangular region bounded by the lines $y = x + .5$, $x = 0$, and $y = 1$. This region of integration is shown in Figure 5.4-6. Thus,

$$\begin{aligned} P(X + .5 \le Y) &= \int_{y=.5}^{1} \int_{x=0}^{y-.5} 9x^2y^2\, dx\, dy \\ &= \int_{y=.5}^{1} 3x^3y^2 \Big|_{x=0}^{y-.5} dy \\ &= \int_{y=.5}^{1} 3(y-.5)^3 y^2\, dy \\ &= \int_{u=0}^{.5} 3u^3(u+.5)^2\, du, \qquad \text{for } u = y - .5 \\ &= \int_{u=0}^{.5} 3\left(u^5 + u^4 + \frac{u^3}{4}\right) du, \\ &= 3\left(\frac{u^6}{6} + \frac{u^5}{5} + \frac{u^4}{16}\right)\Big|_{0}^{.5} \\ &= .038. \end{aligned}$$

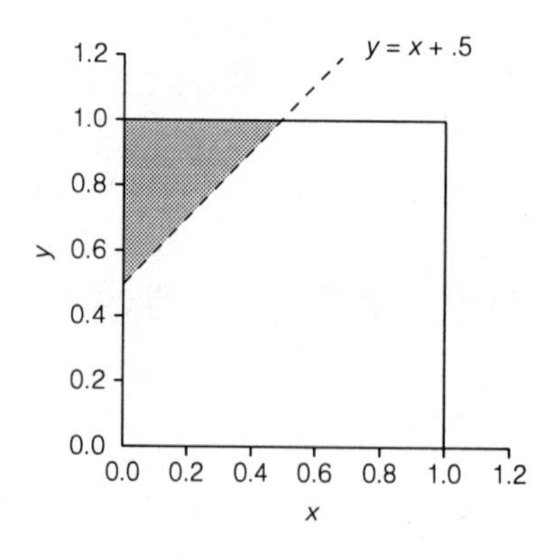

FIGURE **5.4-6** Region of Integration for Example 5.3-6

Theorem 2.7-2, which pertains to the expected value of linear combinations of two random variables, and Theorems 2.7-4 and 2.7-5, which pertain to the product of two independent random variables and the variance of a linear combination of independent random variables, carry over to the continuous case.

EXAMPLE 5.4-8 Referring to Example 5.4-7, let $T = X + Y$ denote the total time the two students spend solving the arithmetic problem. Theorems 2.7-2 and 2.7-5 can be applied to find $E(T)$ and $\text{VAR}(T)$. From the density function for X and Y, we compute $E(X) = E(Y) = 3/4$, $E(X^2) = E(Y^2) = 3/5$. Thus, $\text{VAR}(X) = \text{VAR}(Y) = 3/5 - (3/4)^2 = .0375$. It follows that $E(T) = E(X) + E(Y) = 3/2$, $\text{VAR}(T) = \text{VAR}(X) + \text{VAR}(Y) = .075$, and $\text{STD}(T) = \sqrt{.075} = .274$.

The concept of a joint probability density function and the attendant computations, definitions, and theorems can be extended to several continuous random variables. Of particular importance are Theorems 2.7-3 and 2.7-6, an application of which is given next.

EXAMPLE 5.4-9 A cheese machine in a pizza factory is supposed to put 16 ounces on each pizza. However, the actual amount X is a random variable with a uniform distribution on the interval [15 ounces, 17 ounces]. According to the computations in Example 5.2-3, $E(X) = 16.0$ ounces and $\text{STD}(X) = \sqrt{1/3} = .577$ ounces. Suppose 400 pizzas are made in an hour. Let X_i denote the amount of cheese put on the ith pizza. The total amount of cheese used in an hour is $T = X_1 + X_2 + \cdots + X_{400}$. Applying Theorem 2.7-3, we find $E(T) = 400(16.0) = 6400$ ounces. If the amounts are independent, then an application of Theorem 2.7-6 shows $\text{VAR}(T) = 400(1/3) = 133.33$ and $\text{STD}(T) = 11.5$ ounces. Thus, the amount of cheese used in an hour is likely to vary between the limits $6400 \pm 2(11.5)$ or about 6377 to 6423 ounces. That is, the deviation from the target amount of 6400 ounces will be typically no more than about 23 ounces in an hour.

The univariate variable transformation method described in Section 5.2 extends naturally to the bivariate situation. This extension will be given without proof in the following discussion. Let X_1 and X_2 be jointly continuous random variables with joint probability density function $f(x_1, x_2)$. Suppose $Y_1 = \Phi_1(X_1, X_2)$ and $Y_2 = \Phi_2(X_1, X_2)$ defines a one-to-one transformation. That is, X_1 and X_2 can be expressed uniquely in terms of Y_1 and Y_2. Then

$$g(y_1, y_2) = f(x_1, x_2)|J|,$$

where

$$J = \det \begin{bmatrix} \frac{\partial x_1}{\partial y_1} & \frac{\partial x_1}{\partial y_2} \\ \frac{\partial x_2}{\partial y_1} & \frac{\partial x_2}{\partial y_2} \end{bmatrix}$$

and x_1 and x_2 are expressed in terms of y_1 and y_2 by solving $y_1 = \Phi_1(x_1, x_2)$ and $y_2 = \Phi_2(x_1, x_2)$ for x_1 and x_2. The determinant, J, of the partial derivative matrix is commonly called the *Jacobian* of the transformation.

EXAMPLE 5.4-10 Suppose X_1, X_2 are jointly distributed random variables such that

$$f(x_1, x_2) = \begin{cases} 4x_1x_2, & 0 \le x_1 \le 1, \ 0 \le x_2 \le 1, \\ 0, & \text{otherwise.} \end{cases}$$

Let $Y_1 = X_1 + X_2$ and $Y_2 = 2X_2$. First we determine the region in the (y_1, y_2) plane that contains a nonzero probability. For $0 \le x_1 \le 1$, the variable y_1 ranges over $x_2 \le y_1 \le x_2 + 1$. Since $Y_2 = 2X_2$, this interval becomes $y_2/2 \le y_1 \le y_2/2 + 1$, for $0 \le y_2 \le 2$. The domain of positive probability for $g(y_1, y_2)$ is shown in Figure 5.4-7.

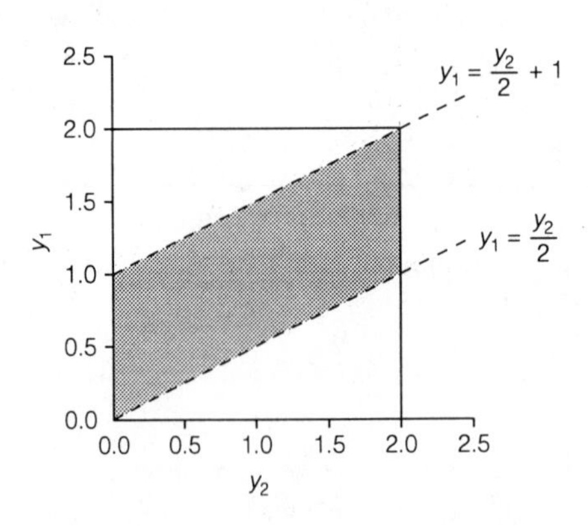

FIGURE 5.4-7 Domain of Positive Probability for $g(y_1, y_2)$ in Example 5.4-10

Now, $X_1 = Y_1 - Y_2/2$, $X_2 = Y_2/2$,

$$J = \det\begin{bmatrix} 1 & -1/2 \\ 0 & 1/2 \end{bmatrix} = \frac{1}{2},$$

and

$$\begin{aligned} g(y_1, y_2) &= f(x_1, x_2)\,|J| \\ &= 4\left(y_1 - \frac{y_2}{2}\right)\left(\frac{y_2}{2}\right)\left|\frac{1}{2}\right| \\ &= \left(y_1 - \frac{y_2}{2}\right)y_2 \end{aligned}$$

for $y_2/2 \le y_1 \le y_2/2 + 1$, $0 \le y_2 \le 2$ and $g(y_1, y_2) = 0$ outside this region.

EXAMPLE 5.4-11 Suppose X_1 and X_2 have joint probability density function

$$f(x_1, x_2) = \begin{cases} e^{-x_1 - x_2}, & x_1 > 0, \ x_2 > 0, \\ 0, & \text{otherwise.} \end{cases}$$

Let $Y_1 = X_1^2$ and $Y_2 = X_2^2$. The joint domain of positive probability for $g(y_1, y_2)$ is $y_1 > 0$ and $y_2 > 0$. Now, $X_1 = \sqrt{Y_1}$, $X_2 = \sqrt{Y_2}$,

$$J = \det\begin{bmatrix} 1/2\sqrt{y_1} & 0 \\ 0 & 1/2\sqrt{y_2} \end{bmatrix} = \frac{1}{4\sqrt{y_1 y_2}},$$

and

$$g(y_1, y_2) = \begin{cases} e^{-\sqrt{y_1} - \sqrt{y_2}} \dfrac{1}{4\sqrt{y_1 y_2}}, & y_1 > 0, \ y_2 > 0, \\ 0, & \text{otherwise.} \end{cases}$$

Exercises 5.4

5.4-1 Let X and Y be continuous random variables with joint probability density function

$$f(x, y) = \begin{cases} k, & 0 \le x \le 1,\ 0 \le y \le 1, \\ 0, & \text{otherwise.} \end{cases}$$

a Sketch the region for which $f(x, y) > 0$.
b Find k.
c Find $P(X < .5,\ Y > .5)$.
d Find the marginal probability density functions $f_X(x)$ and $f_Y(y)$.
e Are X and Y independent random variables?

5.4-2 Let X and Y be continuous random variables with joint probability density function

$$f(x, y) = \begin{cases} k, & 0 \le x \le y \le 1, \\ 0, & \text{otherwise.} \end{cases}$$

a Sketch the region for which $f(x, y) > 0$.
b Find k.
c Find $P(X < .5,\ Y > .5)$.
d Find the marginal probability density functions $f_X(x)$ and $f_Y(y)$.
e Are X and Y independent random variables?

5.4-3 Let X and Y be continuous random variables with joint probability density function

$$f(x, y) = \begin{cases} \frac{1}{2}, & 0 \le x \le 1,\ 0 \le y \le 1,\ 0 \le x + y \le 1, \\ 0, & \text{otherwise.} \end{cases}$$

a Find $P(X < .5,\ Y > .5)$.
b Find $P(X > .75)$.
c Find $P(X + Y < .25)$.

5.4-4 Let X and Y be continuous random variables with joint probability density function

$$f(x, y) = \begin{cases} 4xy, & 0 \le x \le 1,\ 0 \le y \le 1, \\ 0, & \text{otherwise.} \end{cases}$$

Show that X and Y are independent random variables.

5.4-5 Let X and Y be continuous random variables with joint probability density function

$$f(x, y) = \begin{cases} \frac{1}{3}(x+y), & 0 \le x \le 1,\ 0 \le y \le 2, \\ 0, & \text{otherwise.} \end{cases}$$

a Find $P(X < .5,\ Y > 1)$.

b Find $P(X < Y)$.

5.4-6 Let the joint probability density function for X and Y be

$$f(x, y) = \begin{cases} \frac{1}{\pi}, & x^2 + y^2 \le 1, \\ 0, & \text{otherwise.} \end{cases}$$

a Find $P(Y \le .5 \mid X = .75)$.

b Find $E(Y \mid X = .75)$.

c Find $\mathrm{VAR}(Y \mid X = .75)$.

5.4-7 Let X and Y be continuous random variables with joint probability density function

$$f(x, y) = \begin{cases} 6x^2y, & 0 \le x \le 1,\ 0 \le y \le 1, \\ 0, & \text{otherwise.} \end{cases}$$

a Find $E(XY)$, $E(X)$, and $E(Y)$. Does $E(XY) = E(X)E(Y)$?

b Find $P(X \le .5)$ and $P(X \le .5 \mid Y = .25)$. Are these two probabilities equal? What does this imply about the independence of X and Y?

c Find $E(Y \mid X = .5)$.

5.4-8 Let X and Y be independent random variables with probability density functions

$$f(x) = \begin{cases} \frac{x}{2}, & 0 \le x \le 2, \\ 0, & \text{otherwise} \end{cases}$$

and

$$f(y) = \begin{cases} \frac{2-y}{2}, & 0 \le y \le 2, \\ 0, & \text{otherwise.} \end{cases}$$

a Find $f(x, y)$.

b Find $E(XY)$, $E(X)$, and $E(Y)$. Does $E(XY) = E(X)E(Y)$?

c Find $P(X \le 1)$ and $P(X \le 1 \mid Y = 1)$. Are these two probabilities equal?

5.4-9 Suppose an observer is placed at the origin of the (x, y) plane and has a field of view that is the interior of an arc defined by $y = |x|$ and $x^2 + y^2 = 3$. A target is placed at random within the observer's field of view. Let the random variable L denote the straight-line distance from the target to the observer.

a Find the probability density function for L. Note that $L = \sqrt{X^2 + Y^2}$.

b Within what range would you expect L to fall most of the time?

5.4-10 Suppose X_1 and X_2 are independent uniform $[0, 1]$ random variables.

a Find $f(x_1, x_2)$.

b Let $Y_1 = X_1 + X_2$ and $Y_2 = X_1 - X_2$. Find the joint domain of positive probability for $g(y_1, y_2)$.

c Find $g(y_1, y_2)$.

d Find $g_{Y_2}(y_2)$.

5.4-11 Let the random variables Y_1 and Y_2 have the joint probability density function derived in Example 5.4-11.

a Find $g_{Y_1}(y_1)$ and $g_{Y_2}(y_2)$.

b Are Y_1 and Y_2 independent?

5.4-12 Let X and Y be defined as in Exercise 5.4-4. Let $U = X^2$ and $V = Y^2$.

a Find the joint domain of positive probability for $g(u, v)$.

b Find $g(u, v)$.

c Are U and V independent?

5.4-13 Let X_1 be the rate of water flowing into a large reservoir, and let X_2 be the rate at which water is being let out of the reservoir. Assume X_1 and X_2 are independently distributed as

$$f(x_1) = \begin{cases} \dfrac{1}{10{,}000} e^{-x_1/10{,}000}, & x_1 > 0 \text{ cubic feet/second,} \\ 0, & \text{otherwise} \end{cases}$$

and

$$f(x_2) = \begin{cases} \dfrac{1}{20{,}000}, & 0 \le x_2 \le 20{,}000 \text{ cubic feet/second,} \\ 0, \ \text{otherwise.} \end{cases}$$

Let $Y = X_1 - X_2$ = rate of change in the volume of the reservoir.

a Simulate 1000 values for X_1 and X_2 and use this data to find an empirical probability density function for Y.

b Estimate a likely interval for the mean of Y based on the simulated data in part (a).

c Find $g(y)$ and $E(Y) \pm 2$ STD(Y). Compare these values with the empirical results in parts (a) and (b).

5.4-14 Let X_1 be the time a customer waits at an auto repair shop for service to begin on his or her automobile. Let X_2 be the time it takes to complete the service. Assume X_1 and X_2 are independently distributed as

$$f(x_1) = \begin{cases} 2x_1, & 0 \le x_1 \le 1 \text{ hour}, \\ 0, & \text{otherwise} \end{cases}$$

and

$$f(x_2) = \begin{cases} \frac{2}{9}(3 - x_2), & 0 \le x_2 \le 3 \text{ hours}, \\ 0, & \text{otherwise.} \end{cases}$$

Let $Y_1 = X_1 + X_2$ = total time spent at the repair shop and $Y_2 = \$30.00(X_2)$ = cost of labor.

a Simulate 5000 values of X_1 and X_2. For each simulated pair, form Y_1 and Y_2. Estimate the mean and standard deviation of Y_1 and the mean and standard deviation of Y_2 from these simulated values.

b Find $g(y_1, y_2)$, $E(Y_1)$, $E(Y_2)$, $\text{VAR}(Y_1)$, and $\text{VAR}(Y_2)$. Compare this distribution and these means and variances to the simulated values in part (a).

5.4-15 Let X_1 be the total miles a rental car is driven in one year. Let X_2 be the miles per gallon of the rental car. Assume X_1 and X_2 are independently distributed as

$$f(x_1) = \begin{cases} \dfrac{2(20{,}000 - x_1)}{15{,}000^2}, & 5000 \le x_1 \le 20{,}000 \text{ miles}, \\ 0, & \text{otherwise} \end{cases}$$

and

$$f(x_2) = \begin{cases} \dfrac{2x_2}{35^2}, & 0 \le x_2 \le 35 \text{ miles/gallon}, \\ 0, & \text{otherwise.} \end{cases}$$

Let $Y = X_1/X_2$ = total number of gallons of gas consumed for the year.

a Based on 1000 simulated values of X_1 and X_2 estimate a likely interval for Y.

b Compute the sample correlation for X_2 and Y from the simulated data in part (a). Is the result what you expected?

6

Special Continuous Random Variables

SECTION

6.1 Exponential Random Variable

The exponential distribution is used for such random variables as waiting time between successive arrivals at a service station, time between successive emissions of particles from a radioactive substance, and time to failure of an electronic component.

Definition 6.1-1

A random variable X is said to have an *exponential distribution* if the probability density function is given by

$$f(x) = \begin{cases} \lambda e^{-\lambda x}, & \lambda > 0, x > 0, \\ 0, & \text{otherwise.} \end{cases}$$

A graph of the exponential distribution is given in Figure 6.1-1.

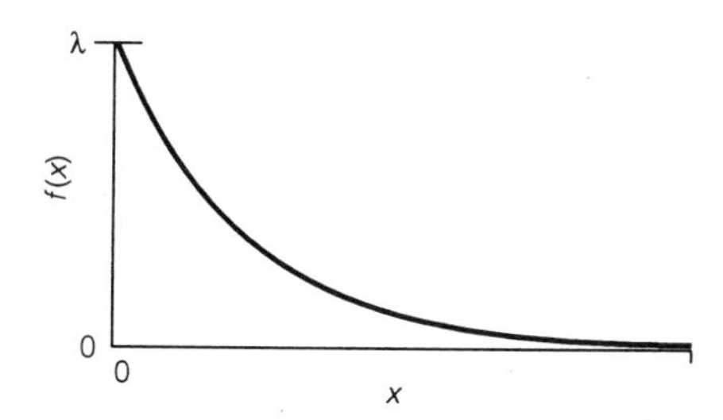

FIGURE **6.1-1**
Graph of Exponential Probability Density Function

We will find the following result from calculus to be useful in dealing with the exponential distribution. For any positive integer n,

$$\int_0^\infty x^n \lambda e^{-\lambda x}\, dx = \frac{n!}{\lambda^n}. \tag{6.1-1}$$

The verification of this result is by induction by using repeated application of integration by parts (see Exercise 6.1-9). This result will be used to derive the expected value and variance of the exponential distribution.

THEOREM **6.1-1** The mean, variance, and standard deviation of an exponential random variable X are

$$E(X) = \frac{1}{\lambda},$$

$$\text{VAR}(X) = \frac{1}{\lambda^2},$$

$$\text{STD}(X) = \frac{1}{\lambda}.$$

The verification of Theorem 6.1-1 is a direct application of Equation (6.1-1). For $n = 1$ and $n = 2$, we have $E(X) = 1/\lambda$ and $E(X^2) = 2/\lambda^2$. Thus, $\text{VAR}(X) = 2/\lambda^2 - (1/\lambda)^2 = 1/\lambda^2$, and $\text{STD}(X) = 1/\lambda$. Notice that the standard deviation and the mean of the exponential random variable are the same.

The cumulative distribution function of an exponential random variable is

$$F(X) = \int_0^x \lambda e^{-\lambda u}\, du = 1 - e^{-\lambda x}, \quad x \geq 0.$$

Since the probability density function of the exponential random variable is zero for $x < 0$, $F(x) = 0$ for $x < 0$.

EXAMPLE 6.1-1 Assume the amount of time, X, it takes to process an order at a fast-food restaurant is an exponential random variable with a mean of 1.5 minutes. The standard deviation of this random variable is also 1.5 minutes. Now, $\lambda = 1/E(X) = 1/1.5 = 2/3$. The probability that it takes more than 3 minutes to process an order is

$$\begin{aligned} P(X > 3) &= 1 - P(X \le 3) \\ &= 1 - F(3) \\ &= 1 - (1 - e^{-(2/3)3}) \\ &= .135. \end{aligned}$$

An exponential random variable with $\mu = 1/\lambda$ may be simulated by using the cumulative distribution function method described in Section 5.3 in the following way.

Simulation of an Exponential Random Variable

1 Generate a unit random number u.

2 Set $u = 1 - e^{-\lambda x}$.

3 Solve for x: $e^{-\lambda x} = 1 - u$

$$\begin{aligned} -\lambda x &= \ln(1-u) \\ x &= -\frac{1}{\lambda}\ln(1-u) \\ &= -\mu \ln(1-u). \end{aligned}$$

EXAMPLE 6.1-2 Suppose we wish to simulate an exponential random variable with $\mu = 1/2$ (i.e., $\lambda = 2$). If $u = .92015$, then $x = -(1/2)\ln(1 - .92015) = 1.26$ is a simulated value of this exponential random variable.

If u is a unit random number, it can be shown that $1 - u$ is also a unit random number (Exercise 6.1-7). Therefore, we may also use the formula $x = -\mu \ln(u)$ to simulate observed values of an exponential random variable in step 3.

The exponential distribution has a curious property called the *lack-of-memory property*. This property can be motivated by the following. Suppose a customer has been waiting in line x units of time for service. The customer would like to know what is the probability that he or she will have to wait an additional t units of time. It turns out that if the time spent in line is an exponential random variable, then this probability does not depend on x, the time already spent in line.

THEOREM 6.1-2 Lack-of-Memory Property. If a random variable X has an exponential distribution, then for all $t > 0$ and $x > 0$

$$P(X > t + x \mid X > x) = P(X > t).$$

Theorem 6.1-2 is verified with the following steps:

$$\begin{aligned} P(X > t + x \mid X > x) &= \frac{P(X > t + x,\ X > x)}{P(X > x)} \\ &= \frac{P(X > t + x)}{P(X > x)} \\ &= \frac{e^{-\lambda(t + x)}}{e^{-\lambda x}} \\ &= e^{-\lambda t} \\ &= P(X > t). \end{aligned}$$

Consider the following intuitive interpretation of the lack-of-memory property. Think of two identical serving lines. In one line, we observe a person who has been in line for x units of time. In another line, we observe a person who has just arrived. If the time in line is an exponential random variable, then both people have the same chance of getting through the line (or not getting through the line) in the next t units of time. That is, the line does not "remember" how long a person has been there. If this result seems contrary to your intuition, then the exponential distribution may be the wrong model for the amount of time in line. On the other hand, if you have ever felt frustrated waiting in a line to be served while latecomers are going through another line more quickly than you, perhaps it is the lack-of-memory property of the exponential distribution at work.

Exercises 6.1

6.1-1 A random variable X has an exponential distribution with density function

$$f(x) = \begin{cases} 2e^{-2x}, & x > 0, \\ 0, & \text{otherwise.} \end{cases}$$

a Find $E(X)$.
b Find $STD(X)$.
c Find $F(X)$.

6.1-2 The lifetime of a light bulb is an exponential random variable with a mean of 1000 hours. If the light bulb is guaranteed to last at least 900 hours, find the probability that it will satisfy the guarantee.

6.1-3 Let X and Y be independent exponential random variables with $E(X) = 2.5$ and $E(Y) = 3.5$.

a Find $E(X + Y)$ and $VAR(X + Y)$.
b Find the joint probability distribution $f(x, y)$.
c Find $P(X + Y > 15)$.

6.1-4 Simulate 100 values of an exponential random variable with a mean of $\mu = 2$. Make a probability histogram of the simulated data to verify that it has the characteristic shape of the exponential distribution.

6.1-5 Find the probability that an exponential random variable falls within two standard deviations of its mean.

6.1-6 The time it takes to complete a service call to a mail-order computer company has an exponential distribution with a mean of 4 minutes. Within what time will 90% of the calls be completed?

6.1-7 Let U be a uniform $[0, 1]$ random variable and let $W = 1 - U$. Verify that W is also a uniform $[0, 1]$ random variable.

6.1-8 Suppose that U has a uniform $[0, 1]$ distribution. Verify that the cumulative distribution function of $X = -\mu \ln(U)$ is that of an exponential random variable with mean μ.

6.1-9 Use integration by parts to show

$$\int_0^\infty y^n e^{-y}\, dy = n!.$$

Then let $y = \lambda x$ to verify Equation (6.1-1). (*Hint*: To do integration by parts, let $u = y^n$, $dv = e^{-y}\, dy$. Repeat this procedure n times.)

SECTION

6.2 Normal Random Variable

Some examples of random variables that often are assumed to have normal distributions are the experimental errors in scientific measurements, test scores on achievement tests, and heights of individuals selected at random from a population. The normal distribution is often used to approximate other distributions. For instance, even if heights of individuals do not have the normal distribution, the probabilities provided by the normal distribution may be good enough for practical purposes.

Definition 6.2-1

A random variable X has the *normal distribution* if the probability density function is

$$f(x) = \frac{1}{\sqrt{2\pi}\,\sigma} e^{-(x-\mu)^2/2\sigma^2}, \qquad -\infty < x < \infty.$$

The normal distribution, which is frequently called bell- or mound-shaped, is shown in Figure 6.2-1. It can be shown that the constants μ and σ are the expected value and standard deviation of the random variable X. These derivations are given in Section 6.5. The normal distribution is symmetric about μ, with half the probability lying above and half lying below μ. It can be shown that about 68% of the probability of the normal distribution falls within one standard deviation of the mean and just over 95% of the probability falls within two standard deviations of the mean.

FIGURE **6.2-1**
The Normal Distribution

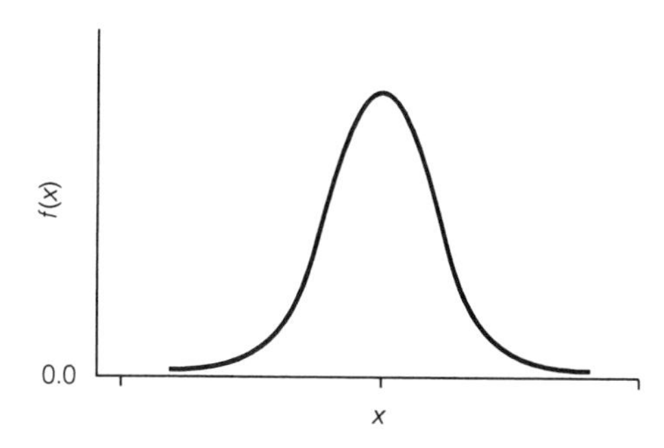

Unlike the exponential distribution, probabilities pertaining to the normal distribution cannot be expressed in terms of closed-form functions. That is, the integral of the normal density function cannot be given in terms of such functions as polynomials and exponentials. Numerical integration techniques have been used to evaluate the normal cumulative distribution function, resulting in tables of the probability values. Table 4 in Appendix A gives values of the cumulative distribution function of the normal distribution with a mean of zero and a standard deviation of one; that is, $\mu = 0$, $\sigma = 1$. We call this distribution the *standard normal*, and denote the corresponding random variable as Z. The following example will illustrate the use of the table.

EXAMPLE 6.2-1 Let Z be a standard normal random variable. Then $P(Z \leq 1.55)$ is found as the entry in the table in the row labeled 1.50 and the column headed by .05. That is,

$$P(Z \leq 1.55) = F(1.55) = .9394.$$

Other examples are

$$\begin{aligned} P(Z > 2.0) &= 1 - P(Z \leq 2.0) \\ &= 1 - F(2.00) \\ &= 1 - .9772 \\ &= .0228, \end{aligned}$$

$$\begin{aligned} P(-1.0 < Z < 1.0) &= P(Z < 1.0) - P(Z \leq -1.0) \\ &= F(1.00) - F(-1.00) \\ &= .8413 - .1587 \\ &= .6826. \end{aligned}$$

To find probabilities for other normal distributions, we will use the following result.

THEOREM 6.2-1 If X has a normal distribution with mean μ and standard deviation σ, then the random variable $Z = (X - \mu)/\sigma$ has a standard normal distribution.

To verify this result, we need to show that the cumulative distribution function of $Z = (X - \mu)/\sigma$ is that of the standard normal. That is,

$$\begin{aligned}
P(Z \le z) &= P\left(\frac{(X-\mu)}{\sigma} \le z\right) \\
&= P(X \le \sigma z + \mu) \\
&= \int_{x=-\infty}^{\sigma z+\mu} \frac{1}{\sqrt{2\pi}\,\sigma} e^{-(x-\mu)^2/2\sigma^2}\, dx \\
&= \int_{u=-\infty}^{z} \frac{1}{\sqrt{2\pi}} e^{-u^2/2}\, du, \qquad \text{for } u = \frac{x-\mu}{\sigma}.
\end{aligned}$$

The latter integral is recognized as the integral form of the cumulative distribution function of a normal random variable with mean $\mu = 0$ and standard deviation $\sigma = 1$. Thus, $Z = (X - \mu)/\sigma$ has a standard normal distribution.

The importance of Theorem 6.2-1 is that it gives a way to use the standard normal distribution to find probabilities for all normal distributions. If X has a normal distribution with mean μ and standard deviation σ, then, for any constant c,

$$\begin{aligned}
P(X \le c) &= P\left(\frac{X-\mu}{\sigma} \le \frac{c-\mu}{\sigma}\right) \\
&= P\left(Z \le \frac{c-\mu}{\sigma}\right),
\end{aligned}$$

where Z is a standard normal random variable. Thus, the probability that X falls below c is the same as the probability that the standard normal variable falls below $(c - \mu)/\sigma$.

EXAMPLE 6.2-2 Let $\mu = 10$ and $\sigma = 2$. To find $P(X \le 12.5)$, we have

$$\begin{aligned}
P(X \le 12.5) &= P\left(Z \le \frac{12.5-10}{2}\right) \\
&= P(Z \le 1.25) \\
&= .8944.
\end{aligned}$$

EXAMPLE 6.2-3 Test scores on an examination have a normal distribution with $\mu = 50$ and $\sigma = 5$. The probability that the scores are between 45 and 60 is

$$\begin{aligned}
P(45 \le X < 60) &= P(X \le 60) - P(X \le 45) \\
&= P\left(Z \le \frac{60-50}{5}\right) - P\left(Z < \frac{45-50}{5}\right)
\end{aligned}$$

$$= P(Z \le 2.00) - P(Z < -1.00)$$
$$= .9772 - .1587$$
$$= .8185.$$

Note that we have treated test scores as if they were continuous variables, although in practice they would generally be recorded as integer values. Moreover, since we have treated the scores as continuous variables, it would not have made any difference in our probability calculations whether we had included or excluded the endpoints of 45 and 60.

EXAMPLE 6.2-4 Let X denote the time it takes a worker to assemble an electronic device. Suppose that X has a normal distribution with $\mu = 4.5$ minutes and $\sigma = .2$ minutes. The probability that it will take the worker more than 5 minutes to do the work is

$$P(X > 5) = 1 - P(X \le 5) = 1 - P\left(Z \le \frac{5 - 4.5}{.2}\right)$$
$$= 1 - P(Z \le 2.50)$$
$$= 1 - .9938$$
$$= .0062.$$

EXAMPLE 6.2-5 Let T denote the number of hours a light bulb will last before burning out. Suppose that $X = \ln(T)$ has a normal distribution with $\mu = 8.0$ and $\sigma = 4$. The probability that the light bulb fails in less than 900 hours is

$$P(T < 900) = P(\ln(T) < \ln(900))$$
$$= P(X < 6.80)$$
$$= P\left(Z < \frac{6.80 - 8.00}{4}\right)$$
$$= P(Z < -.30)$$
$$= .3821.$$

EXAMPLE 6.2-6 Assume the heights of a certain group of individuals are normally distributed with a mean of 70 inches and a standard deviation of 3 inches. Let us find the height that defines the tallest 10% of the individuals. We first find the upper 10% point of the standard normal distribution. From Table 4 in Appendix A the desired point is 1.28 since $P(Z > 1.28) = .10$. It follows that the height x that corresponds to the tallest 10% of the group must satisfy the relationship $(x - 70)/3 = 1.28$. Therefore, $x = 70 + 3(1.28) = 73.8$ inches.

The next theorem is a more general version of Theorem 6.2-1. It pertains to linear functions of a normal random variable.

THEOREM 6.2-2 If X has a normal distribution with mean μ and standard deviation σ, then $Y = a + bX$ has a normal distribution with mean $\mu_Y = a + b\mu$ and standard deviation $\sigma_Y = |b|\sigma$.

The verification of this result is similar to that of Theorem 6.2-1 and is left as an exercise (Exercise 6.2-12).

EXAMPLE 6.2-7 If X denotes temperatures on the Celsius scale and Y denotes temperatures on the Fahrenheit scale, then $Y = (9/5)X + 32$. If daily temperatures during September have a normal distribution with a mean of 15 degrees Celsius and a standard deviation of 5 degrees Celsius, then the Fahrenheit temperatures are normally distributed with a mean of 59 degrees and a standard deviation of 9 degrees.

An important use of the normal distribution is as an approximation to the binomial distribution. For large values of n, the binomial distribution has the same shape as normal distribution. For example, consider the probability histogram of the binomial probabilities for $n = 20$ and $p = .5$ shown in Figure 6.2-2. On top of this probability histogram, a normal distribution with mean $\mu = np = 10$ and $\sigma = \sqrt{np(1-p)} = \sqrt{5} = 2.24$ has been graphed. It is easy to see that areas under the normal curve will be approximately equal to the corresponding rectangular areas of this binomial distribution.

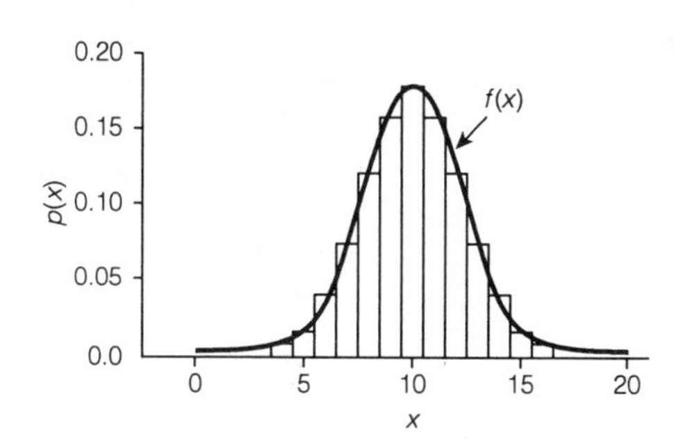

FIGURE 6.2-2 Normal Approximation to a Binomial Distribution

THEOREM 6.2-3 Let X have a binomial distribution with parameters n and p. For large n, X will have an approximate normal distribution with mean $\mu = np$ and standard deviation $\sigma = \sqrt{np(1-p)}$.

This result tells us that probabilities pertaining to a binomial random variable with a large value of n may be approximated by corresponding probabilities of a normal random variable whose mean and standard deviation are the same as the binomial random variable. For a fixed value of p, the approximation improves with increasing n. For a given value of n, the approximation will be better for values of p near .5 than for values of p near 0 or 1. Theorem 6.2-3 will be verified in Section 8.2. The next two examples illustrate the approximation.

EXAMPLE 6.2-8 Let $n = 20$ and $p = .5$. From Table 2 of Appendix A, we find $P(X \le 12) = .8684$. The mean and standard deviation of this binomial random variable are 10 and 2.24, respectively. The corresponding normal probability is

$$\begin{aligned} P(X \le 12) &= P\left(Z \le \frac{12-10}{2.24}\right) \\ &= P(Z \le .89) \\ &= .8133. \end{aligned}$$

Notice that the approximate probability is somewhat too low because we have approximated a discrete distribution by a continuous one. The discrete distribution has positive probability at $X = 12$, whereas the continuous distribution assigns 0 probability to this point. To "capture" this lost probability, we can go .5 unit beyond 12 in computing the probability under the normal curve. That is, we can improve the approximation by finding the probability under the normal curve up to 12.5. If we do this, we obtain

$$\begin{aligned} P(X \le 12.5) &= P\left(Z \le \frac{12.5-10}{2.24}\right) \\ &= P(Z \le 1.12) \\ &= .8686. \end{aligned}$$

The adjustment discussed in Example 6.2-8 is called the *continuity correction*.

In the remainder of what we do, we will not use the continuity correction. Actually, the continuity correction becomes less important as the value of n increases. For smaller values of n, binomial probabilities can be readily found with a computer, making the approximation unnecessary in these cases.

EXAMPLE 6.2-9 A 400-point multiple choice test has four possible responses for each question, one of which is correct. The problem is to compute the probability that the score will be more than 110, based on the assumption that a student guesses at every question. If X denotes the number of correct answers, then X has a binomial distribution with $n = 400$ and $p = .25$. The mean and standard deviation are $\mu = 100$ and $\sigma = \sqrt{75} = 8.66$. Applying the normal approximation gives

$$\begin{aligned} P(X > 110) &= P\left(Z > \frac{110 - 100}{8.66}\right) \\ &= P(Z > 1.15) \\ &= 1 - .8749 \\ &= .1251. \end{aligned}$$

The simulation of normal random variables using the distribution function technique of Section 5.3 presents a problem because a closed-form expression for $F(x)$ is not available. Fortunately, we only need to find a way to generate standard normal random variables, because general normal random variables can be obtained from standard normal variables by application of Theorem 6.2-2. Specifically, if Z is a standard normal random variable, then $X = \mu + \sigma Z$ has a normal distribution with mean μ and standard deviation σ. Standard normal random number generators are available in various programming and simulation languages, but a discussion of these generators is beyond the scope of this book. We consider just one of many possible ways to generate standard normal random numbers. The method is based on the *Box–Mueller* transformation, which is stated in the following theorem.

THEOREM 6.2-4 Let U_1 and U_2 be independent, uniform $[0, 1]$ random variables. Then the random variables

$$Z_1 = \sqrt{-2\ln(U_1)}\,\cos(2\pi U_2), \qquad Z_2 = \sqrt{-2\ln(U_1)}\,\sin(2\pi U_2)$$

are independent standard normal random variables.

Using Theorem 6.2-4, we can obtain pairs of standard normal random numbers from pairs of unit random numbers as illustrated in the following example. Theorem 6.2-4 can be verified by using the variable transformation technique discussed in Section 5.4 and is left as an exercise.

EXAMPLE 6.2-10 Let $u_1 = .10577$ and $u_2 = .83445$ be two unit random numbers. The following numbers are simulated values of two independent, standard normal random variables.

$$Z_1 = \sqrt{-2\ln(.10577)}\ \cos(2\pi(.83445))$$
$$= 1.07269;$$
$$Z_2 = \sqrt{-2\ln(.10577)}\ \sin(2\pi(.83445))$$
$$= -1.82820.$$

It is sometimes convenient to use just half of the Box–Mueller pair, say Z_1. If a single standard normal random variable is needed, it is still necessary to generate two unit random numbers to apply the Box–Mueller transformation. For instance, if we wish to obtain 100 standard normal variables, we could use 200 unit random numbers to obtain 100 values of Z_1.

EXAMPLE 6.2-11 Heights of a certain population have a normal distribution with a mean of 71 inches and a standard deviation of 3 inches. Let X denote the height of a person from this population. The computations illustrate the generation of five values of X by using the first half of the Box–Mueller pair.

Unit Random Numbers		Simulated Standard Normal Random Variable	Height
u_1	u_2	z_1	$x = 71 + 3z_1$
.83661	.50210	-0.59726	69.2
.26754	.87301	1.13381	74.4
.78872	.53273	-0.67445	69.0
.76476	.86402	0.48096	72.4
.61769	.88984	0.75573	73.3

Exercises 6.2

6.2-1 Let Z be a standard normal random variable. Use Table 4 in Appendix A to find

a $P(Z \le -.52)$

b $P(Z \le .52)$

c $P(-.52 \le Z \le .52)$

d $P(Z > 1.91)$

e $P(Z \le -1.91)$

f $P(.35 \le Z \le 1.75)$

6.2-2 If X is a normal random variable, what is the probability it will take on a value within two standard deviations of its mean?

6.2-3 Let X be a normal random variable with $\mu = 8$ and $\sigma = 3.5$. Find

a $P(X < 0)$

b $P(-2 \le X \le 2)$

c $P(X > 15)$

6.2-4 The time spent by students working on a project is a normal random variable with a mean of 12 hours and a standard deviation of 4 hours.

a Find the probability that the amount of time spent on the project is less than 14 hours.

b Find the probability that the amount of time spent on the project is greater than 8 hours.

6.2-5 A bottle cap manufacturing machine produces bottle caps with a standard deviation of .001 inch. At what nominal (mean) diameter should the machine be set to produce caps for which no more than 10% will exceed 1.24 inches in diameter?

6.2-6 On average, 3% of the watches manufactured by a particular company will be returned for warranty repair. Use the normal approximation to the binomial to find the probability that among 1000 watches sold, 40 or fewer will be returned.

6.2-7 An occupational physical therapist feels that 5% of a production plant's accidents are caused by employees returning to their jobs before completion of their therapy. If the therapist is correct, find the approximate probability that among 75 accidents more than 5 were due to employees returning to their jobs prematurely.

6.2-8 A certain fireworks display will burn for an average of 7.5 seconds with a standard deviation of .5 second before exploding. Find a time t_0 so that the fireworks has a probability of .99 exploding after t_0. Assume the burn time is normally distributed.

6.2-9 A certain type of product must be packaged manually for distribution. Let N be the number of such packages produced in a day. Suppose $\ln(N)$ is normally distributed with $\mu = 5.3$ and $\sigma = 1.6$. Find the probability that at least 500 packages will be produced in a day.

6.2-10 The time it takes to code an algorithm depends on its complexity. Suppose there is a fixed start-up time of 3.5 hours. The additional time to code an algorithm with baseline complexity is normally distributed with $\mu = 2$ hours and $\sigma = 1$ hour. Assume an algorithm that is k times as complex as the baseline will take k times the baseline's additional hours.

a Find the distribution of the total time to code an algorithm with baseline complexity.

b Find the distribution of the total time to code an algorithm that is three times as complex as the baseline.

c Find the distribution of the total time to code an algorithm that is half as complex as the baseline.

6.2-11. Use both Z_1 and Z_2 from the Box–Mueller transformation to simulate 5000 normal random variables with $\mu = 1.7$ and $\sigma = .475$. Find the sample average and sample standard deviation. Make a histogram of the empirical probability density function, using 20 evenly spaced intervals.

6.2-12. Verify Theorem 6.2-2.

6.2-13 Verify Theorem 6.2-4. Hint: $z_1^2 + z_2^2 = -2\ln(U_1)$ and $z_2/z_1 = \tan(2\pi U_2)$.

a Use the bivariate random variable transformation technique discussed in Section 5.4 to find $f(z_1, z_2)$ from the joint distribution of U_1 and U_2.

b Verify that Z_1 and Z_2 are independent standard normal random variables.

6.2-14. Collect data on the heights of 50 to 100 randomly selected college-age males. Make a histogram of the data. Does it appear that the normal distribution is a good probability distribution for these heights?

SECTION

6.3 Gamma Random Variable

An important function in mathematics is the *gamma function*. It is defined as the integral

$$\Gamma(\alpha) = \int_0^\infty y^{\alpha-1} e^{-y}\, dy.$$

Through integration by parts the following properties about the gamma function can be derived (see Exercise 6.3-8):

i $\Gamma(1) = 1.$

ii $\Gamma(\alpha) = (\alpha - 1)\Gamma(\alpha - 1)$ for $\alpha > 1.$

iii $\Gamma(n) = (n-1)!$ for any positive integer n.

Definition 6.3-1

A random variable X is said to have the *gamma distribution* if the probability density function is

$$f(x) = \begin{cases} \dfrac{x^{\alpha-1}e^{-x/\beta}}{\beta^{\alpha}\Gamma(\alpha)}, & \alpha, \beta > 0, \quad x > 0, \\ 0, & \text{otherwise.} \end{cases}$$

Graphs of the gamma distribution for various values of α and $\beta = 1$ are shown in Figure 6.3-1.

FIGURE **6.3-1** Gamma Probability Density Function

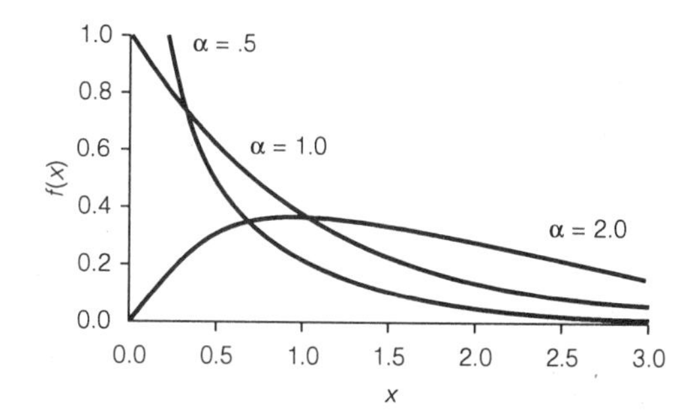

We have already considered a special case of a gamma random variable. The exponential random variable is a gamma random variable with $\alpha = 1$. Another special gamma random variable that plays an important role in statistics is a *chi-square random variable.*

Definition 6.3-2

A gamma random variable with $\alpha = n/2$, n a positive integer, and $\beta = 2$ is called a chi-square random variable. The probability density function is

$$f(x) = \begin{cases} \dfrac{x^{n/2-1}\,e^{-x/2}}{2^{n/2}\Gamma(n/2)}, & n = 1, 2, \ldots; \quad x > 0, \\ 0, & \text{otherwise.} \end{cases}$$

The parameter n of a chi-square random variable is called the *degrees of freedom.* Figure 6.3-2 shows graphs of the chi-square distribution for various values of n. It can be shown that the square of a standard normal random variable has a chi-square distribution with one degree of freedom (Exercise 6.3-10).

FIGURE 6.3-2 Chi-Square Probability Density Functions

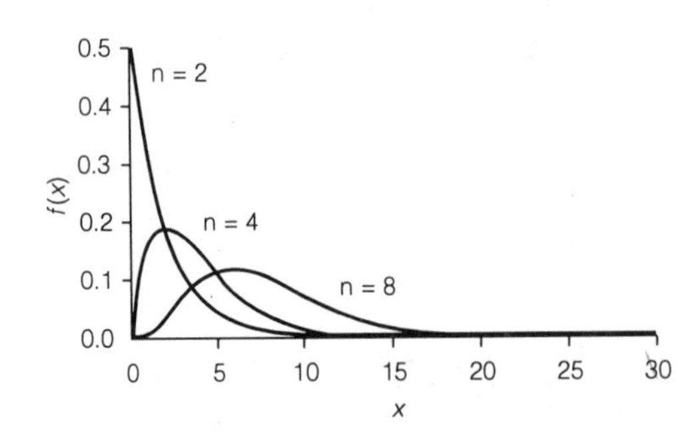

THEOREM 6.3-1 The mean and variance of a gamma random variable with parameters α and β are

$$\mu = \alpha\beta \quad \text{and} \quad \sigma^2 = \alpha\beta^2.$$

Verification of Theorem 6.3-1 is given in Exercise 6.3-9. Given the mean and variance of a gamma random variable, the parameters α and β can be determined as

$$\beta = \frac{\sigma^2}{\mu} \quad \text{and} \quad \alpha = \frac{\mu^2}{\sigma^2}.$$

EXAMPLE 6.3-1 Let X be a gamma random variable with mean .5 and variance .1. Then $\beta = .1/.5 = .2$ and $\alpha = .25/.1 = 2.5$.

When $\beta = 1$, the cumulative distribution function for the random variable of Definition 6.3-1 is called the *incomplete gamma function*. Table 5 in Appendix A gives values of the incomplete gamma function for $\alpha = 1, 2, \ldots, 10$ and $x = 1, 2, \ldots, 15$. This table can be used to evaluate probabilities for gamma random variables.

EXAMPLE 6.3-2 Let X be a gamma random variable with $\beta = 1$ and $\mu = \alpha = 4$. The following probabilities are found using Table 5 in Appendix A with $\alpha = 4$.

$$\begin{aligned} P(X \le 6) &= .849, \\ P(X > 3) &= 1 - P(X \le 3) \\ &= 1 - .353 \\ &= .647. \end{aligned}$$

The incomplete gamma function can also be used to find probabilities for other gamma random variables when $\beta \neq 1$. The following theorem provides a relationship between a gamma random variable with arbitrary β and the incomplete gamma function.

THEOREM **6.3-2** If X is a gamma random variable with parameters α and β, then the cumulative distribution function for the random variable $Y = X/\beta$ is an incomplete gamma function with parameter α.

To verify this result, we simply need to show that the cumulative distribution function of $Y = X/\beta$ is an incomplete gamma function. We have

$$\begin{aligned} P(Y \le y) &= P\left(\frac{X}{\beta} \le y\right) \\ &= P(X \le y\beta) \\ &= \int_{x=0}^{y\beta} \frac{x^{\alpha-1}e^{-x/\beta}}{\beta^{\alpha}\Gamma(\alpha)}\,dx \\ &= \int_{u=0}^{y} \frac{(u\beta)^{\alpha-1}e^{-u}}{\beta^{\alpha}\Gamma(\alpha)}\beta\,du, \qquad \left(\text{for } u = \frac{x}{\beta}\right) \\ &= \int_{u=0}^{y} \frac{u^{\alpha-1}e^{-u}}{\Gamma(\alpha)}\,du. \end{aligned}$$

The latter integral is recognized as the incomplete gamma function. If X is gamma with parameters β and α, then for any constant c

$$\begin{aligned} P(X \le c) &= P\left(\frac{X}{\beta} \le \frac{c}{\beta}\right) \\ &= P\left(Y < \frac{c}{\beta}\right), \end{aligned}$$

where the cumulative distribution function of Y is an incomplete gamma.

EXAMPLE 6.3-3 Let X be a gamma random variable with $\mu = 12$ and $\sigma^2 = 48$. This gives $\beta = 4$ and $\alpha = 3$. Using Theorem 6.3-1 and Table 6 in Appendix A, we obtain

$$\begin{aligned} P(X \le 20) &= P\left(Y \le \frac{20}{4}\right) \\ &= .875 \qquad \text{(incomplete gamma with } \alpha = 3\text{).} \end{aligned}$$

EXAMPLE 6.3-4 The time, T, to unload a truck at the loading dock is a gamma random variable with $\mu = 80$ minutes and $\sigma = 40$ minutes. The probability that it will take more than 2 hours to unload a truck is

$$\begin{aligned} P(T > 120) &= P(Y > 6) \qquad \text{(incomplete gamma with } \alpha = 4\text{)} \\ &= 1 - .849 \\ &= .151. \end{aligned}$$

Exercises 6.3

6.3-1 Evaluate the following.

- **a** $\Gamma(5)$.
- **b** Given $\Gamma(1/2) = \sqrt{\pi}$, find $\Gamma(3/2)$.

6.3-2 Let X be a gamma random variable with $\beta = 1$ and $\alpha = 5$.

- **a** Find $P(X \le 6)$.
- **b** Find $P(X > 4)$.
- **c** Find $P(4 \le X \le 6)$.
- **d** Find the mean and variance of X.

6.3-3 Let X be a chi-square random variable with $n = 8$.

- **a** Find $P(X \le 8)$.
- **b** Find the mean and variance of X.
- **c** Do you think X will have a symmetric distributional shape? Why or why not?

6.3-4 Let X be a gamma random variable with $\mu = 21$ and $\sigma^2 = 63$.

- **a** Find α and β.
- **b** Find $P(18 \le X \le 30)$.

6.3-5 Let X be a chi-square random variable with $\mu = 7$.

- **a** Find σ^2.
- **b** Find $P(\mu - 2\sigma \le X \le \mu + 2\sigma)$. How does this compare to the Chebychev inequality?

6.3-6 Suppose that the CPU time for an execution of a particular software package has a gamma distribution with mean 5 seconds and standard deviation of 2.5 seconds.

a Find α and β.

b Find the probability that it will take more than 10 seconds for an execution of this software.

6.3-7 The minimum time for fire fighters at a certain station to reach the scene of a fire is 3 minutes. The actual time is $3 + X$, where the random variable X has a gamma distribution with $\mu = 4$ minutes and $\sigma = 2$ minutes. Find the probability that it takes more than 9 minutes for the fire fighters to reach a fire.

6.3-8 Use integration by parts to verify

i $\Gamma(1) = 1$;

ii $\Gamma(\alpha) = (\alpha - 1)\Gamma(\alpha - 1), \ \alpha > 1$;

iii $\Gamma(n) = (n-1)!, \ n = 1, 2, \ldots$.

6.3-9 For a gamma random variable X, verify that $\mu = \alpha\beta$ and $\sigma^2 = \alpha\beta^2$. (*Hint*: Rewrite the integrals for $E(X)$ and $E(X^2)$ as constants multiplied by integrals involving gamma density functions with parameters $(\alpha + 1, \beta)$ and $(\alpha + 2, \beta)$, respectively.)

6.3-10 If Z has a standard normal distribution, show that $Y = Z^2$ has a chi-square distribution with 1 degree of freedom.

SECTION

6.4 The Weibull Random Variable

In this section we study the *Weibull* distribution, named after the physicist Waloddi Weibull, who suggested its use.

Definition 6.4-1

A random variable X is said to have the *Weibull distribution* if its cumulative distribution is

$$F(x) = \begin{cases} 1 - e^{-(x/\alpha)^\beta}, & \alpha, \beta > 0, \ x \geq 0, \\ 0, & x < 0. \end{cases}$$

The probability density function of the Weibull is

$$f(x) = \begin{cases} \left(\frac{\beta}{\alpha}\right)\left(\frac{x}{\alpha}\right)^{\beta-1} e^{-(x/\alpha)^\beta}, & \alpha, \beta > 0, \ x \geq 0, \\ 0, & x < 0. \end{cases}$$

Graphs of the Weibull probability density function are shown in Figure 6.4-1.

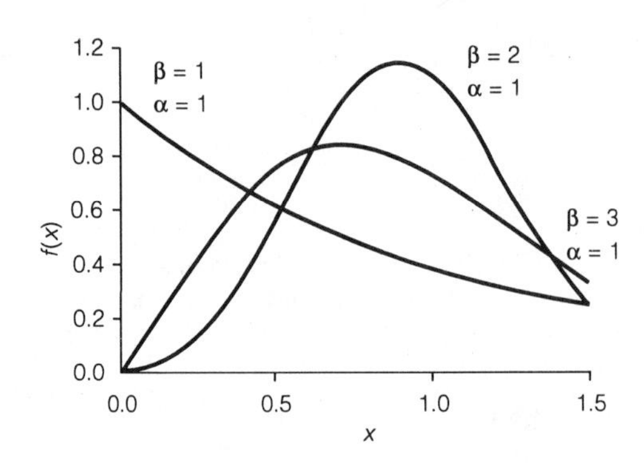

FIGURE **6.4-1**
Plots of the Weibull Probability Density Function

The mean and standard deviation of the Weibull distribution involve rather complicated functions of the parameters α and β. It can be shown (see Exercise 6.4-5) that the expected value and variance of the Weibull random variable are

$$\mu = \alpha\Gamma\left(1 + \frac{1}{\beta}\right),$$

$$\sigma^2 = \alpha^2\left(\Gamma\left(1 + \frac{2}{\beta}\right) - \left[\Gamma\left(1 + \frac{1}{\beta}\right)\right]^2\right),$$

where $\Gamma(x)$ denotes the gamma function defined in Section 6.3. If the mean and variance of the Weibull distribution are specified, these expressions may be used to obtain α and β. First,

$$\frac{\mu^2}{\mu^2 + \sigma^2} = \frac{[\Gamma(1 + 1/\beta)]^2}{\Gamma(1 + 2/\beta)}. \qquad (6.4\text{-}1)$$

Equation (6.4-1) can be solved by evaluating the right-hand side for various values of β until one is found in which the right-hand side matches the left-hand side. Having so determined β, we find

$$\alpha = \frac{\mu}{\Gamma(1 + 1/\beta)}. \qquad (6.4\text{-}2)$$

Table 6.4-1 contains values of $[\Gamma(1 + 1/\beta)]^2/\Gamma(1 + 2/\beta)$ and $\Gamma(1 + 1/\beta)$ for β ranging from .2 to 4.0 in increments of .2. Example 6.4-1 illustrates the use of this table to determine α and β from μ and σ.

TABLE **6.4-1**

β	$\dfrac{[\Gamma(1+1/\beta)]^2}{\Gamma(1+2/\beta)}$	$\Gamma(1+1/\beta)$
0.200	0.004	120
0.400	0.092	3.323
0.600	0.244	1.505
0.800	0.386	1.133
1.000	0.500	1.000
1.200	0.588	0.940
1.400	0.656	0.911
1.600	0.709	0.896
1.800	0.752	0.889
2.000	0.785	0.886
2.200	0.813	0.886
2.400	0.835	0.886
2.600	0.854	0.888
2.800	0.870	0.890
3.000	0.883	0.893
3.200	0.895	0.896
3.400	0.905	0.898
3.600	0.913	0.901
3.800	0.921	0.904
4.000	0.927	0.906

EXAMPLE **6.4-1** A Weibull distribution has mean $\mu = 4$ and $\sigma = 2$. Thus, $\mu^2/(\mu^2+\sigma^2) = .800$. From Table 6.4-1 we see that β falls between 2.00 and 2.20. The interpolated value is $\beta = 2.11$. Also from Table 6.4-1, we find $\Gamma(1 + 1/\beta) = .886$ for $\beta = 2.11$, so $\alpha = 4/.886 = 4.51$.

Another way of determining the parameters of the Weibull distribution is based on specifying two values of the cumulative distribution function. Since $1 - F(x) = e^{-(x/\alpha)^\beta}$, we have

$$\ln[-\ln(1-F(x))] = \beta[\ln(x) - \ln(\alpha)]. \tag{6.4-3}$$

Suppose we know two values of X, $x_{.05}$ and $x_{.95}$, that correspond to the 5% and 95% values of $F(x)$. That is, $F(x_{.05}) = .05$ and $F(x_{.95}) = .95$. (Note that any two values of x could be used.) Find $\ln[-\ln(1-F(x_{.05}))] = \ln[-\ln(.95)] = -2.970$, and likewise, $\ln[-\ln(1 - F(x_{.95}))] = \ln[-\ln(.05)] = 1.097$. Substituting these values into Equation (6.4-3), we obtain two equations in two unknowns that may be readily solved for α and β. Thus,

$$\begin{aligned} -2.970 &= \beta[\ln(x_{.05}) - \ln(\alpha)], \\ 1.097 &= \beta[\ln(x_{.95}) - \ln(\alpha)]. \end{aligned} \tag{6.4-4}$$

Subtracting the first equation from the second in Equation (6.4-4) and simplifying gives us the solution for β. Dividing the first equation by the second and simplifying gives the solution for $\ln(\alpha)$ from which the value of α may be obtained. We have

$$\begin{aligned} \beta &= \frac{4.067}{\ln(x_{.95}) - \ln(x_{.05})}, \\ \ln(\alpha) &= \frac{1.097\ln(x_{.05}) + 2.970\ln(x_{.95})}{4.067}. \end{aligned} \tag{6.4-5}$$

EXAMPLE 6.4-2 The time to failure of a rechargeable battery has a Weibull distribution. It has a 5% chance of failing in less than 2 hours and a 95% chance of failing in less than 10 hours. With this information, we find

$$\begin{aligned} \beta &= \frac{4.067}{\ln(10) - \ln(2)} = 2.53, \\ \ln(\alpha) &= \frac{1.097\ln(2) + 2.970\ln(10)}{4.067} = 1.87, \\ \alpha &= e^{1.87} = 6.49. \end{aligned}$$

An observed value of a Weibull random variable may be simulated in the following way.

Simulation of a Weibull Random Variable

1 Generate a unit random number u.

2 Set $u = 1 - e^{-(x/\alpha)^\beta}$.

3 Solve for x: $e^{-(x/\alpha)^\beta} = 1 - u$

$$\begin{aligned} (x/\alpha)^\beta &= -\ln(1-u) \\ x &= \alpha[-\ln(1-u)]^{1/\beta}. \end{aligned}$$

Since $1 - u$ is also a unit random number (see Exercise 6.1-7), we may simulate values of the Weibull random variable by using $x = \alpha[-\ln(u)]^{1/\beta}$ in step 3.

Exercises 6.4

6.4-1 The CPU time T, in seconds, to execute a piece of software changes based on the input parameters. Suppose the CPU time follows a Weibull distribution with parameters $\alpha = .05$ and $\beta = .25$.

a Find $f(t)$.

b What is $E(T)$ and $VAR(T)$?

c What is the probability that a certain set of input parameters will cause the software to take longer than 1 second to execute?

6.4-2 The diameters of pebbles selected randomly from a riverbed have a Weibull distribution with a mean of 5 millimeters with a standard deviation of 2 millimeters.

a Find α and β.

b Find the probability that a pebble has a diameter of more than 7 millimeters.

6.4-3 A physical education instructor studied the vertical distances that students in a ninth grade class could jump. The instructor found that these distances follow a Weibull distribution with a mean of 10 inches and a standard deviation of 3 inches. What range of distances constitutes the middle 50% of the probability density?

6.4-4 It has been observed that 5% of the students who take a certain examination will finish in less than 20 minutes and 95% will finish in less than 90 minutes. Assume that the time to finish an examination is a Weibull random variable.

a Find α, β, μ, and σ.

b Find the probability that a student will finish the examination in less than 1 hour.

6.4-5 Verify that $E(X) = \alpha\Gamma(1 + 1/\beta)$ and $E(X^2) = \alpha^2\Gamma(1 + 2/\beta)$, from which σ^2 may be determined. (*Hint*: Use the variable substitution $u = (x/\alpha)^\beta$ and the gamma function.)

SECTION

6.5 Moment-Generating Functions

The concepts of moments and moment-generating functions carry over from the discrete to the continuous case. Computation is simply a matter of replacing summations with integrals. (See Section 3.6 for moment-generating functions for discrete random variables.)

EXAMPLE 6.5-1 The kth moment of a uniform $[c, d]$ random variable X is

$$E(X^k) = \int_{x=c}^{d} x^k \frac{1}{d-c}\, dx = \frac{d^{k+1} - c^{k+1}}{(d-c)(k+1)}.$$

The moment-generating function of X is

$$\begin{aligned} M_X(t) = E(e^{Xt}) &= \int_{x=c}^{d} e^{xt} \frac{1}{d-c}\, dx \\ &= \frac{1}{t(d-c)} e^{xt} \Big|_c^d \\ &= \frac{e^{dt} - e^{ct}}{t(d-c)}. \end{aligned}$$

As with discrete random variables, it is sometimes useful to use the moment-generating function to compute the moments of a continuous random variable. The next two examples use the moments of normal random variables to find their means and variances.

EXAMPLE 6.5-2 Let Z be a standard normal random variable. Then

$$\begin{aligned} M_Z(t) = E(e^{Zt}) &= \int_{-\infty}^{\infty} e^{zt} \frac{1}{\sqrt{2\pi}} e^{-z^2/2}\, dz \\ &= \int_{-\infty}^{\infty} \frac{1}{\sqrt{2\pi}} e^{-(z^2-2zt)/2}\, dz \\ &= \int_{-\infty}^{\infty} \frac{1}{\sqrt{2\pi}} e^{-(z^2-2zt+t^2)/2} e^{t^2/2}\, dz \\ &= e^{t^2/2} \int_{-\infty}^{\infty} \frac{1}{\sqrt{2\pi}} e^{-(z-t)^2/2}\, dz \\ &= e^{t^2/2}. \end{aligned}$$

The first and second derivatives of this moment-generating function are

$$M_Z^{(1)}(t) = te^{t^2/2} \quad \text{and} \quad M_Z^{(2)}(t) = (1+t^2)e^{t^2/2}.$$

Evaluating these derivatives at $t = 0$ (Theorem 3.6-1), we find

$$\mu = M_Z^{(1)}(0) = 0 \quad \text{and} \quad \sigma^2 = M_Z^{(2)}(0) - \left[M_Z^{(1)}(0)\right]^2 = 1.$$

EXAMPLE 6.5-3 Let the random variable X have a normal distribution with parameters μ and σ. We will find the moment-generating function for X. From Theorem 6.2-1, if Z is a standard normal random variable, we may express X as $X = \mu + \sigma Z$. Thus,

$$M_X(t) = E(e^{Xt}) = E(e^{(\mu+\sigma Z)t}) = e^{\mu t}E(e^{Z\sigma t})$$
$$= e^{\mu t}M_Z(\sigma t) = e^{\mu t}e^{\sigma^2 t^2/2} = e^{\mu t + \sigma^2 t^2/2}.$$

We use moment-generating functions to verify that the parameters μ and σ are the mean and standard deviation of the normal random variable. Now,

$$M_X^{(1)}(t) = (\mu + \sigma^2 t)e^{\mu t + \sigma^2 t^2/2}$$

and

$$M_X^{(2)}(t) = [\sigma^2 + (\mu + \sigma^2 t)^2]e^{\mu t + \sigma^2 t^2/2}.$$

Evaluating these derivatives at zero gives

$$M_X^{(1)}(0) = \mu \quad \text{and} \quad M_X^{(2)}(0) = \sigma^2 + \mu^2.$$

It follows that $E(X) = \mu$ and $\mathrm{VAR}(X) = \sigma^2$.

Exercises 6.5

6.5-1 Find the fifth moment of the random variable X with probability density function

$$f(x) = \begin{cases} 2x, & 0 \le x \le 1, \\ 0, & \text{otherwise.} \end{cases}$$

6.5-2 Find the moments for the random variable X with probability density function

$$f(x) = \begin{cases} 2(1-x), & 0 \le x \le 1, \\ 0, & \text{otherwise.} \end{cases}$$

6.5-3 Find the moment-generating function of an exponential random variable and use it to verify that $\mu = \sigma$.

6.5-4 Let X be a uniform $[0, 1]$ random variable. Find the moment-generating function of $Y = 5X/3$. Verify that $M_Y(t) = M_X(5t/3)$.

6.5-5 Let X be a gamma random variable with parameters α and β.

a Show that the moment-generating function of a gamma random variable with parameters α and β is $M_X(t) = 1/(1 - \beta t)^{\alpha}$, $t < 1/\beta$.

b Use this moment-generating function to verify that $\mu = \alpha\beta$ and $\sigma^2 = \alpha\beta^2$.

6.5-6 Let X have a Weibull distribution. Find the moment-generating function of $Y = (X/\alpha)^{\beta}$. Show that this is the moment-generating function of an exponential random variable with mean $\mu = 1$.

SECTION

6.6 Method of Moments Estimation

The probability distribution and its associated parameter values are what distinguishes one random variable from another. Often it is the case that for some random phenomenon the general distributional shape of the random variable (normal, uniform, gamma, Weibull, etc.) might be known, but the corresponding parameters might be unknown. Given a random sample, the parameters can frequently be estimated by using a technique based on the *sample moments* of the random variable.

Definition 6.6-1

Given a random sample $\{X_1, X_2, \ldots, X_n\}$, the *kth sample moment* is

$$M_k = \frac{1}{n}\sum_{i=1}^{n} X_i^k.$$

We have already noticed through multiple examples that the moments of a random variable, $E(X^k)$, $k = 1, 2, \ldots$, are functions of the probability distribution parameters. A parameter estimation technique called *method of moments* equates the sample moments M_k to $E(X^k)$, forming a simultaneous system of equations. The parameter estimates are then the solution to this system of equations.

Method of Moments Parameter Estimation

Let $\{\theta_1, \theta_2, \ldots, \theta_t\}$ be the parameters of some distribution. Given a random sample $\{X_1, X_2, \ldots, X_n\}$, the parameter estimates are the solution for $\{\theta_1, \theta_2, \ldots, \theta_t\}$ in the system of t equations defined by

$$M_k = E(X^k),$$

$k = 1, 2, \ldots, t$. This solution is denoted as $\{\hat{\theta}_1, \hat{\theta}_2, \ldots, \hat{\theta}_t\}$.

EXAMPLE **6.6-1** Let X be a uniform $[0, \theta]$ random variable. Since the distribution of X has only one parameter, the first moment and first sample moment will suffice for the estimation of θ. Now

$$E(X) = \mu = \frac{\theta}{2}.$$

The method of moments estimator for θ is found by solving

$$M_1 = \bar{X} = \frac{\theta}{2}.$$

This solution (or estimator for θ) is

$$\hat{\theta} = 2\bar{X}.$$

Given the data $\{1.2, 1.5, 4.0, 3.5, 5.0, 2.7, 3.2, 4.9, 0.1, 4.1, 2.3, 2.6, 3.9, 0.4, 1.4\}$ from a random sample of size 15 for X yields $\hat{\theta} = 2(2.72) = 5.44$.

EXAMPLE **6.6-2** The first two moments of a gamma random variable are

$$E(X) = \alpha\beta, \qquad E(X^2) = \alpha\beta^2 + \alpha^2\beta^2.$$

The method of moments estimators for α and β are found by solving the system of equations

$$M_1 = \alpha\beta, \qquad M_2 = \alpha\beta^2 + \alpha^2\beta^2.$$

The solution is

$$\hat{\alpha} = \frac{(M_1)^2}{M_2 - (M_1)^2} \quad \text{and} \quad \hat{\beta} = \frac{M_1}{\hat{\alpha}}.$$

The following data are from a random sample of X:

$$\{2.63, 9.30, 13.41, 8.22, 11.86, 8.55, 5.21, 11.38, 15.70, 6.08, 7.13, 19.22, 8.43, 10.77, 10.73, 3.51, 7.00, 7.11, 9.94, 11.22\}$$

Based on the data, the sample moments are

$$M_1 = \overline{X} = 9.37$$

and

$$M_2 = \frac{1}{20}\sum_{i=1}^{20} X_i^2 = 102.68.$$

From these sample moments

$$\hat{\alpha} = \frac{(9.37)^2}{102.68 - (9.37)^2} = 5.90$$

and

$$\hat{\beta} = \frac{9.37}{5.90} = 1.59.$$

The method of moments estimation techniques can be used for discrete random variables as well as continuous random variables.

EXAMPLE 6.6-3 Let the random variable X have a binomial distribution with m trials and probability of success p. Recall from Example 3.6-1 that the first moment is

$$E(X) = mp.$$

The method of moments estimator for p is found by solving for p in

$$M_1 = mp,$$

where $M_1 = \overline{X}$ is the sample average of observed binomial random variables. The solution is

$$\hat{p} = \frac{M_1}{m} = \frac{\overline{X}}{m}.$$

The following data are a random sample of size 25 of a binomial random variable with 10 trials and $p = .85$.

$$\{10, 8, 7, 9, 10, 10, 8, 8, 7, 8, 10, 10, 8, 9, 8, 10, 9, 6, 8, 10, 8, 8, 9, 9, 9\}$$

Based on this data, the first sample moment, or the sample average, is $\overline{X} = 8.64$, and the method of moments estimator for p is $\hat{p} = .864$. Notice that $\hat{p}$ is relatively close to the true value of p.

EXAMPLE 6.6-4 Let the random variable X have a geometric distribution with probability p of a success on each Bernoulli trial. Recall that $E(X) = 1/p$. Therefore, the method of moments estimator for p is

$$\hat{p} = \frac{1}{\overline{X}}.$$

Exercises 6.6

6.6-1 Show that the method of moments estimators for the parameters of a normal random variable are $\hat{\mu} = \overline{X}$ and $\sigma^2 = S^2$.

6.6-2 Verify that $\hat{\alpha} = (M_1)^2/(M_2 - (M_1)^2)$ and $\hat{\beta} = M_1/\hat{\alpha}$ from Example 6.5-5 are the method of moments estimators for α and β.

6.6-3 Suppose X is an exponential random variable with parameter λ.

a Find the method of moments estimator for λ.

b Use the estimator to estimate λ and σ^2 from the following data:

$$\{0.107, 0.279, 1.277, 0.656, 0.051, 1.337, 0.175, 0.197, 0.119, 0.029, 0.103, 1.962, 0.020, 0.051, 0.474, 0.048, 1.451, 0.251, 0.371, 0.049\}$$

6.6-4 Let X be a Poisson random variable with parameter λ. Derive the method of moments estimator for λ.

6.6-5 The times in 1000 hours it takes nine light bulbs to fail are

$$0.75, 1.23, 1.85, 0.82, 1.57, 2.05, 0.45, 1.35, 0.24.$$

a Assuming that the data come from a Weibull distribution, use the sample mean and sample standard deviation to estimate α and β.

b Estimate the probability that a light bulb will fail in less than 800 hours.

6.6-6 Suppose that the following data are the pounds per square inch that caused each of 20 samples of plastic to break. Assume that the data are a random sample of a Weibull random variable, the values having been arranged in order from smallest to largest.

$$\{11.5, 13.9, 15.4, 16.1, 16.8, 19.1, 20.4, 20.5, 21.4, 21.5,$$
$$22.8, 26.4, 28.2, 29.1, 35.3, 38.4, 39.9, 44.4, 54.9, 64.0\}.$$

a Use Equations (6.4-1) and (6.4-2) and Table 6.4-1 to find the method of moments estimators for α and β from the Weibull distribution.

b Since 5% of the observed data values are 11.5 or less, $P(X \le 11.5) \approx .05$, or $x_{.05} \approx 11.5$. Similarly $P(X \le 54.9) \approx .95$, or $x_{.95} \approx 54.9$. Estimate α and β based on these sample values, using the technique described with Equations (6.4-3) through (6.4-5).

6.6-7 Simulate five random samples of size 20 from a Weibull distribution with $\alpha = 1$ and $\beta = 2$.

a For each simulated sample, estimate α and β by using the method of moments (see Exercise 6.6-5a).

b For each simulated sample, estimate α and β by using $x_{.05}$ and $x_{.95}$ (see Exercise 6.6-5b).

c The mean-squared error of the estimates of a parameter α for a given method of estimation is computed as $(1/r)\sum_{i=1}^{r}(\hat{\alpha}_i - \alpha)^2$, where $\hat{\alpha}_i$ is the estimated value of α from the ith random sample and r is the total number of samples. Of course, $r = 5$ and $\alpha = 1$ in this case. Compute the mean-squared error for each of the two methods of estimation in parts (a) and (b) to determine which method did a better job of estimating α.

d Repeat part (c) for the estimates of β.

e What would you expect to happen to the magnitudes of the mean-squared errors in parts (c) and (d) if each data set had contained 40 values instead of 20?

6.6-8 Suppose X has probability density function

$$f(x) = \begin{cases} \theta x^{\theta-1}, & 0 \le x \le 1, \quad \theta > 0, \\ 0, & \text{otherwise.} \end{cases}$$

a Derive the method of moments estimate for θ.

b Let $\theta = 2$. Simulate 50 random samples of sizes 10, 30, 100, and 1000 for X. For these four sample sizes, estimate θ by using the method of moments estimate in part (a). How do these estimates compare to the true value of $\theta = 2$? Use the mean-squared error defined in Exercise 6.6-7 to do the comparison.

6.6-9 The following two sets of data are random samples of size 30 of a geometric random variable with $p = .2$ and of a binomial random variable with 20 Bernoulli trials and $p = .35$. Which data set do you believe is for which random variable? Why?

Data set 1: 8, 2, 5, 4, 9, 3, 6, 4, 5, 6,
9, 9, 6, 5, 8, 14, 9, 6, 5, 8,
9, 7, 5, 6, 7, 10, 7, 4, 10, 6

Data set 2: 4, 3, 1, 1, 3, 5, 15, 13, 5, 12,
1, 2, 8, 8, 3, 1, 9, 8, 2, 2,
9, 3, 6, 1, 3, 2, 5, 6, 6, 5

7

Markov Counting and Queuing Processes

SECTION 7.1 Bernoulli Counting Process

Counting processes arise naturally in dealing with the number of outcomes of various kinds that occur across time. Some examples of counting processes are the number of failures of a computer operating system, the number of phone calls that come into a switchboard, and the number of traffic accidents at an intersection. A counting process is often a part of a model of a larger system. For example, the number of failures of an operating system might be part of a model for a computing network. A formal definition of a counting process follows.

Definition 7.1-1

A stochastic process is called a *counting process* if

i The possible states are the nonnegative integers.

ii For each state i the only possible transitions are

$$i \to i, i \to i+1, i \to i+2, \ldots.$$

If the process has also the Markov property, it is called a *Markov counting process*. As with the binomial random variable, we refer to the outcomes being counted as *successes*.

In this section we study a particular discrete-time counting process called the *Bernoulli counting process*. Assume that we have divided a continuous-time interval into discrete disjoint subintervals of equal length. We call these subintervals *frames*. Additionally, assume that the frame size is selected such that at most one success can occur in each frame. For example, if we were counting the number of calls completed by a telephone sales person, a frame of 10 seconds would probably be small enough so that at most one completion would occur in each frame. Similarly, if we were to count the number of ships arriving at a loading dock, a frame of 30 minutes might be small enough for at most one ship to arrive. If the successes from frame to frame are independent and identical Bernoulli trials, then this counting process is a Bernoulli counting process.

Definition 7.1-2

A counting process is said to be a *Bernoulli counting process* if

i The number of successes that can occur in each frame is either 0 or 1.

ii The probability, p, that a success occurs during any frame is the same for all frames.

iii Successes in nonoverlapping frames are independent of one another.

THEOREM 7.1-1 Let $X(n)$ denote the total number of successes in a Bernoulli counting process at the end of the nth frame, $n = 1, 2, \ldots$. Let the initial state be $X(0) = 0$. The probability distribution of $X(n)$ is the binomial

$$P(X(n) = x) = \binom{n}{x} p^x (1-p)^{n-x}.$$

To verify Theorem 7.1-1, simply note that $X(n)$ is the sum of the outcomes of n independent and identical Bernoulli trials.

The one-step transition probabilities for the Bernoulli counting process are

$$P(i \to i+1) = p, \quad P(i \to i) = 1-p.$$

The one-step transition matrix is

$$P = \begin{array}{c} \\ 0 \\ 1 \\ 2 \\ 3 \\ \vdots \end{array} \begin{array}{c} \begin{array}{ccccc} 0 & 1 & 2 & 3 & \cdots \end{array} \\ \left[\begin{array}{ccccc} 1-p & p & 0 & 0 & \cdots \\ 0 & 1-p & p & 0 & \cdots \\ 0 & 0 & 1-p & p & \cdots \\ 0 & 0 & 0 & 1-p & \cdots \\ \vdots & \vdots & \vdots & \vdots & \cdots \end{array}\right] \end{array}.$$

The Bernoulli counting process can be simulated frame-by-frame with the following algorithm.

Simulation of a Bernoulli Counting Process

Initialize $n = 0$
$X(0) = 0$

Repeat the following steps once for each frame.

1. Generate a unit random number u.
2. If $u < p$, then set $X(n+1) = X(n) + 1$;
   else set $X(n+1) = X(n)$.
3. Increment $n = n + 1$.

EXAMPLE 7.1-1 Figure 7.1-1 shows the simulation of a Bernoulli counting process for 60 frames where the probability of a success in each frame is .07. A unit random number u was generated for each frame, and a success was declared to occur if $u < .07$. Successes occurred in the 4th, 34th, 36th, and 46th frames. The step function shows the increment in counts for the successes that occurred at those frames.

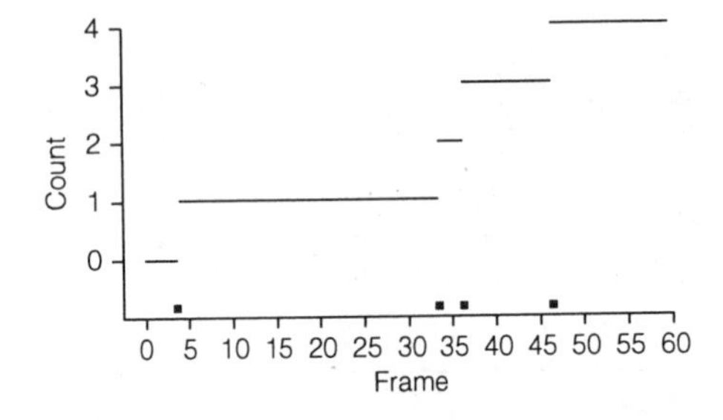

FIGURE 7.1-1 Simulation of a Bernoulli Counting Process, $p = .07$

We must consider two practical aspects when using the Bernoulli counting process as a model for a real process: the length of a frame and the probability that a success occurs during a frame. The choice of the length of a frame is usually just a matter of practical judgment. The physical context of the modeling problem should suggest a frame length so that at most one success will occur during each frame. For instance, in counting the number of cars that pass a point on a highway, a frame of a second or less might be required for city traffic, but a longer frame might suffice for rural traffic.

Having chosen the length of the frame, we can determine the probability that a success will occur during a frame. To do this, we relate the probability of a success in a frame to the expected number of successes in some predetermined unit of time. For example, suppose that the unit of time is chosen to be an hour and the expected number of successes in an hour is 72. Also suppose that the frame length is chosen to be 1 second. Since there are 3600 frames in an hour, the expected hourly count is $np = 3600p = 72$. Therefore, $p = 72/3600 = .02$. If λ denotes the expected number of successes in a unit of time, and n is the number of frames in this unit of time, then

$$p = \frac{\lambda}{n}.$$

The constant λ is called the *rate* of success. In practice it is estimated as

$$\hat{\lambda} = \frac{\text{Number of successes in } t \text{ units of time}}{t}.$$

For instance, if we observe 144 successes in 8 hours, the estimated rate is $144/8 = 18$ per hour. If there are n frames in the unit of time and we let the frame length, denoted by Δ, be expressed in the same unit of time as λ (seconds, minutes, hours, etc.), then $\Delta = 1/n$. A useful relationship between probability, rate, and frame length in a Bernoulli counting process is

$$p = \lambda\Delta.$$

This equation tells us that the probability of success is proportional to the length of the frame where the constant of proportionality is the rate.

EXAMPLE 7.1-2 A company has a toll-free number for taking orders. It has been observed that the company receives on average $\lambda = 3$ calls per hour. Suppose that the number of calls can be modeled by a Bernoulli counting process with 1-minute frames. Since λ is in hours, we have $\Delta = 1/60$ hour. Thus, the probability of a call during a 1-minute frame is $p = 3\,(1/60) = .05$.

EXAMPLE 7.1-3 Instead of 1-minute frames, suppose that frames of length 5 minutes had been chosen for Example 7.1-2. In this case $\Delta = 5/60 = 1/12$ hour, so the probability would be $p = 3\,(1/12) = .25$. The expected number of successes per hour does not change as the result of the change in frame size. However, the standard deviation does change. For instance, with 1-minute frames the standard deviation of the number of counts in 60 minutes is $\sqrt{60(.05)(.95)} = 1.69$ counts per hour, whereas with 5-minute frames the standard deviation is $\sqrt{12(.25)(.75)} = 1.50$

counts per hour. Generally speaking, small changes in frame length will not critically affect the behavior of the process. In this example, increasing the length of frame by a factor of 5 caused only a relatively small change in the standard deviation of hourly counts.

The Bernoulli counting process is time homogeneous because the probability of a success does not change with time. Nonhomogeneous processes can be modeled by dividing them into homogeneous segments. For example, consider counting the number of accidents on a highway. When viewed on a 24-hour basis, this process would be nonhomogeneous due to changing traffic intensities throughout the day. However, when viewed during a short period of time, say a mid-morning hour, the probability of an accident would not change much because the intensity of traffic would be relatively constant throughout the hour. Hence, during that time, the process could be considered to be homogeneous.

EXAMPLE 7.1-4 Example 7.1-2 can be modified to account for changing rates of calls. Suppose that the company averages three calls per hour in the morning but only one and one half calls per hour in the afternoon. The probability of a call during a 1-minute frame changes from .05 in the morning to .025 in the afternoon. If the calls are to be simulated over an entire day, the computer simulation would be different for the morning frames and for the afternoon frames. Given a unit random number u, a simulated call occurs in a morning frame if $u < .05$, whereas one occurs in an afternoon frame if $u < .025$. Counting the number of calls across time is just a matter of bookkeeping.

We now derive the probability distribution for the number of frames it takes to go from one success to the next. We denote this quantity by Y. For instance, if a success has just occurred and it is followed by the sequence 0 0 0 0 1, then $Y = 5$. The possible values of Y are $1, 2, 3, \ldots$. Moreover, Y will have the value y if and only if there are $y - 1$ successive frames in which the outcome of interest does not occur before it occurs again. That is, the probability that $Y = y$ is just the probability of $(y - 1)$ 0's in a row followed by a 1. This leads to the following theorem.

THEOREM 7.1-2 Let Y denote the number of frames from one success to the next. The probability mass function of Y is

$$P(Y > y) = (1 - p)^{y-1}p, \quad y = 1, 2, 3, \ldots.$$

We recognize this distribution as the geometric distribution discussed in Section 3.2 with $\mu = 1/p$ and $\sigma^2 = (1-p)/p^2$.

EXAMPLE 7.1-5 In reference to Example 7.1-2, what is the probability that there is more than 1 hour from one call to the next? Since there are 60 frames in an hour, we wish to find $P(Y > 60)$. In general,

$$P(Y > y) = \sum_{k=y+1}^{\infty} (1-p)^{k-1} p$$

$$= (1-p)^y.$$

Since $p = .05$ in our example, $P(Y > 60) = (.95)^{60} = .046$. The mean, variance, and standard deviation are

$$\mu = \frac{1}{.05} = 20 \text{ frames},$$

$$\sigma^2 = \frac{.95}{(.05)^2} = 380,$$

$$\sigma = \sqrt{380} = 19.5 \text{ frames}.$$

A natural quantity of interest is the amount of time that passes from one success to the next. If T is the time it takes to go from one success to the next and Y is the number of frames from one success to the next, then $T = Y\Delta$. Therefore,

$$E(T) = E(Y)\Delta$$

$$= \frac{1}{p}\Delta = \frac{1}{\lambda\Delta}\Delta = \frac{1}{\lambda}$$

and

$$\text{VAR}(T) = \text{VAR}(Y)\Delta^2$$

$$= \left[\frac{1-p}{p^2}\right]\Delta^2$$

$$= \left[\frac{1-\lambda\Delta}{(\lambda\Delta)^2}\right]\Delta^2$$

$$= \frac{1-\lambda\Delta}{\lambda^2}.$$

This result is summarized as follows.

THEOREM 7.1-3 If T is the time from one success to the next in a Bernoulli counting process, then

$$E(T) = \frac{1}{\lambda}, \quad \text{VAR}(T) = \frac{1-\lambda\Delta}{\lambda^2}.$$

For small Δ, $\text{VAR}(T) \approx 1/\lambda^2$.

EXAMPLE 7.1-6 Suppose that the number of freight trucks arriving at a terminal is a Bernoulli counting process with a rate of $\lambda = 5$ per hour and the frame length is $\Delta = 1/12$ hour. The expected time from the arrival of one truck to the next is the reciprocal of the rate; that is, 1/5 hour = 12 minutes and $\text{STD}(Y) = \sqrt{[1-5(1/12)]/5^2} = .15$ hour or 9.2 minutes.

Exercises 7.1

7.1-1. If each of the following processes can be modeled as a Bernoulli counting process, what length of frame do you think could be used so that at most one occurrence of the outcome of interest will occur during each frame. Choose the largest frame that will do. Base your answers on your experiences with these kinds of processes.

- **a** The number of cars that come in for service at a fast-food restaurant during noon hour.
- **b** The number of calls to a fire station requesting emergency assistance.
- **c** The number of students that come to a computing laboratory during evening hours.

7.1-2. Suppose we wish to determine the probability of an outcome for the frames in parts (a) through (c) of Exercise 7.1-1. Describe how you would obtain the data to do this.

7.1-3 Suppose that we wish to use the Bernoulli counting process to model the arrival of customers at a self-service gas station. It has been observed that customers come in at the rate of 20 per hour. How many frames per hour should we choose, and what should be the length of each frame if the probability of an arrival during a frame is to be .05?

7.1-4. Two researchers each model the number of customers entering a grocery store during an hour's time using Bernoulli counting processes. It has been observed that there are an average of 36 customers per hour. Researcher A uses frames of 1 second in length, whereas researcher B uses frames of 1 minute in length.

- **a** Find the probability of an arrival during a frame for both A and B.
- **b** Give the means and standard deviations for the number of customers arriving in an hour for the two frame lengths in part (a).
- **c** Researcher A points out that 1 minute is too long a time for a frame because more than

one customer can arrive in that time. Nevertheless, B goes ahead with a simulation using 1-minute frames. In what way will B's simulation give an incorrect picture of the arrival of customers at the store?

7.1-5 Suppose that a number of defects coming from an assembly line can be modeled as a Bernoulli counting process with frames of one-half-minute length and probability $p = .02$ of a defect during each frame.

a Find the probability of going more than 20 minutes without a defect.

b Determine the rate λ in units of defects per hour.

c If the process is stopped for inspection each time a defect is found, on average how long will the process run until it is stopped?

d Give a likely range for the number of minutes between stops for inspections.

7.1-6 In reference to Exercise 7.1-5, suppose that we have the following procedure for stopping the process. Whenever a defect occurs, the record is checked for the preceding hour. If in that time there was another defect, the process is stopped for inspection and then started again. Otherwise, it is allowed to continue. Simulate 8 hours of the process and determine the number of times it is stopped. Repeat the simulation 20 times and use your results to estimate the mean and standard deviation of the number of times the process is stopped.

7.1-7 The number of requests for reserved library material will vary throughout the day. Suppose from 8 A.M. to 12 noon the rate of requests is three per hour, from 12 noon to 5 P.M., the rate is six per hour, and from 5 P.M. to 8 P.M., the rate is two per hour.

a Assuming a frame length of 2 minutes, determine the probability of reserved material being requested for frames in each of the three time periods.

b Using a table of random numbers, simulate the number of requests for reserved library material from 11:30 A.M. to 12:30 P.M.

?**c** What is the probability that there will be no requests for reserved material from 11:30 A.M. to 12:30 P.M.?

d What is the expected amount of time from one request to the next in each of these three time periods?

e Write a computer program to simulate the number of requests from 8 A.M. to 8 P.M. Obtain a random sample of 100 and display the data graphically.

7.1-8 Suppose that we count the number of weeks in a year that the weekly state Lotto has a winner. Assume that this process is a Bernoulli counting process with frame size one week and probability of a winner $p = .70$.

a What is the probability that the Lotto will go more than three weeks in a row without a winner?

b What are the mean and standard deviation for the number of weeks between successive winners?

c Suppose the prize is $1,000,000 the first week and doubles by this amount each week that there is no winner. Simulate the Lotto over three years' time and determine the largest amount of prize money available during any given week.

7.1-9 Suppose a Bernoulli counting process can be used to model customers purchasing soda from a vending machine with frames of 2 minutes and $p = .15$ of a soda purchase each frame.

a What is the daily rate, λ, of soda purchases? Assume an 8-hour day.

?**b** Give a likely range for the number of sodas purchased each day.

c Give a likely range for the number of minutes between purchases.

7.1-10 Refer to Exercise 7.1-3. Find the expected time and the standard deviation of the time between arrivals of customers at the gas station.

SECTION

7.2 The Poisson Process

We now discuss a continuous-time counting process called the *Poisson process*. It is obtained from the Bernoulli counting process by letting the frame length become arbitrarily small. First, we will show how the Poisson distribution arises naturally in conjunction with the Bernoulli counting process. Suppose that a Bernoulli counting process has n frames in the interval $[0, t]$ each with length Δ. Moreover, suppose that there are at least 20 frames and p is .05 or less. As shown in Section 3.5, the probability distribution of the number of successes in $[0, t]$, while binomial, can be approximated by a Poisson distribution. That is,

$$P(X(n) = x) = \binom{n}{x} p^x (1-p)^{n-x}$$
$$\approx e^{-\mu} \frac{\mu^x}{x!},$$

where $\mu = np$.

The mean of the Poisson distribution can be expressed in terms of the rate λ. Since λ is the expected number of successes in one unit of time, the expected number in $[0, t]$ is $\mu = \lambda t$. Substituting this for μ, we have

$$P(X(n) = x) \approx e^{-\lambda t} \frac{(\lambda t)^x}{x!}.$$

Since the Bernoulli counting process is time homogeneous, this approximation will apply to any interval of length t, not just $[0, t]$.

Now suppose that the frame length of the Bernoulli counting process becomes smaller and smaller while the expected number of successes in the interval $[0, t]$ remains the same, namely λt. We can imagine the existence of a continuous-time counting process, obtained as a limit of the Bernoulli counting process, for which the distribution of the number of successes in an interval of length t is exactly Poisson with mean λt. This limiting process is called the *Poisson process*.

Definition 7.2-1

Let $N(t)$ denote the number of successes in the interval $[0, t]$. Assume that $N(t)$ is a continuous-time counting process and that the count begins at zero.

$N(t)$ is said to be a *Poisson process* if

i Successes in nonoverlapping intervals occur independently of one another.

ii The probability distribution of the number of successes depends only on the length of the interval and not on the starting point of the interval.

iii The probability of x successes in an interval of length t is

$$P(N(t) = x) = e^{-\lambda t}\frac{(\lambda t)^x}{x!}, \qquad x = 0, 1, 2, \ldots,$$

where λ is the expected number of successes per unit of time.

We can clearly see that the defining properties of the Poisson process are analogous to the properties of the Bernoulli counting process with small frame size. Property i is analogous to the independence property of the Bernoulli counting process. Property ii, which is the time-homogeneous property of the Poisson process, implies that the rate λ is unchanging across time. This is analogous to the unchanging value of p in the Bernoulli counting process. Property iii follows from the Poisson approximation to the binomial distribution.

EXAMPLE 7.2-1 Suppose transmission errors in sending data over a telephone line is a Poisson process with a rate of $\lambda = 1.2$ errors per minute. The probability of having no errors in 4 seconds of transmission time is computed as follows. Since the time in question is $t = 1/15$ minute, the expected number of errors is $\lambda t = 1.2/15 = .08$. Therefore,

4/60

$$P(N(1/15) = 0) = e^{-.08}\frac{(.08)^0}{0!} = .92.$$

To compute the probability of two or fewer errors in 5 minutes, we note that the expected number of errors is $\lambda t = (1.2)(5) = 6$, so

$$P(N(5) \le 2) = \sum_{x=0}^{2} e^{-6}\frac{6^x}{x!}$$

$$= .062.$$

EXAMPLE 7.2-2 The number of requests for computer searches at a certain library during the evening hours is a Poisson process with $\lambda = 12$ requests per hour. To compute the probability that there are 10 or more request for service in 30 minutes, note that the mean is $\lambda t = (12)(.5) = 6$ requests. From Table 3 of Appendix A,

$$\begin{aligned} P(N(.5) \geq 10) &= 1 - P(N(.5) \leq 9) \\ &= 1 - .916 \\ &= .084. \end{aligned}$$

The mean number of requests in 2 hours is $(12)(2) = 24$. Since the standard deviation is the square root of the mean for the Poisson distribution, the standard deviation for 2 hours is 4.9 requests. A likely range of requests for the 2-hour period is 24 $\pm$ 2(4.9), which in rounded numbers is 14 to 34 requests.

Next consider the question of how long it will take for the first success to occur in a Poisson process. Let T be the random variable denoting this time. The time for the first success to occur will be larger than t if and only if there are no successes in the interval $[0, t]$. Thus,

$$P(T > t) = P(N(t) = 0) = e^{-\lambda t}.$$

If $F_T(t)$ denotes the cumulative distribution function of T, then

$$F_T(t) = 1 - e^{-\lambda t}, \qquad t > 0.$$

This is just the cumulative distribution function of an exponential random variable. Moreover, since the Poisson process is time homogeneous, the time between any two consecutive successes has an exponential distribution. This result is summarized in the following theorem.

THEOREM 7.2-1 Let T be the time between consecutive successes in a Poisson process with rate λ. Then T has an exponential distribution with probability density function

$$f_T(t) = \lambda e^{-\lambda t}, \qquad t > 0,$$

and

$$E(T) = \frac{1}{\lambda}, \quad \text{VAR}(T) = \frac{1}{\lambda^2}.$$

The expressions for $E(T)$ and $VAR(T)$ are the same as the corresponding expressions in Theorem 7.1-3 for the Bernoulli counting process when $\Delta \to 0$.

EXAMPLE 7.2-3 The number of private planes taking off from a municipal airport is a Poisson process with an expected value of 24 in an 8-hour business day. We will find the probability that there are 30 minutes or less between consecutive takeoffs. If time is measured in hours, then $\lambda = 3$ per hour and $t = .5$ hours. Thus,

$$\begin{aligned} P(T \leq .5) &= 1 - e^{-(3)(.5)} \\ &= .78. \end{aligned}$$

Since the private planes land at the rate of $\lambda = 3$ per hour, the expected time between landings is 1/3 hour or 20 minutes. The standard deviation is also 20 minutes.

Exercises 7.2

7.2-1 The number of baseball games rained out in Mudville is a Poisson process with $\lambda = 5$ per 30 days.

a Find the probability that there are more than five rained-out games in 15 days.

b Find the probability that there are no rained-out games in seven days.

7.2-2 Suppose that the number of customers entering a hardware store is a Poisson process. Explain how to determine a reasonable value for λ.

7.2-3 Shipments of paper arrive at a printing shop according to a Poisson process with a mean of .5 shipments per day.

a Find the probability that the printing shop receives more than two shipments in a day.

b If there are more than four days between shipments, all the paper will be used up and the presses will be idle. What is the probability that this will happen?

7.2-4 The number of times a piece of military hardware must be serviced in an arctic environment is a Poisson process with an average of one service every 200 hours of operation.

a If a mission involving the use of this hardware takes 24 hours to complete, what is the probability that it will be completed without service being required on the hardware?

b If the probability is .95 that no service on this hardware is required during a mission, what is the mission time?

7.2-5 The number of customers visiting a real-estate sales booth at a mall is a Poisson process with $\lambda = 4$ customers per hour. Give a likely range of values for the number of customers visiting the booth in 8 hours.

7.2-6 Suppose that the process in Example 7.2-3 is modeled as a Bernoulli counting process with rate $\lambda = 3$ per hour and frame length of 1 minute. Simulate the number of takeoffs in 8 hours. Repeat this simulation 10 times. Use the Poisson process to obtain a likely range for the number of takeoffs in 8 hours. Most of your simulated results for the number of takeoffs should fall in this range because the Bernoulli counting process behaves like a Poisson process when the frame length and probability of success in each frame are small.

7.2-7 Information obtained by a data base supervisor indicates that the number of times a particular type of user accesses the data base is a Poisson process with an average of nine times per week. Over a two-week period of time the supervisor noted 30 accesses by one of these users. Does this level of activity indicate anomalous behavior?

7.2-8 In an assembly line, parts leave a particular station according to a Poisson process at the rate of six per minute. A quality control engineer notices that only 300 parts have left this station in the last hour. Is there evidence that the assembly line may be out of control—that is, producing too few parts?

7.2-9 In a Bernoulli counting process, the probability that there are more than n frames from one outcome to the next is $P(N > n) = (1-p)^n$. If T is the time from one outcome to the next, then $T = N\Delta$ where Δ is the frame length. Thus, $P(T > t) = P(N\Delta > t) = P(N > t/\Delta) = (1-p)^{t/\Delta}$. Substitute $p = \lambda\Delta$ into this expression and take the limit as $\Delta \to 0$. Interpret this limiting value in terms of the Poisson process.

SECTION

7.3 Exponential Random Variables and the Poisson Process

The time T between two successes in the Poisson process is called an *interarrival* time. In Theorem 7.2-1, we showed that T has an exponential distribution. Since successes in nonoverlapping intervals of a Poisson process are independent of one another, the interarrival times are independent random variables.

THEOREM 7.3-1 Let T_i, $i = 1, 2, \ldots,$ denote the interarrival times in a Poisson process with rate λ. The T_i's are independent exponential random variables each with mean $\mu = 1/\lambda$.

This theorem is useful in simulating a Poisson process. The interarrival times $T_1, T_2, \ldots, T_n$ are simulated as n exponential random variables. The first n successes occur at times $T_1, T_1 + T_2, \ldots, T_1 + T_2 + \cdots + T_n$.

EXAMPLE 7.3-1 Suppose that a Poisson process has rate $\lambda = 2$. Table 7.3-1 shows the simulated values of five interarrival times and the corresponding times at which five successes occur. For instance, we see that the first three interarrival times are .1123, .8723, and .0273, so the first time at which $N(t) = 3$ is $t = .1123 + .8723 + .0273 = 1.0119$.

Similarly the first time at which $N(t) = 4$ is $t = 1.1847$. Hence, if $1.0119 \leq t < 1.1847$, $N(t) = 3$.

TABLE 7.3-1 Simulation of Five Interarrival and Success Times for a Poisson Process with Rate $\lambda = 2$

i	u	t_i	t	$N(t)$
1	.2021	.1123	.1123	1
2	.8235	.8723	.9846	2
3	.0531	.0273	1.0119	3
4	.2922	.1728	1.1847	4
5	.7845	.7674	1.9521	5

Next we derive the probability distribution of the time that it takes to observe n successes in a Poisson process. This time is called a *waiting time* and is denoted W_n. The derivation of the probability distribution of W_n depends on the following relationship between waiting times and the number of successes. It takes more than t units of time for the nth success to occur if and only if there are $n-1$ or fewer successes in $[0, t]$. That is,

$$P(W_n > t) = P(N(t) \leq n-1). \tag{7.3-1}$$

Now let $F_n(t)$ denote the cumulative distribution function of W_n. From Equation (7.3-1) we have

$$\begin{aligned} 1 - F_n(t) &= P(W_n > t) \\ &= P(N(t) \leq n-1) \\ &= \sum_{x=0}^{n-1} e^{-\lambda t} \frac{(\lambda t)^x}{x!}. \end{aligned}$$

THEOREM 7.3-2 The cumulative distribution function of the waiting time W_n is

$$F_n(t) = 1 - \sum_{x=0}^{n-1} e^{-\lambda t} \frac{(\lambda t)^x}{x!}.$$

The distribution in Theorem 7.3-2 is called an *Erlang* distribution. The probability density function of the Erlang distribution can be found by differentiating the expression in Theorem 7.3-2. With a little algebra, it can be shown that

$$f_n(t) = \frac{\lambda^n t^{n-1}}{(n-1)!} e^{-\lambda t}.$$

The Erlang distribution is a special case of the gamma distribution with $\alpha = n$ and $\beta = 1/\lambda$.

EXAMPLE 7.3-2 The number of thunderstorms occurring in a city during the summer is a Poisson process with rate $\lambda = .8$ per week. The probability that the waiting time for two storms is more than five days is just the probability that there are zero or one storms in five days. Since the number of storms in five days has a Poisson distribution with a mean of $.8(5/7) = .57$, we have

$$\begin{aligned} P(W_2 > 5) &= P(N(5) \le 1) \\ &= e^{-.57} + e^{-.57}(.57) \\ &= .888. \end{aligned}$$

That is, there is about an 89% chance that we will have to wait more than five days for two storms.

The waiting time for the nth success can be expressed as

$$W_n = T_1 + T_2 + \cdots + T_n,$$

where the T_i's are the interarrival times. Since $E(T_i) = 1/\lambda$ and $\mathrm{VAR}(T_i) = 1/\lambda^2$ and by applying Theorems 2.7-3 and 2.7-5, we have the following result for the mean and variance of W_n.

THEOREM 7.3-3 The mean and variance of the waiting time random variable W_n are

$$\mu = \frac{n}{\lambda}, \quad \sigma_n^2 = \frac{n}{\lambda^2}.$$

EXAMPLE **7.3-3** Referring to Example 7.3-2, we find that the mean and standard deviation of the time to observe two thunderstorms are

$$\mu = \frac{2}{.8} = 2.5 \text{ weeks,}$$

$$\sigma = \frac{\sqrt{2}}{.8} = 1.8 \text{ weeks.}$$

Exercises 7.3

7.3-1 Military vehicles arrive at a service facility according to a Poisson process at the rate of five per hour. Simulate 2 hours of vehicles arriving at the service facility by using the method of simulation discussed in Section 7.3. Repeat this simulation 100 times and compute the sample average and sample standard deviation for the number of arrivals in 2 hours.

7.3-2 The washing machine repair shop receives calls at the rate of two per day. Assume an 8-hour day. Simulate five days of calls by using the method of simulation in Section 7.3. Repeat this simulation 500 times. If the shop can handle 13 calls a week, is it likely that there will be a backlog at the end of the week? Answer this question based on your simulation results.

7.3-3 Let $N(t)$ be a Poisson process with rate $\lambda = 1$. What are the probability distribution, mean, and standard deviation of the waiting time W_5?

7.3-4 The number of floods to hit a certain location on the Mississippi River is a Poisson process with a rate of one every four years. Find the probability that there are more than two floods in a year.

7.3-5 Instantaneous surges of electrical current occur on a certain line according to a Poisson process with a rate of $\lambda = .2$ per hour. Suppose it takes four surges for the electrical system to fail. Find the probability that the system will last more than 6 hours.

SECTION

7.4 The Bernoulli Single-Server Queuing Process

A *queuing process* is a stochastic process in which the states of the process are the number of persons or objects in the system. There are two competing processes that "drive" a queuing process, the *arrivals* and *services*. The arrivals act to increase the number in the queue. These occur as customers (i.e., persons or objects) come to a facility for service. The services act to decrease the number in the queue as customers leave when service is completed. Characteristics of a queuing process are governed by such factors as how frequently the arrivals occur, how fast the services occur, and the number of servers.

Some queues are limited in the number of persons or objects that can be waiting for service. For instance, the number of customers waiting for a haircut at the barbershop might be limited to a few because of lack of waiting space. On the other hand, the length of some queues might grow without bound. The number of cars waiting to get through the toll plaza for the Golden Gate Bridge is an example where a lengthy queue is possible. If a queue is limited in the number it can hold at any time, it is called a *finite* or *limited-capacity queue*. Otherwise, it is called an *unlimited-capacity queue*.

The time scale for observing a queuing process may be discrete or continuous. Our development of the queuing processes will follow the pattern that we used in dealing with the Bernoulli counting process and the Poisson process. Continuous-time processes will be obtained from the discrete-time processes by letting frame sizes become small. To simplify the presentation, we consider the single-server process before considering multiple-server processes. A system that could be modeled as a single-server queuing process would be the checkout station at a small convenience store that has one server to handle transactions.

To describe the behavior of a single-server queuing process, we need to describe the arrival process and the service process. Here we have several choices. For instance, arrivals could occur at regular intervals of time or randomly. Likewise, services could take a fixed amount of time to complete, or their time could be random. If there is a random component, then we must select the mechanism for generating the random events. The processes that we consider will have randomness in both arrivals and services.

Definition 7.4-1

A queuing process is said to be a *single-server Bernoulli queuing process*, with unlimited capacity, if the following three conditions are satisfied.

i The arrivals occur according to the rules of a Bernoulli counting process. The probability of an arrival during any frame is denoted P_A.

ii When the server is busy, completions of service occur according to the rules of a Bernoulli counting process. The probability that service is completed by the busy server during a frame is denoted P_S.

iii Arrivals occur independently of services.

We adopt the convention that the earliest a customer can complete service is one frame after arriving. When referring to the number in the queuing system, we include those being served and those in line that have not yet begun service. The states of the system are $0, 1, 2, \ldots,$ representing the number of customers in the system. Since the arrival and service probabilities remain unchanged throughout time, we say that such processes are *time homogeneous*.

The Bernoulli queuing process can be easily visualized in terms of coin tossing. Imagine tossing two biased coins each frame—a service coin and an arrival coin. If the server is busy, we toss the service coin. If the service coin comes up heads, a service has been completed. Next, we toss the arrival coin. If the arrival coin comes up heads, an arrival has occurred. This order of coin tossing assures us that the earliest a customer can complete service is one frame after arrival.

We will now derive transition probabilities for the single-server Bernoulli queuing process. First consider transitions from state 0. Since arrivals follow the Bernoulli counting process, there can be at most one arrival during a frame. The transition $0 \to 0$ will occur if and only if there is no arrival during the frame. Similarly, the transition $0 \to 1$ will occur if and only if there is one arrival during the frame. Such possibilities as two arrivals and one service in the same frame are ruled out by the assumptions. Thus,

$$P(0 \to 0) = 1 - P_A, \qquad P(0 \to 1) = P_A. \tag{7.4-1}$$

Now suppose that the process is in state $i \neq 0$. Assumptions i and ii from Definition 7.4-1 allow at most one arrival and at most one service during a frame. Thus, the possible transitions that can be made are $i \to (i-1)$, $i \to i$, and $i \to (i+1)$. For the $i \to (i-1)$ transition, there must be one service and no arrival. Using the independence of arrivals and services, we have

$$P(i \to i-1) = P_S(1 - P_A). \tag{7.4-2}$$

The transition $i \to i$ can occur if there are no arrivals and no services or one arrival and one service. In this case, one arrival and one service during a frame involve different customers. Thus,

$$P(i \to i) = (1 - P_A)(1 - P_S) + P_A P_S. \tag{7.4-3}$$

Since the transition $i \to (i+1)$ only occurs when there is one arrival and no service, we have

$$P(i \to i+1) = P_A(1 - P_S). \tag{7.4-4}$$

EXAMPLE 7.4-1 Let a single-server queue have probability of arrival $P_A = .10$ and probability of service $P_S = .15$. The one-step transition matrix is

$$P = \begin{array}{c} \\ 0 \\ 1 \\ 2 \\ 3 \\ \vdots \end{array} \begin{array}{c} \begin{array}{cccccc} 0 & 1 & 2 & 3 & 4 & \cdots \end{array} \\ \begin{bmatrix} .90 & .10 & 0 & 0 & 0 & \cdots \\ .135 & .780 & .085 & 0 & 0 & \cdots \\ 0 & .135 & .780 & .085 & 0 & \cdots \\ 0 & 0 & .135 & .780 & .085 & \cdots \\ \ddots & \ddots & \ddots & \ddots & \ddots & \ddots \end{bmatrix} \end{array}.$$

Simulation of the number of customers in a single-server Bernoulli queuing process may be carried out with the following algorithm.

Simulation of Single-Server Bernoulli Queuing Process

Variables:

FRAME	Frame number
$Q(\bullet)$	An array consisting of the number of customers in the queue at the end of each frame
IA	Bernoulli variable indicating whether an arrival has occurred in the present frame. If an arrival has occurred, IA = 1; otherwise, IA = 0.
IS	Bernoulli variable indicating whether a service has occurred in the present frame. If a service has occurred, IS = 1; otherwise, IS = 0.

Initialize:

FRAME = 0

$Q(0) = 0$

Repeat the following steps for a predetermined number of frames:

1 Set IA = 0 and IS = 0.
Increment FRAME = FRAME + 1.

2a If $Q(\text{FRAME} - 1) > 0$ (i.e., a customer from the previous frame is waiting to complete service), generate a unit random number u_S.

2b If $u_S < P_S$ (i.e., a service has occurred), then IS = 1.

3 Generate a unit random number u_A.
If $u_A < P_A$ (i.e., an arrival has occurred), then IA = 1.

4 Update the number in the queue as follows:

$Q(\text{FRAME}) = Q(\text{FRAME} - 1) + \text{IA} - \text{IS}$ (add the arrival when it occurs and subtract the service when it occurs).

Simulation of eight frames of a single-server queue using this method is shown in Table 7.4-1. Here $P_A = .10$ and $P_S = .15$.

TABLE 7.4-1 Simulation of a Single-Server Queue, $P_A = .10,\ P_S = .15$

FRAME	Uniform Random Numbers		Outcome		Number in Queue
	u_S	u_A	IS	IA	$Q(\cdot)$
0	*	*	0	0	0
1	*	.09236	0	1	1
2	.28740	.00231	0	1	2
3	.14583	.89357	1	0	1
4	.48912	.74591	0	0	1
5	.10335	.08711	1	1	1
6	.05342	.56721	1	0	0
7	*	.02321	0	1	1
8	.78341	.67381	0	0	1

* Indicates no random number was generated.

An alternative method of simulating the number in the queue is to use the one-step transition matrix. The procedure is the same as that in Section 4.5 for simulating the behavior of Markov chains. The procedure just described is attractive because it focuses directly on the underlying mechanisms that govern the behavior of the Bernoulli queuing process.

Since the arrivals occur according to a Bernoulli counting process, the number of frames from the arrival of one customer to the next has the same distribution as the number of trials it takes to obtain a success in a sequence of Bernoulli trials, which is a geometric distribution with probability P_A. Similarly, the number of frames it takes to service a customer has a geometric distribution with probability P_S.

In modeling a queuing process, we need the quantities P_A and P_S. A value of P_A can be determined just as we determined p in the Bernoulli counting processes. Let λ_A denote the rate of arrival, let n be the number of frames in a unit of time, and let Δ denote frame length. Then

$$P_A = \frac{\lambda_A}{n} = \lambda_A \Delta. \tag{7.4-5}$$

The service rate λ_S is expressed in terms of the expected number of services that are completed in a unit of time when there are customers to be served. The service probability is

$$P_S = \frac{\lambda_S}{n} = \lambda_S \Delta. \tag{7.4-6}$$

It is sometimes convenient and natural to express service rate in terms of the expected amount of time μ_S that it takes to complete service for one customer. From Theorem 7.1-3, we have $\mu_S = 1/\lambda_S$, which gives $P_S = (1/\mu_S)\Delta$. For instance, if it takes a server 20 minutes on average to complete a service and the frame size is 3 minutes, then $P_S = (1/20)3 = .15$.

It is easy to modify the frame-by-frame method of simulating a Bernoulli queuing process to accommodate varying arrival and service probabilities. Such processes are called *nonhomogeneous* Bernoulli queuing processes. The following example illustrates the simulation of one such process.

EXAMPLE 7.4-2 Suppose we model a single-server car wash with frame sizes $\Delta = 1/2$ minute. Assume the arrival probabilities follow a cyclical pattern defined by

$$P_A(n) = .05 + .05 \sin\left(\frac{2\pi n}{2880}\right),$$

where n is the frame number. Let $n = 0$ denote 8 A.M. The argument of the sine function for $P_A(n)$ is chosen so that a cycle is completed in 2880 frames, which is the number of frames in 24 hours. The maximum arrival probability is .10, which occurs at 2 P.M., $n = 720$ frames. This corresponds to a rate of 12 customers per hour. The minimum arrival probability is zero, and this occurs at 2 A.M., $n = 2160$ frames. The service probability is chosen to be constant throughout the 24-hour cycle with $P_S = .05$, which corresponds to a mean service time of 10 minutes. The results of simulating the process for one 48-hour period are summarized in Figure 7.4-1. The plotted data are the hourly averages of the number of customers in the system. As expected, we see a cyclical-like behavior of the number in this system. The randomness is also apparent in the difference between the second 24 hours and the first 24 hours.

FIGURE 7.4-1 Simulation of Single-Server Queuing Process with Variable Arrival Rate

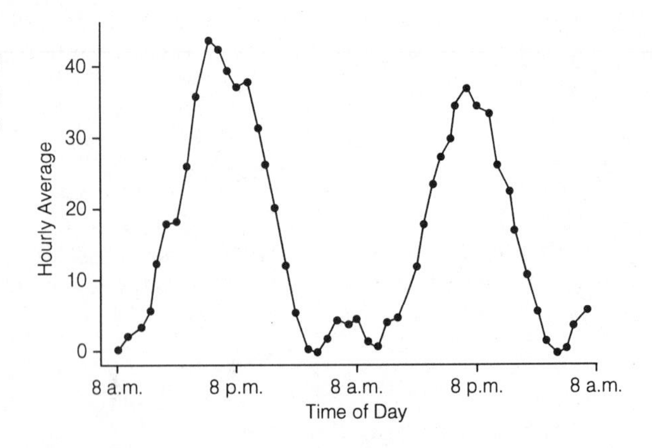

The Bernoulli queuing process may be modified so that only a limited number of customers can become part of the system. Suppose the system has a capacity of C customers. If there are fewer than C customers in the system, transitions occur as if the process were a Bernoulli queuing process with unlimited capacity. If the system is full, we will assume that potential customers continue to arrive, but they do not join the system unless a service has occurred. If the system is full, the number of customers in the system stays the same under the following conditions:

1 No potential customer arrives and no service takes place.

2 One potential customer arrives and one service takes place.

3 One potential customer arrives and no service takes place.

The transition probability from state C to state C is

$$\begin{aligned} P(C \to C) &= P(\text{condition } 1) + P(\text{condition } 2) + P(\text{condition } 3) \\ &= (1 - P_A)(1 - P_S) + P_A P_S + P_A(1 - P_S) \\ &= 1 - P_S(1 - P_A). \end{aligned}$$

The number in the system decreases by 1 if no potential customer arrives and one service takes place, or

$$P(C \to C - 1) = P_S(1 - P_A).$$

For $i < C$ the possible transitions from state i occur as in the unlimited-capacity queuing process.

EXAMPLE 7.4-3 A telephone has the capability of keeping one caller on hold while another is talking. If a call to the telephone is attempted while a caller is on hold, a busy signal is received and the incoming call is lost. Thus, at most two callers can be in the system at any time. Suppose the probability that a call arrives during a frame of 1 minute is $P_A = .10$ and the probability that a call is completed during a frame is $P_S = .15$. The one-step transition matrix for this system with limited-capacity Bernoulli queuing process is

$$P = \begin{array}{c} \\ 0 \\ 1 \\ 2 \end{array} \begin{array}{c} \begin{array}{ccc} 0 & 1 & 2 \end{array} \\ \begin{bmatrix} .90 & .10 & 0 \\ .135 & .780 & .085 \\ 0 & .135 & .865 \end{bmatrix} \end{array}.$$

The steady-state probability vector for this limited-capacity queuing process can be found with the techniques of Section 4.6. We find $\pi = (.45, .34, .21)$. Thus, 45%

of the time the telephone will be idle, and approximately 21% of the time it will be at capacity.

Exercises 7.4

7.4-1 Television sets arrive at a repair shop at the rate of four per day. Assume an 8-hour working day. The expected time to complete service on a television set is 1.25 hours. Model this process as a single-server Bernoulli queuing process with 15-minute frames.

- **a** Find the service rate λ_S expressed as the expected number of television sets that can be repaired per day.
- **b** Find the arrival and service probabilities P_A and P_S.
- **c** Use a table of random numbers to simulate the operation of the repair shop for two days. Start with no televisions in the shop. Compute the mean number of televisions in the shop each hour and plot the results.

7.4-2 Use a computer to simulate the number of television sets in the repair shop described in Exercise 7.4-1. Simulate 100 two-day periods where 8 hours make up one day.

- **a** Make an empirical distribution for the number in the shop at the end of the first hour, the end of the first day, and the end of the second day.
- **b** Compute the sample mean and sample standard deviation of the number in the shop for each of the times in part (a). Comment on the changes you observe across time.

7.4-3 The drive-up window at a Burger Barn can be modeled as a single-server Bernoulli queuing process with a frame length of 10 seconds. The arrival rate is one customer every 3 minutes. The expected time to serve a customer is 2 minutes.

- **a** Express the arrival rate and service rate in terms of number of customers per minute.
- **b** Obtain the arrival and service probabilities.
- **c** Simulate the operation of this system for 1 hour. Repeat 100 times. Find the sample mean and sample standard deviation for the number in the system at the end of the hour.

7.4-4 In a Bernoulli single-server queuing process compare the behavior of the system when

- **a** $P_A = 1$ and $P_S < 1$.
- **b** $P_A < 1$ and $P_S = 1$.

7.4-5 If the arrival probability is greater than the service probability, one would expect the length of the queue to grow with time. What do you suppose would happen if the arrival and service probabilities are the same? Investigate this question simulating the system with $P_A = P_S = .1$ for 2000 frames.

7.4-6 Cars arrive at a parking garage according to a Bernoulli counting process with frames of 15 seconds and arrival probability $P_A = .6$. Suppose cars are serviced (i.e., pass through the ticket gate) each frame with $P_S = .8$.

- **a** Give the hourly arrival and service rates.
- **b** Using the frame-by-frame method, simulate 1 hour of activity at this parking garage, assuming this system has unlimited capacity. Repeat this simulation 100 times. Compute

the sample mean, sample standard deviation, minimum, and maximum number of cars in the system at the end of 1 hour.

c Suppose the street leading into the garage entrance can only accommodate four cars waiting to pass through the ticket gate. Simulate 1 hour of activity of this limited capacity system 100 times. Compute the sample mean, sample standard deviation, and minimum and maximum number of cars in the system at the end of 1 hour. Compare your results with those obtained in part (b).

d Find the steady-state probability vector for the limited capacity system in part (c). Interpret these probabilities.

7.4-7 A barbershop has one barber and two chairs for waiting. The expected time for the barber to cut a customer's hair is 15 minutes. Customers arrive at the rate of two per hour provided the barbershop is not full. However, if the barbershop is full (three customers), potential customers go elsewhere. Assume that the barbershop can be modeled as single-server Bernoulli queuing process with limited capacity. Use frame sizes of 3 minutes.

a Derive the one-step matrix for this process.

b Find steady-state probabilities and interpret them.

7.4-8 Repeat Exercise 7.4-7, using a frame size of 6 minutes. Comment on the effect of frame size for these two exercises.

7.4-9 Families move to a certain community at the rate of six per month during the months of June, July, and August. During other times of the year the arrival rate is three per month. Families move away from this community at the rate of five per month during June, July, and August, and at the rate of three per month during other times of the year.

a Suppose there are 600 families in the community on January 1. Simulate the increases and decreases in the number of families in this community due to family moves for two years. Treat the moves into and out of the community as we would the arrivals and services in a Bernoulli queue. Use frame sizes of one day, and assume for simplicity that each month has 30 days.

b In a deterministic model for movement, there would be no random variability. For instance, exactly six families would move to the community in the months of June, July, and August, and exactly five would move away. In the other months the number of families in the community remains constant with exactly three families moving in and exactly three families moving out. Compare changes in the number of families in this community using the random model in part (a) to changes given by the deterministic model. Which model do you think is more appropriate?

7.4-10 The demand for a copy machine in an office is variable, with the highest demand being in the last 10 minutes of each hour. For the first 50 minutes of each hour, the arrival rate is one job per 5 minutes. For the last 10 minutes of each hour, the arrival rate is two jobs per 5 minutes. The service rate also varies. For the first 50 minutes of each hour, the expected time to complete a job is 4 minutes. For the last 10 minutes, the expected time to complete a job is 3 minutes. Simulate the number in the system for 8 hours, using frame sizes of 1 minute. Compute the average number in the system during each 10-minute interval. Plot these averages against time of day.

SECTION

7.5 The M/M/1 Queuing Process

In this section we derive a continuous-time queuing process by letting the frame size in the single-server Bernoulli queuing process become small. We will call the continuous-time process an *M/M/1 queuing process*. The first and second letters in this notation denote the type of arrival and service processes, respectively. The letter *M* indicates that these processes have the Markov property. The letters *G* and *D* are also used in this notation; *G* denotes a general distribution for either arrivals or services and *D* denotes deterministic (nonrandom) arrivals and services. The number denotes the number of servers.

To begin the development, we consider the transition probabilities of the Bernoulli single-server queuing process when Δ, the frame size, is small. From Equations (7.4-5) and (7.4-6),

$$P_A = \lambda_A \Delta \quad \text{and} \quad P_S = \lambda_S \Delta,$$

where Δ is the length of a frame. Substitute these expressions into the transition probabilities in Equations (7.4-1) through (7.4-4), and drop terms that involve $P_A P_S = \lambda_A \lambda_S \Delta^2$ because they are negligible in comparison with terms involving Δ when Δ is small. The result is

$$P(0 \to 0) = 1 - \lambda_A \Delta, \qquad P(0 \to 1) = \lambda_A \Delta, \tag{7.5-1}$$

and for $i > 0$,

$$P(i \to i-1) \approx \lambda_S \Delta, \tag{7.5-2}$$

$$P(i \to 1) \approx 1 - \lambda_A \Delta - \lambda_S \Delta, \tag{7.5-3}$$

$$P(i \to i+1) \approx \lambda_A \Delta. \tag{7.5-4}$$

Equations (7.5-1) through (7.5-4) represent approximate transition probabilities for the M/M/1 queuing process in a small time interval of length Δ. These approximations tell us that the M/M/1 queuing process essentially can have just two or three possible outcomes in a small frame of length Δ. If there is no customer to be served, then there can be one arrival with probability $\lambda_A \Delta$ or no arrival with probability $1 - \lambda_A \Delta$. If there is at least one customer to be served, there can be one arrival with probability $\lambda_A \Delta$, one service with probability $\lambda_S \Delta$, or neither with probability $1 - \lambda_A \Delta - \lambda_S \Delta$. All other transitions have negligible probability when the frame size is small. This discussion leads to the following definition.

Definition 7.5-1

A continuous-time, single-server queuing process is said to be an *M/M/1 queuing process* if i For sufficiently small time intervals of length Δ, transition probabilities can be approximated by Equations (7.5-1) through (7.5-4) where the errors of approximation are negligible.

ii Transitions occurring in nonoverlapping intervals are independent of one another.

The arrivals and services of an M/M/1 queuing process are limits of Bernoulli counting processes as the frame size becomes small. Thus, the arrivals in an M/M/1 queuing process behave as a Poisson process with rate λ_A. Likewise if there are customers to be served, the service process behaves as a Poisson process with rate λ_S. The following theorem is a consequence of Theorem 7.3-1 for the Poisson process.

THEOREM 7.5-1 The time between the arrivals of customers in an M/M/1 queuing process is an exponential random variable with mean $\mu_A = 1/\lambda_A$ and, given that a customer is being served, the time to complete service is an exponential random variable with mean $\mu_S = 1/\lambda_S$.

We now derive the steady-state distribution for the number of customers in an M/M/1 queuing system. Imagine that we have divided a continuous interval of time into infinitesimally small intervals of length Δ. We treat the approximations in Equations (7.5-1) through (7.5-4) as if they were exact, the logic being that the frame lengths can be chosen small enough to make the errors of approximation negligible. The derivation expresses the number in the system at time $t + \Delta$ in terms of the number at time t.

If there are $i > 0$ customers at time $t + \Delta$, there could have been only $i - 1$, $i + 1$, or i customers at time t. The reason for this is that in the small time interval between t and $t + \Delta$, there could have been only one of three outcomes: one arrival, one service, or neither. This is shown in Table 7.5-1.

TABLE 7.5-1 Transitions Leading to i Customers in System at Time $t + \Delta$

State at time t	Activity in $(t, t + \Delta)$	State at time $t + \Delta$
$i - 1$ customers	1 arrival	i customers
$i + 1$ customers	1 service	i customers
i customers	0 arrivals, 0 services	i customers

$$\begin{aligned} P(i \text{ customers at time } t+\Delta) &= P(i-1 \text{ customers at time } t)\,P(\text{arrival in frame of length } \Delta) \\ &+ P(i+1 \text{ customers at time } t)\,P(\text{service in frame of length } \Delta) \\ &+ P(i \text{ customers at time } t)\,P(\text{neither arrival nor service in frame of length } \Delta)\,. \end{aligned}$$

Let π_i denote the steady-state probability that there are $i > 0$ customers in the system. If the system is in the steady-state condition, then for all states i,

$$\begin{aligned} P(i \text{ customers at time } t+\Delta) &= P(i \text{ customers at time } t) \\ &= \pi_i . \end{aligned}$$

For $i > 0$ the steady-state forms of the equations in Table 7.5-1 are

$$\pi_i = \pi_{i-1}\lambda_A\Delta + \pi_{i+1}\lambda_S\Delta + \pi_i(1 - \lambda_A\Delta - \lambda_S\Delta). \tag{7.5-5}$$

Similarly, it can be shown, for $i = 0$,

$$\pi_0 = \pi_0(1 - \lambda_A\Delta) + \pi_1\lambda_S\Delta. \tag{7.5-6}$$

Simplifying Equation (7.5-6) yields

$$\pi_0\lambda_A = \pi_1\lambda_S.$$

For $i = 1$, Equation (7.5-5) becomes

$$\pi_1(\lambda_A + \lambda_S) = \pi_0\lambda_A + \pi_2\lambda_S.$$

which simplifies to

$$\pi_1\lambda_A = \pi_2\lambda_S.$$

This pattern continues, producing what is known as the system of *balance equations* defined by

$$\pi_i\lambda_A = \pi_{i+1}\lambda_S, \qquad i = 0, 1, \ldots . \tag{7.5-7}$$

The term on the left of the equals sign in Equation (7.5-7) can be thought of as the amount of probability mass that leaves state i for state $i + 1$ at each transition, and the term on the right can be thought of as the amount that leaves state $i + 1$ for state i. For the steady-state condition to occur, these masses must "balance," hence the term *balance equation.*

The solution of the balance equations is readily obtained in terms of π_0. We have

$$\pi_1 = \pi_0\left(\frac{\lambda_A}{\lambda_S}\right),$$

$$\pi_2 = \pi_1\left(\frac{\lambda_A}{\lambda_S}\right),$$

$$= \pi_0\left(\frac{\lambda_A}{\lambda_S}\right)^2$$

$$\vdots$$

$$\pi_i = \pi_0\left(\frac{\lambda_A}{\lambda_S}\right)^i, \qquad i > 0.$$

Since the sum of the probabilities must be 1,

$$1 = \sum_{i=0}^{\infty} \pi_i$$

$$= \pi_0 \sum_{i=0}^{\infty}\left(\frac{\lambda_A}{\lambda_S}\right)^i.$$

For this infinite series above to converge, the condition $\lambda_A/\lambda_S < 1$ must be satisfied. Thus, the probabilities will sum to 1, and the steady-state probabilities will exist if and only if $\lambda_A/\lambda_S < 1$. If $\lambda_A/\lambda_S < 1$, then

$$\sum_{i=0}^{\infty}\left(\frac{\lambda_A}{\lambda_S}\right)^i = \frac{1}{1-\lambda_A/\lambda_S}$$

and

$$\pi_0 = 1 - \frac{\lambda_A}{\lambda_S}.$$

Since the other probabilities π_i, $i > 0$, are expressed in terms of π_0,

$$\pi_i = \left(1 - \frac{\lambda_A}{\lambda_S}\right)\left(\frac{\lambda_A}{\lambda_S}\right)^i.$$

THEOREM **7.5-2** Let $r = \lambda_A/\lambda_S$ denote the *arrival/service ratio* of an M/M/1 queuing process. The steady-state distribution exists if and only if $r < 1$ and the steady-state probability mass function for the number in the system is

$$\pi_i = (1-r)r^i, \qquad i = 0, 1, 2, \ldots.$$

The probability mass function from Theorem 7.5-2 can be used to compute various probabilities of interest. For example, if X is the number of customers in the system in steady state, then

$$P(X > x) = \sum_{i=x+1}^{\infty} (1-r)r^i$$
$$= r^{x+1}$$

and

$$P(X \leq x) = 1 - P(X > x)$$
$$= 1 - r^{x+1}.$$

EXAMPLE 7.5-1 Orders are called into a company at the rate of 30 per day and are filled sequentially by a single service facility on a first-come, first-served basis. Servicing an order takes an average of 12 minutes, so the service rate in an 8-hour day is 40 orders per day. Assume that this process can be regarded as an M/M/1 queuing process. Since the arrival/service ratio is $r = 30/40 = .75$, the steady-state probability that there is no order to be filled is $\pi_0 = 1 - r = .25$. This tells us that in the long run the service facility is idle 25% of the time. The probability that there are two or fewer orders in the system in steady state is

$$P(X \leq 2) = 1 - (.75)^3 = .58.$$

The probability of 10 or more orders in the system in steady state is

$$P(X > 9) = (.75)^{10} = .056.$$

The steady-state distribution of the M/M/1 queuing process is related to the geometric distribution defined in Section 2.8. Recall that the geometric distribution is of the form

$$p(x) = p(1-p)^{x-1}, \quad x = 1, 2, \ldots.$$

Make the substitutions $Y = X - 1$ and $r = 1 - p$. Then Y has the same distribution as the steady-state distribution of the number of customers in the M/M/1 queuing system. We can use this fact to find the mean and standard deviation of the number of customers in the M/M/1 queuing system in steady state. We have

$$\begin{aligned} E(Y) &= E(X) - 1 \\ &= \frac{1}{1-r} - 1 \\ &= \frac{r}{1-r} \\ \text{STD}(Y) &= \text{STD}(X) = \frac{\sqrt{r}}{1-r}. \end{aligned}$$

THEOREM 7.5-3 The steady-state mean and standard deviation of the number of customers in an M/M/1 queuing system with arrival/service ratio $r < 1$ are

$$\mu = \frac{r}{1-r}, \qquad \sigma = \frac{\sqrt{r}}{1-r}.$$

EXAMPLE 7.5-2 Consider the service facility in Example 7.5-1. In steady state, the expected number of orders in the system after a long time is $.75/(1-.75) = 3$, and the standard deviation is $\sqrt{.75}/(1-.75) = 3.5$.

Note that the mean and standard deviation of the number of customers in the system in Theorem 7.5-3 approaches infinity as r approaches 1. This shows that the steady-state queue length will be long in queues where r is less than 1 but near 1. If $r \geq 1$, the steady-state distribution does not exist. Rather the queue length will grow without bound since services cannot keep up with arrivals.

Consider the expected amount of time that a customer spends in the system. If there are N customers ahead of a newly arriving customer, the newly arriving customer must wait for N services to occur before beginning service, and then must complete service in order to leave the queue. The time of each service is an exponential random variable with mean $1/\lambda_S$. Thus, a customer who arrives when N customers are in the system must spend an average of $(N+1)(1/\lambda_S)$ units of time in the system. Since the expected number of customers in the system when the customer arrives is $r/(1-r)$, the mean time spent in the system is

$$\left(\frac{r}{1-r} + 1\right)\frac{1}{\lambda_S} = \frac{1}{(1-r)\lambda_S}.$$

THEOREM 7.5-4 Assume the M/M/1 queuing system is in steady state. Let T denote the amount of time a customer spends in the system and $r = \lambda_A/\lambda_S$. Then

$$E(T) = \frac{1}{(1-r)\lambda_S}.$$

EXAMPLE 7.5-3 Referring to Example 7.5-1, we find that the expected amount of time that an order spends in the system from the time it is called in until it has been filled is $1/(1-.75)40 = .1$ day $= 48$ minutes (assuming an 8-hour work day).

EXAMPLE 7.5-4 A company loses \$600 a day for each day a certain piece of equipment is broken down. There are two possible repair shops where the equipment can be fixed. Shop 1, which charges \$400 to fix the equipment, has requests for service from an average of four customers per day and can service at the rate of five customers per day. Shop 2, which charges \$700 to fix the equipment, also has requests from an average of four customers per day but can service at the rate of eight customers per day. Assuming that both shops can be modeled as M/M/1 queuing processes, we will determine which shop would be the least expensive in the long run for the company to use to repair its equipment.

The average time a piece of equipment spends in shop 1 is $1/(1-.8)5 = 1.00$ day, which gives an average loss, due to downtime, to the company of \$600. Add to this the \$400 repair bill, and the total cost to have repairs done at shop 1 is, on average, \$1000. The average time a piece of equipment spends in shop 2 is $1/(1-.5)8 = .25$ day. Thus, the downtime cost to the company at shop 2 is only $(.25)(600) =$ \$150, which when added to the \$700 repair bill gives an average cost of \$850. Thus, shop 2 is preferable.

Exercises 7.5

For the following problems, the assumptions of the M/M/1 queuing process apply and the systems have reached steady state.

7.5-1 Plot the expected value and the standard deviation of the number in the system as functions of the arrival/service ratio r.

7.5-2 The probability that a queuing process is idle in steady state is .3. Find the expected number in the system.

7.5-3 The arrival rate is $\lambda_A = 10$ customers in a unit of time. What must the service rate be so that the probability that there are two or fewer customers in the system is .90?

7.5-4 A system with a service rate λ_S is capable of serving an average of λ_S customers in a unit of time when the server is busy, but the server is not always busy. What is the expected number of customers actually served in a steady-state M/M/1 queuing process in a unit of time?

7.5-5 A walk-in clinic provides medical services to patients. The expected time to treat a patient is 10 minutes. Patients come to the clinic at the rate of five per hour.

- **a** What is the expected number of patients in the clinic at any time?
- **b** What is the probability that there is at least one patient in the clinic?
- **c** Ninety percent of the time the number in this system will be less than or equal to what value?
- **d** How many minutes, on average, will a patient spend in the clinic?

7.5-6 A small branch bank has a single teller to handle transactions. The expected time it takes the teller to handle a transaction is 2 minutes. Customers come to the bank at the rate of one every 3 minutes.

- **a** Find the expected value and standard deviation of the number of customers in the bank at any time.
- **b** On average how long does a customer spend in the bank?
- **c** What fraction of time is the teller idle?

7.5-7 A salvage company receives wrecked automobiles at the rate of 10 per day. These can be processed at the rate of 11 per day.

- **a** On average, how many cars are on the salvage lot at any time?
- **b** What would the service rate have to be to reduce the expected number of cars on the lot to one-half what it is at present?

7.5-8 Jobs sent to a printer are held in a buffer until they can be printed. Jobs are printed sequentially on a first-come, first-served basis. Jobs arrive at the printer at the rate of four per minute. The average time to print a job is 10 seconds.

- **a** Find the expected value and standard deviation of the number of jobs in this system at any time.
- **b** If a person submits a job to be printed, what is the probability it will begin printing immediately?

7.5-9 The judicial system can be viewed as a Markov queuing process where the judges are the servers. Consider a night court with one judge. People arrive for arraignment at the rate of 8 per hour. The mean time to complete an arraignment is 6 minutes. What is the probability that three or more cases will be waiting for arraignment in the courtroom?

SECTION

7.6 *k*-Server Queuing Process

In this section we examine the behavior of queuing processes in which there is more than one server. If there is an available server when a customer arrives, the customer begins service immediately. Otherwise, the customer joins the end of a single line. Service will begin with the first available server when the customer reaches the head of line. This setup is shown in Figure 7.6-1. As in our treatment of single-server queuing processes, we consider the k-server Bernoulli queuing process and its continuous-time analogue, the M/M/k queuing process.

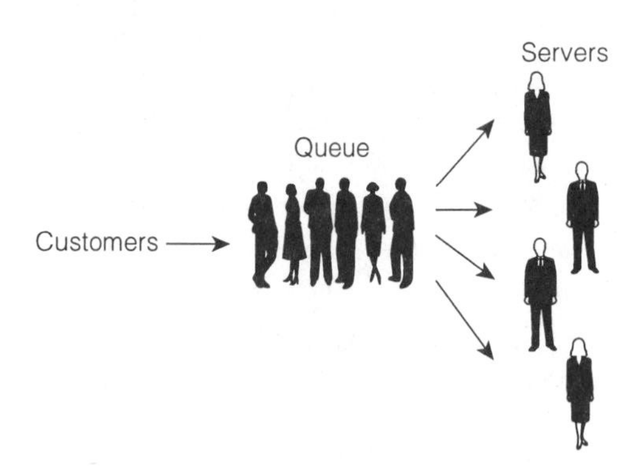

FIGURE **7.6-1**
Diagram for the k-Server Queuing Process

The k-server Bernoulli queuing process is a straightforward extension of the single-server Bernoulli queuing process, and can be described in terms of a biased coin-tossing experiment analogous to the single-server queuing process. The only difference is there are more coins to toss. If there are k servers, then up to $k + 1$ coins will be tossed each frame to determine arrivals and services. The first k coins, one for each server, determine how many services have been completed. If all servers are busy, then all k of the server coins are tossed, and those servers whose coins turn up heads have services completed while the others do not. If there are fewer than k busy servers, then coins are tossed only for those servers that are busy. The server coins are assumed to be identical, and the probability of heads for each server coin is denoted P_S. The remaining coin determines whether or not an arrival occurs. The probability of heads for the arrival coin is denoted P_A.

With this intuitive description of the k-server Bernoulli queuing process as a background, we have the following definition.

Definition 7.6-1

A process with k servers and unlimited capacity is said to be a *k-server Bernoulli queuing process* if the following three conditions are satisfied.

i The arrivals occur according to the rules of a Bernoulli counting process. The probability of an arrival during any frame is denoted P_A.

ii For each busy server, the completions of service occur according to the rules of a Bernoulli counting process. The probability that service is completed by a busy server during a frame is denoted P_S, which is the same for all servers.

iii Arrivals occur independently of services, and the servers, when busy, function independently of one another.

As with the single-server queuing process, we assume that the earliest a customer can complete service is one frame after arriving. The number in the system includes those being served and those in line that have not yet begun service.

To obtain the transition probabilities for the k-server queuing process, note that the probability of one or more services being completed in any frame will depend on the number of busy servers. If there are n busy servers, $n \le k$, then the probability of j services being completed has a binomial probability distribution with probability of success P_S. Therefore,

$$P(j \text{ services} \mid n \text{ busy servers}) = \binom{n}{j} P_S^j (1 - P_S)^{n-j}, \quad j \le n.$$

Transition probabilities for the k-server Bernoulli queuing process can be derived from the fact that the arrival probability is P_A and that arrivals and services are independent. We illustrate the method in the case of the two-server queuing process.

EXAMPLE 7.6-1 When there are no busy servers or just one busy server, the two-server queuing process behaves just as a single-server queue in terms of arrival and service probabilities. Therefore, the transition probabilities from states 0 and 1 are identical to those of the single-server queue in Section 7.4. When there are two or more customers in the queue, both servers are busy, and the probabilities of zero, one, or two services are binomial probabilities given by $(1 - P_S)^2$, $2(1 - P_S)P_S$, and P_S^2, respectively. Therefore, the transition probabilities are

$$\begin{aligned}
P(0 \to 0) &= 1 - P_A, \\
P(0 \to 1) &= P_A, \\
P(1 \to 0) &= (1 - P_A)P_S, \\
P(1 \to 1) &= (1 - P_A)(1 - P_S) + P_A P_S, \\
P(1 \to 2) &= (1 - P_S)P_A,
\end{aligned}$$

and for $i \ge 2$,

$$\begin{aligned}
P(i \to i+1) &= P(\text{one arrival, no service}) \\
&= P_A(1 - P_S)^2, \\
P(i \to i) &= P(\text{no arrival, no service}) + P(\text{one arrival, one service}) \\
&= (1 - P_A)(1 - P_S)^2 + P_A\,[2(1 - P_S)P_S] \\
P(i \to i-1) &= P(\text{no arrival, one service}) + P(\text{one arrival, two services}) \\
&= (1 - P_A)[2(1 - P_S)P_S] + P_A P_S^2, \\
P(i \to i-2) &= P(\text{no arrival, two services}) \\
&= (1 - P_A)P_S^2.
\end{aligned}$$

When $P_A = .10$ and $P_S = .15$, the one-step transition matrix is

$$P = \begin{array}{c} \\ 0 \\ 1 \\ 2 \\ 3 \\ \vdots \end{array} \begin{array}{c} \begin{array}{cccccc} 0 & 1 & 2 & 3 & 4 & \cdots \end{array} \\ \left[\begin{array}{cccccc} .90 & .10 & 0 & 0 & 0 & \cdots \\ .135 & .780 & .085 & 0 & 0 & \cdots \\ .020 & .232 & .676 & .072 & 0 & \cdots \\ 0 & .020 & .232 & .676 & .072 & \cdots \\ \ddots & \ddots & \ddots & \ddots & \ddots & \ddots \end{array}\right] \end{array}.$$

Simulating the number of customers in a k-server queuing process is simple. We can use the one-step transition matrix to simulate the states of the system from one frame to the next, or we can use the biased coin-tossing analogy to simulate the process. The latter method is defined by the following algorithm.

Simulation of k-Server Bernoulli Queuing Process

Variables:

FRAME	Frame number
$Q(\bullet)$	An array consisting of the number of customers in the system at the end of each frame
IA	Bernoulli variable indicating whether an arrival has occurred in the present frame if an arrival has occurred (IA = 1; otherwise, IA = 0).
TS	Binomial random variable indicating the number of completed services that have occurred in the present frame. (TS $= 0, 1, \ldots, n$, where n is the number of busy servers at the start of the present frame and $n = \min[k, Q(\text{FRAME} - 1)]$.)

Initialize:

FRAME $= 0$

$Q(0) = 0$

Repeat the following steps for a predetermined number of frames:

1. Set IA $= 0$ and TS $= 0$.
   Increment FRAME $=$ FRAME $+ 1$.
2. If $Q(\text{FRAME} - 1) > 0$, then generate $\min[k, Q(\text{FRAME} - 1)]$ unit random numbers (i.e., one for each busy server).
   Compare each u_i to P_S and set TS $=$ number of u_i's $<$ PS.

3 Generate a unit random number u_A.
If $u_A < P_A$, then IA = 1.

4 Update the number in the queue as follows:
$Q(\text{FRAME}) = Q(\text{FRAME} - 1) + \text{IA} - \text{TS}$

EXAMPLE 7.6-2 We will illustrate the simulation of the two-server queuing process from Example 7.6-1. That is, $P_A = .10$ and $P_S = .15$. The results of 10 frames are shown in Table 7.6-1.

TABLE 7.6-1 Simulation of a 2-Server Queuing Process, $P_A = .10$, $P_S = .15$

Frame	Unit Random Numbers u_1	u_2	u_A	Outcome $u_1 < P_S$	$u_2 < P_S$	$u_A < P_A$	Number in System
—	*	*	*	*	*	*	0
1	*	*	.09877	*	*	yes	1
2	.33453	*	.00612	no	*	yes	2
3	.45112	.66124	.78799	no	no	no	2
4	.77931	.10552	.69001	no	yes	no	1
5	.78190	*	.04335	no	*	yes	2
6	.33556	.99121	.01334	no	no	yes	3
7	.08771	.00211	.94534	yes	yes	no	1
8	.45642	*	.88341	no	*	no	1
9	.13131	*	.46778	yes	*	no	0
10	*	*	.88343	*	*	no	0

* Indicates no random number was generated.

Now consider the continuous-time M/M/k queuing process. It is obtained from the k-server Bernoulli queuing process by letting frame size become small. Let λ_A and λ_S denote the rate of arrival and rate of service, respectively. Approximate transition probabilities for small frame sizes can be obtained by substituting $P_A = \lambda_A \Delta$ and $P_S = \lambda_S \Delta$ into the expressions for the transition probabilities of the Bernoulli k-server queuing process. These expressions are expanded in powers of Δ, and terms involving Δ^j for $j \geq 2$ are ignored since they are negligible in comparison with terms involving Δ when Δ is small. For instance, the transition probability $P(i \to i-1)$ in the two-server queuing process in Example 7.6-1 for $i \geq 2$ can be approximated by

$$\begin{aligned} P(i \to i-1) &= (1-P_A)[2(1-P_S)] + P_A P_S^2 \\ &= (1-\lambda_A\Delta)\,[2(1-\lambda_S\Delta)\lambda_S\Delta] + \lambda_A\Delta(\lambda_S\Delta)^2 \\ &= 2\lambda_S\Delta + \text{terms involving } \Delta^j \text{ for } j \geq 2 \\ &\approx 2\lambda_S\Delta. \end{aligned}$$

Now consider the probability of j services given n busy servers. This is

$$\begin{aligned} P(j \text{ services} \mid n \text{ busy servers}) &= \binom{n}{j} P_S^j (1-P_S)^{n-j} \\ &= \binom{n}{j} (\lambda_S\Delta)^j (1-\lambda_S\Delta)^{n-j}. \end{aligned}$$

If $j \geq 2$, the powers of Δ are all 2 or greater; hence, this probability is negligible for small Δ. If $j = 1$, we have

$$\begin{aligned} P(1 \text{ service} \mid n \text{ busy servers}) &= n\lambda_S\Delta + (\text{terms involving } \Delta^j \text{ for } j \geq 2) \\ &\approx n\lambda_S\Delta. \end{aligned}$$

For $j = 0$, $P(\text{no service} \mid n \text{ busy servers}) \approx 1 - n\lambda_S\Delta$.

As this discussion shows, there can be essentially only zero or one services in a frame of small length Δ. Likewise there can essentially be only zero or one arrivals. All other possibilities have negligible probabilities. Moreover, the probability of both an arrival and a service occurring in the same frame is negligible since it would involve terms Δ^j for $j \geq 2$. These considerations lead to the following defining properties of an M/M/k queuing process.

Definition 7.6-2

A *continuous-time process* is an M/M/k queuing process with unlimited capacity if it satisfies the following conditions.

i When there are n busy servers and i customers in the system, the following approximations apply to the possible transitions in a sufficiently small frame of length Δ:

$$\begin{aligned} P(i \to i+1) &\approx \lambda_A\Delta; \\ P(i \to i-1) &\approx n\lambda_S\Delta; \\ P(i \to i) &\approx 1 - \lambda_A\Delta - n\lambda_S\Delta. \end{aligned}$$

ii Transitions in nonoverlapping frames are independent of one another.

The term $n\lambda_S$ can be interpreted as the *system service rate*— that is, the rate at which the system services customers when there are n busy servers. For instance, if a single busy server can serve at the rate of $\lambda_S = 10$ customers per hour, then three busy servers can serve at the rate of $3\lambda_S = 30$ customers per hour. If n = 0, there are no busy servers; so $P(i \to i-1) = 0$. As with the M/M/1 queuing process, the time between arrivals of customers is an exponential random variable with mean $\mu_A = 1/\lambda_A$, and the time for a customer to complete service once the customer has begun service is an exponential random variable with mean $\mu_S = 1/\lambda_S$.

In modeling real queuing systems, it is necessary to determine the arrival rate λ_A and the service rate λ_S. If $N(t)$ denotes the number of arrivals in a time period of length t, we note that $E(N(t)/t) = \lambda_A$. Thus, the method of moments estimate of λ_A is obtained by observing the process for a sufficiently long time and computing

$$\hat{\lambda}_A = \frac{\text{Number of arrivals}}{\text{Amount of time spent observing the process}}.$$

However, the service rate is usually determined differently. Since the rate can be computed only when there are customers to be served, what is typically done is to obtain the amount of time each of a number of customers spends being served. Since the mean time to complete service is $1/\lambda_S$ for an exponential service time, the method of moments estimate of λ_S is determined by the equation

$$\frac{1}{\hat{\lambda}_S} = \text{Sample mean of the time for } n \text{ customers to complete service.}$$

Thus, λ_S can be estimated as $1/\bar{T}$ where $\bar{T}$ is the sample mean of the observed service times.

EXAMPLE **7.6-3** In a certain queuing system, there were 10 arrivals in 8 hours. The arrival rate is estimated as $\hat{\lambda}_A = 10/8 = 1.25$ customers per hour. The times (in hours) that it took to service these 10 arrivals are $0.2, 0.7, 0.5, 0.3, 1.0, 2.1, 0.7, 0.5, 0.9, 0.1$. The sample mean is $\bar{T} = 0.7$ hour. Thus, the service rate is estimated as $\hat{\lambda}_S = 1/.7 = 1.4$ services per hour.

Exercises 7.6

7.6-1 In a three-server Bernoulli queuing process, the probability that an arrival occurs during each frame is .10, and the probability that a busy server completes service is .15. Compute the transition probabilities for the system.

7.6-2 Partly assembled items come to a production facility where they may be sent to one of three stations for completion of the assembly process. Items arrive at the rate of four per minute. The average amount of time it takes to complete each item is 3 minutes. Assume that the system can be modeled as an M/M/3 queuing process. What is the system service rate when there is one busy station? two busy stations? three busy stations? Express these rates in terms of the number of items completed per minute.

7.6-3 To purchase tickets for an amusement park, customers form a single line and go to the first available ticket booth when they are at the head of the line. The entrance has two ticket booths. The expected number of people arriving is 3 per minute, and the expected time to purchase a ticket is 30 seconds. Assuming that the system can be modeled as a Bernoulli queuing process, with frame size 1 second, simulate the number of people in the system for 1 hour.

7.6-4 To go through customs at an international airport, passengers form a single line and go to the first available customs agent when their time for service comes. There are five available customs agents. Passengers arrive at the rate of 80 per hour, and the expected service time is 5 minutes. Assume that this system can be modeled as an M/M/5 queuing process. By choosing an appropriate frame size, approximate this system with a Bernoulli queuing process and simulate for 1 hour. Compute the number in the system at the end of this time. Repeat this simulation 100 times and find the sample mean and the sample standard deviation of the number of passengers in the system at the end of 1 hour.

7.6-5 Suppose in a k-server queuing process, k lines may be formed, one for each server. When a customer arrives, he or she goes to the shortest line. How would the number in the system compare to the single-line k-server queuing process? Consider two situations.

a Customers may change lines.

b Customers do not change lines.

7.6-6 At the checkout desk at the library, 25 people were observed checking out books between 8:30 A.M. and 10:00 A.M. The times in minutes for the librarian to complete the checkout process on each person are

$$\{2.3, 4.5, 1.8, 6.9, 7.0, 3.5, 5.4, 2.7, 8.0, 9.4, 3.8, 6.7, 8.1,$$
$$4.7, 10.0, 9.3, 4.0, 1.2, 3.0, 2.7, 8.4, 11.2, 4.7, 2.8, 9.2\}.$$

From these data estimate λ_A and λ_S assuming that the system is an M/M/k queuing process.

SECTION

7.7 Balance Equations and Steady-State Probabilities

In this section, we find the steady-state probabilities for an M/M/k queuing process when such probabilities exist. It is helpful to see the pattern of the solution for the steady-state probabilities if we first consider a queuing process in which both the rate of arrival of customers and the rate at which the system services customers depend on the number of customers in the system. Such systems are common. For instance, customers may come at a slower rate if there are more customers in the system. Similarly, the rate at which the system services customers may increase as the number of customers increases. This could happen not only because there might be more busy servers as occurs in the M/M/k queuing process, but also because servers might work more quickly.

Definition 7.7-1

Let a_i and s_i denote the arrival rate and system service rate, respectively, of a queuing process with i customers in the system. The process is called a *general Markov queuing process* if the following conditions are met. In a sufficiently small time frame of length Δ we have the following approximations for transition probabilities:

$$\begin{aligned} P(i \to i+1) &\approx a_i\Delta, \\ P(i \to i-1) &\approx s_i\Delta, \\ P(i \to i) &\approx 1 - a_i\Delta - s_i\Delta. \end{aligned}$$

The error of approximation is negligible in comparison with the terms involving Δ when Δ is small. The transitions in nonoverlapping intervals are independent of one another.

THEOREM 7.7-1 Let $a_j, j = 0, 1, 2, \ldots,$ be the arrival rates and let $s_j, j = 1, 2, 3, \ldots,$ be the system service rates of a general Markov queuing process. Let $\pi_j, j = 0, 1, 2, \ldots,$ denote the steady-state probability distribution of the process. The π_j's satisfy the system of balance equations

$$\pi_j a_j = \pi_{j+1} s_{j+1}, \qquad j = 0, 1, 2, \ldots.$$

The verification of this result is similar to the derivation of the balance equations for the M/M/1 queuing process (see Exercise 7.7-9).

THEOREM 7.7-2 The steady-state probability distribution for the general Markov queuing process exists if and only if

$$1 + \frac{a_0}{s_1} + \frac{a_0 a_1}{s_1 s_2} + \frac{a_0 a_1 a_2}{s_1 s_2 s_3} + \cdots < \infty. \tag{7.7-1}$$

The steady-state probability distribution is

$$\pi_j = \pi_0 \frac{a_0 a_1 \cdots a_{j-1}}{s_1 s_2 \cdots s_j}, \qquad j = 1, 2, \ldots, \tag{7.7-2}$$

$$\pi_0 = \left(1 + \frac{a_0}{s_1} + \frac{a_0 a_1}{s_1 s_2} + \frac{a_0 a_1 a_2}{s_1 s_2 s_3} + \cdots\right)^{-1}. \tag{7.7-3}$$

To verify Theorem 7.7-2, we solve the balance equations explicitly. The solution for π_j in terms of π_0 is readily obtained as

$$\pi_1 = \pi_0 \frac{a_0}{s_1},$$

$$\pi_2 = \pi_1 \frac{a_1}{s_2},$$

$$= \pi_0 \frac{a_0 a_1}{s_1 s_2},$$

$$\vdots$$

$$\pi_j = \pi_0 \frac{a_0 a_1 \ldots a_{j-1}}{s_1 s_2 \ldots s_j}.$$

The solution of this system of equations will be a probability distribution if and only if the sum of the probabilities is one. Thus,

$$1 = \sum_{j=0}^{\infty} \pi_j$$

$$= \pi_0 \left(1 + \frac{a_0}{s_1} + \frac{a_0 a_1}{s_1 s_2} + \frac{a_0 a_1 a_2}{s_1 s_2 s_3} + \cdots\right).$$

The equality holds if and only if Equation (7.7-1) from Theorem 7.7-2 is satisfied. In this case,

$$\pi_0 = \left(1 + \frac{a_0}{s_1} + \frac{a_0 a_1}{s_1 s_2} + \frac{a_0 a_1 a_2}{s_1 s_2 s_3} + \cdots\right)^{-1}.$$

The solutions for the other values of π_j, $j > 0$, are obtained by substituting the solution for π_0 from Equation (7.7-3) into Equation (7.7-2).

We now apply the balance equations to the M/M/k queuing process. In the M/M/k process, the arrival rate does not depend on the number in the system, thus

$$a_j = \lambda_A, \quad j = 0, 1, 2, \ldots. \tag{7.7-4}$$

The system service rate of the M/M/k queuing process is proportional to the number of busy servers, which may be as few as zero and as many as k. Therefore,

$$\begin{aligned} s_j &= j\lambda_S, \quad j = 0, 1, \ldots, k-1, \\ s_j &= k\lambda_S, \quad j \geq k. \end{aligned} \tag{7.7-5}$$

Theorem 7.7-2 is satisfied for the M/M/k queuing process if and only if $\lambda_A/\lambda_S < k$.

To obtain the steady-state probabilities we must compute terms of the form

$$C_j = \frac{a_0 a_1 \ldots a_{j-1}}{s_1 s_2 \ldots s_j}, \quad j = 1, 2, \ldots,$$

For the M/M/k queuing process, the pattern for C_j differs, depending on whether $1 \leq j < k$ or $j \geq k$. Substitute the arrival and service rates in Equations (7.7-4) and (7.7-5) into the expression for C_j. Let $r = \lambda_A/\lambda_S$. Then

$$\begin{aligned} C_j &= \frac{\lambda_A^j}{\lambda_S^j j!} \\ &= \frac{r^j}{j!}, \quad 1 \leq j < k, \end{aligned}$$

and

$$\begin{aligned} C_j &= \frac{\lambda_A^j}{\lambda_S^j k! k^{j-k}} \\ &= \frac{r^j}{k! k^{j-k}}, \quad j \geq k. \end{aligned}$$

Using these expressions we obtain

$$\begin{aligned} \pi_0 &= \left[1 + \sum_{j=1}^{k-1} \frac{r^j}{j!} + \sum_{j=k}^{\infty} \frac{r^j}{(k! k^{j-k})}\right]^{-1} \\ &= \left[1 + \sum_{j=1}^{k-1} \frac{r^j}{j!} + \frac{r^k}{k!(1 - r/k)}\right]^{-1}, \end{aligned} \tag{7.7-6}$$

$$\pi_j = \pi_0 \frac{r^j}{j!}, \quad j < k, \tag{7.7-7}$$

$$\begin{aligned} \pi_j &= \pi_0 \frac{r^j}{k!k^{j-k}}, \\ &= \pi_k \left(\frac{r}{k}\right)^{j-k}, \quad j \geq k. \end{aligned} \tag{7.7-8}$$

EXAMPLE 7.7-1 Consider an M/M/2 queuing process with $r = .25$. To three-decimal-place accuracy we have

$$\begin{aligned} \pi_0 &= \left[1 + .25 + \frac{.25^2}{2!\,(1 - .25/2)}\right]^{-1} = .778, \\ \pi_1 &= (.778)(.25) = .195, \\ \pi_2 &= (.778)\frac{.25^2}{2!} = .024, \\ \pi_3 &= (.778)\frac{.25^3}{2!2} = .003, \\ \pi_4 &= (.778)\frac{.25^4}{2!2^2} = .000, \\ \pi_j &= .000 \quad \text{for } j > 4. \end{aligned}$$

Ignoring probabilities for $j \geq 4$, we can approximate the mean, variance, and standard deviation of the number of customers in the system as follows:

$$\begin{aligned} \mu &\approx (0(.778) + 1(.195) + 2(.024) + 3(.003)) \\ &= .252, \\ \sigma^2 &\approx (0 - .252)^2(.778) + (1 - .252)^2(.195) \\ &\quad + (2 - .252)^2(.024) + (3 - .252)^2(.003) \\ &= .254, \\ \sigma &\approx \sqrt{.254} = .504. \end{aligned}$$

EXAMPLE 7.7-2 Consider two different system configurations. One has a single fast server, and the other has two slower servers. The arrival rate is 10 customers per minute for both. The service rate for the fast server is 80 customers per minute, and the service rate for each of the slower servers is 40 customers per minute. Assume that the single-server system can be modeled as an M/M/1 queuing process and the two-server system can be modeled as an M/M/2 queuing process.

For the system with the fast server we have $r = 10/80 = .125$. Thus, the mean number of customers in the system is $.125/(1 - .125) = .143$. For the system with the two slower servers, $r = 10/40 = .25$, so the computations in Example 7.7-1 apply. Thus, the mean number of customers in the system is approximately .252, which shows that the system with two slower servers performs worse than the one with the single fast server in terms of mean number of customers in the system. It is interesting to note that the two systems have the same system service rate of 80 customers per minute when the two servers are both busy. The reason for the difference between the mean number of customers in the system is that the system with the two servers has a service rate of only 40 customers per minute when there is only one customer to be processed as compared to a rate of 80 customers per minute for the single-server system.

EXAMPLE 7.7-3 The checkout line at the student cafeteria presently uses a single cashier. Suppose the line can be modeled as an M/M/1 queuing process with an arrival–service ratio of $r = .9$. In this case, the mean number of customers in the system is $.9/(1 - .9) = 9$. Consider the effect of adding two more cashiers, which are assumed to provide identical service to the first. The steady-state probabilities to three-decimal-place accuracy are

$$\pi_0 = \left[1 + .9 + \frac{.9^2}{2!} + \frac{.9^3}{3!\,(1 - .9/3)}\right]^{-1} = .403,$$

$$\pi_1 = (.403)(.9) = .363,$$

$$\pi_2 = (.403)\frac{.9^2}{2!} = .163,$$

$$\pi_3 = (.403)\frac{.9^3}{3!} = .049,$$

$$\pi_4 = (.403)\frac{.9^4}{3!3} = .015,$$

$$\pi_5 = (.403)\frac{.9^5}{3!3^2} = .004,$$

$$\pi_6 = (.403)\frac{.9^6}{3!3^3} = .001,$$

$$\pi_j = .000 \quad \text{for } j > 6.$$

The mean number of customers in this three-server queuing system is approximately

$$\mu \approx (0(.403) + 1(.363) + 2(.163) + \cdots + 6(.001)) \\ = .922.$$

Thus, with the two additional cashiers, the mean number of customers in the system is reduced from nine customers to less than one.

The following theorem gives a method for finding the expected amount of time a customer spends in the system in terms of the expected number of customers in the system.

THEOREM 7.7-3 Assume that an M/M/k queuing system has reached the steady-state condition. Let T denote the amount of time a customer spends in the system, and let N denote the number of customers in the system. Then

$$\lambda_A E(T) = E(N).$$

Suppose customers arrive at the rate of $\lambda_A = 2$ per minute and the mean number in the system is $E(N) = 10$ customers. Then customers spend an average of $E(T) =$ 5 minutes in the system.

The equation in Theorem 7.7-3 is called *Little's formula*. The plausibility of this result can be seen with the pictorial representation of a deterministic queuing system given in Table 7.7-1. The columns of Table 7.7-1 represent 1-minute time periods, and the rows represent customers. In this system, customers arrive at a nonrandom rate of two per minute and spend exactly 5 minutes in the system. The asterisks show when each customer is in the system. The number of asterisks in each column represent the number of customers in the system at that time. As predicted by Little's formula, there are eventually 10 customers in the system. In fact, there are 10 customers at each time after the first 4 minutes.

TABLE **7.7-1** Representation of Little's Formula

		Time Period							
Customers	**1**	**2**	**3**	**4**	**5**	**6**	**7**	**8**	...
1	*	*	*	*	*				
2	*	*	*	*	*				
3		*	*	*	*	*			
4		*	*	*	*	*			
5			*	*	*	*	*		
6			*	*	*	*	*		
7				*	*	*	*	*	
8				*	*	*	*	*	
9					*	*	*	*	
10					*	*	*	*	
11						*	*	*	
12						*	*	*	
13							*	*	
14							*	*	
15								*	
16								*	
...								...	

* represents time periods when customers are in the system.

EXAMPLE **7.7-4** Consider the single-server and three-server queuing system in Example 7.7-3. Suppose customers arrive at the rate of six per minute. The expected time for a customer to spend in the single-server system is $E(N)/\lambda_A = 9/6$ minutes $= 90$ seconds. The expected time for a customer to spend in a three-server system is $.922/6 = 9.2$ seconds.

Exercises 7.7

Where appropriate, assume that the systems in question can be modeled as M/M/k queuing processes for the indicated value of k and that the systems have reached the steady-state condition.

7.7-1 Compute the steady-state probability distributions for an M/M/4 queuing system with arrival/service ratios of $r = .5$ and $r = .9$. Find the means and standard deviations of the number of customers in these systems.

7.7-2 Write a program to compute the steady-state probabilities of an M/M/k queuing system. The program should have the arrival rate, the service rate, and the value of k as inputs. Stop computing the probabilities after they are less than .0005. Also include computations for the mean and standard deviation of the number in the system and the mean time a customer spends in the system.

7.7-3 Customers arrive at a copy center at the rate of nine per hour. The expected time a customer spends using a copier is 6 minutes.

a Compute the means and standard deviations of the number of customers in the system when one and two copiers are in operation.

b Compute the expected time a customer spends in the system when one and two copiers are in operation.

c Based on these computations, do you think it makes sense to have two copiers?

7.7-4 The tourist information center at a resort area has an information booth with three employees to answer questions that tourists may have. Suppose that tourists arrive at the rate of 20 per hour and each employee spends an average of 8 minutes answering a tourist's questions. What is the probability that there are no tourists at the booth?

7.7-5 Requests for gift wrapping of packages arrive at a service counter at the rate of 15 per hour. The requests are handled by two people each of whom can wrap a package in an average of 6 minutes. On average, how long will each customer have to wait to have his or her package wrapped?

7.7-6 The arrival rate for a certain system is 30 customers per hour. When two servers work, each has a service rate of 25 customers per hour. At what rate would one server have to work in order for the mean number of customers in the one-server system to be the same as that of this two-server system?

7.7-7 Calls arrive at a certain crisis center at the rate of 8 per hour. There are five counselors to take the calls. Each call takes an average of 20 minutes. When a person calls the center, what is the probability that he or she will find a counselor free to take the call?

7.7-8 A formula for the expected number of customers in an M/M/k queuing system that has reached steady state is

$$E(N) = r + \frac{r\pi_k/k}{(1-r/k)^2}.$$

Apply the formula to the M/M/4 queues in Exercise 7.7-1. Also show that the formula reduces to $r/(1-r)$ for the M/M/1 queue. Derive the formula.

7.7-9 The following steps outline an intuitive derivation that the steady-state probabilities of a general Markov queuing process satisfy the system of balance equations in Theorem 7.7-1. There are only three transitions in a small interval $[t, t+\Delta]$ that can lead to $i > 0$ customers being in the queue at time $t+\Delta$. These are $(i-1) \to i$, $(i+1) \to i$, and $i \to i$, and they occur with probabilities $a_{i-1}\Delta$, $s_{i+1}\Delta$, and $(1-a_i\Delta - s_i\Delta)$, respectively (why?). Thus, in the steady-state condition and for $i > 0$

$$\pi_i = \pi_{i-1}(a_{i-1}\Delta) + \pi_{i+1}(s_{i+1}\Delta) + \pi_i(1-a_i\Delta - s_i\Delta).$$

Likewise, for $i = 0$, we have

$$\pi_0 = \pi_1(s_1\Delta) + \pi_0(1-a_0\Delta).$$

Show that simplification of these equations leads to the balance equations.

SECTION

7.8 More Markov Queuing Processes

Consider the steady-state probabilities for the M/M/k queuing process as k becomes large. From Equation (7.7-6), we see

$$\begin{aligned} \lim_{k\to\infty} \pi_0 &= \left[1 + \sum_{j=1}^{\infty} \frac{r^j}{j!}\right]^{-1} \\ &= e^{-r}. \end{aligned}$$

Thus,

$$\lim_{k\to\infty} \pi_j = e^{-r}\frac{r^j}{j!}, \quad j = 1, 2, \ldots.$$

Therefore, if there are infinitely many servers, the number in the system has a Poisson distribution in the steady state with mean $\mu = r$. Since there is no waiting for service with an infinite-server queuing process, the number in the system is the number of busy servers. In practice, the result is applied to systems with numerous servers.

EXAMPLE **7.8-1** Mega-mart has provided a large number of checkout stations to meet the potential demand. Customers arrive at the check-out stations at the rate of four per minute. They spend an average of 1.5 minutes being serviced, which gives a service rate of 2/3 customers per minute. The arrival/service ratio is $r = 4/(2/3) = 6$. Since the mean of the steady-state Poisson distribution is $r = 6$, there is an average of six busy check-out stations at any time. From Table 3 of Appendix A,

$$\begin{aligned} P(\text{more than 10 busy lines}) &= 1 - P(\text{10 or fewer busy lines}) \\ &= 1 - .9574 \\ &= .0426. \end{aligned}$$

Thus, 10 checkout stations would handle the demand most of the time.

We now consider a steady-state example of an M/M/k queuing process with finite capacity. Note that for an M/M/k process with finite capacity c, the arrival rates a_j in the balance equations of Definition 7.7-1 are zero for $j \geq c$, and the service rates s_j need only be defined for $j \leq c$ because at most c customers can be in the system at any time.

EXAMPLE 7.8-2 Super-Lube has two service bays and room for two additional cars to wait for service. Cars come to Super-Lube at the rate of one car every 10 minutes, or .1 car per minute provided there is space for a car. The time to service a car averages 10 minutes, so the service rate for the system is .1 car per minute for one busy server and .2 car per minute for two busy servers. Since the system has a capacity of four, only the arrival rates a_j for $j = 0, 1, 2, 3$ need to be considered since $a_j = 0$ for $j \geq 4$. Likewise, only the service rates s_j for $j = 1, 2, 3, 4$ need to be considered since at most four cars can be in the queue at any one time. Therefore,

$$a_0 = a_1 = a_2 = a_3 = .1,$$

$$s_1 = .1, \quad s_2 = s_3 = s_4 = .2.$$

$$\begin{aligned}
\pi_0 &= \left(1 + \frac{a_0}{s_1} + \frac{a_0 a_1}{s_1 s_2} + \frac{a_0 a_1 a_2}{s_1 s_2 s_3} + \frac{a_0 a_1 a_2 a_3}{s_1 s_2 s_3 s_4}\right)^{-1} \\
&= [1 + 1 + .5 + (.5)^2 + (.5)^3]^{-1} \\
&= .348, \\
\pi_1 &= .348, \\
\pi_2 &= .348(.5) = .174, \\
\pi_3 &= .348(.5)^2 = .087, \\
\pi_4 &= .348(.5)^3 = .043.
\end{aligned}$$

The mean and standard deviation of the number in the system can be computed directly from the foregoing distribution. We find $\mu = 1.13$ cars and $\sigma = 1.11$ cars. There are no customers about 35% of the time, and the system is full about 4% of the time.

In the M/M/k queuing process, the arrival rate is assumed to be constant. However, it sometimes happens that the rate of arrival depends on the number in the system. Customers are less likely to join when there are more people in line than when there are fewer in line. A possible model that has this characteristic is one in which the arrival rate has the form

$$a_i = \frac{\lambda_A}{i + 1}. \tag{7.8-1}$$

Here λ_A denotes the arrival rate when there are no customers in the system, and the arrival rate decreases as $1/(i + 1)$, where i is the number in the system. Suppose that we have a single-server queuing process with variable arrival rate as defined in Equation (7.8-1) and a service rate of $s_i = \lambda_S$. It is easy to verify that the terms involved in the solution of the balance equations have the form

$$\frac{a_0 a_1 \ldots a_{j-1}}{s_1 s_2 \ldots s_j} = \frac{\lambda_A^j}{\lambda_S^j j!} = \frac{r^j}{j!},$$

where $r = \lambda_A / \lambda_S$. It follows that

$$\pi_0 = \left[\sum_{j=0}^{\infty} \frac{r^j}{j!}\right]^{-1} = e^{-r} \quad \text{and} \quad \pi_j = e^{-r} \frac{r^j}{j!}.$$

We have shown that the number in the system under the variable arrival rate assumption has a Poisson distribution with mean $\mu = r$.

EXAMPLE 7.8-3 Suppose the number of customers waiting for service from a pretzel vendor can be modeled as a single-server queuing process with variable arrival rate as defined in Equation (7.8-1). If there are no customers waiting for service, the arrival rate is four customers per minute. It is two customers per minute with one customer in the queue, one and one-third customers per minute with two customers in the queue, and so forth. Customers can be served at the rate of two customers per minute. Thus the number of customers in line after a long period of time has a Poisson distribution with a mean of $r = 2$ customers. From Table 3 in Appendix A, we find that the probability that there are more than five customers in the queue is $1 - .9834 = .0166$. Thus, it is likely that there will be five or fewer customers in line at any one time.

Exercises 7.8

7.8-1 An aircraft engine repair facility can service one engine at a time and has storage space for three other engines. The arrival rate is one engine every five days, and the average service time is four days for each engine. Assume that the system can be modeled as a limited-capacity M/M/1 queuing process.

a Find the steady-state distribution for the number of engines in the system.

b Find the expected value and the standard deviation for the number in the system in steady state.

7.8-2 A large number of personal computers have been made available in a microcomputer laboratory for student use. Students come to the laboratory at the rate of 12 per hour, and the typical student spends 50 minutes using the computer each visit. Suppose this system is modeled as an infinite-server queuing system.

a What is the steady-state distribution for the number of computers in use at any time?

b What is a likely range within which the number of busy computers will fall?

c Based on your computations, how many computers do you think the laboratory should have to handle the demand most of the time?

7.8-3 The sale of lottery tickets at a certain outlet is handled by a single clerk. When there is no one in line, customers arrive at the rate of eight per minute to purchase tickets. If there are customers in line, the arrival rate is determined by Equation (7.8-1). It takes an average of 20 seconds to handle a transaction. Find the steady-state distribution and mean for the number of customers in the system.

7.8-4 The local PBS station has 14 telephones to handle calls during its fund-raising drive. Subscribers call in at the rate of 16 per minute, and each call takes an average of 45 seconds to complete. Callers cannot be put on hold if all phones are busy. Assume that this system can be modeled as a limited-capacity M/M/14 queuing process.

a Find the steady-state distribution for the number of busy telephones.

b What is the probability that a caller will find all phones busy?

7.8-5 A car rental operation at a municipal airport has 15 cars available for rent. Cars are rented for an average duration of three days. Customers arrive to rent cars at a rate of five per day. Assume that this operation can be modeled as an M/M/15 queuing process with finite capacity of 15.

a Find the probability that when a customer arrives there is no car available.

b How many cars does this operation need so that the probability that no car is available when a customer arrives is less than .05?

7.8-6 A laundromat has eight washing machines. Customers arrive at the rate of 10 per hour. Assume that a customer uses only one machine. The mean service time for a load of wash is 1/2 hour. If a potential customer arrives at this laundromat and finds that all the machines are occupied, the person will leave. Suppose this system can be modeled as an M/M/8 limited-capacity queuing process.

a What is the steady-state distribution for the number of busy washing machines?

b What percent of the time are people likely to leave the laundromat due to no available machines?

7.8-7 Compare queuing systems at various fast-food restaurants. Consider both inside and drive-up service. Based on your experience, which queuing systems seem to deliver the best service in terms of minimizing the customer's time in line?

SECTION

7.9 Simulating an M/M/*k* Queuing Process

One way to simulate an M/M/*k* queuing process is to approximate it by a Bernoulli queuing process as described in Section 7.6. Choose frame sizes sufficiently small so that the approximations in Definition 7.6-2 apply, and then let $P_A = \lambda_A \Delta$ and

$P_S = \lambda_S \Delta$. If the probabilities are small enough so that $P_A < .05$ and $kP_S < .05$, then the simulation provides a reasonable approximation to the behavior of the M/M/*k* queuing process. However, such simulations are time consuming because they involve simulations over many frames.

It is more common to simulate the time between arrivals of customers and the time it takes to service each customer once service has begun. Since these times have exponential distributions with means $1/\lambda_A$ and $1/\lambda_S$, respectively, they can be readily simulated, as discussed in Section 6.1. Using these quantities, we can obtain the time each customer arrives for service, the time each spends waiting for service, and the time each spends in the system. We give an algorithm for simulating the arrivals and departures of customers in an M/M/1 queuing system. This algorithm may be used with any distribution of interarrival times and any distribution of service times, not just exponential distributions. Thus, it allows for simulating more general queuing processes than the M/M/1 queue.

Simulation of M/M/1 Queuing Process

Variables:

n	number of the customer
Nmax	maximum number of customers to go through system
$A(n)$	interarrival time (time between arrival of $(n-1)$th and nth customer)
$S(n)$	service time for nth customer
$T(n)$	arrival time of nth customer
$B(n)$	time service begins for nth customer
$D(n)$	time nth customer departs from the system
$W(n)$	time nth customer spends in the system
$L(n)$	time nth customer spends waiting in line before beginning service

Initialize:

$n = 0$

$T(0) = 0$

$D(0) = 0$

$N\text{max} =$ predetermined value

Repeat the following steps for $n = 1$ to Nmax:

1 Generate values for $A(n)$ and $S(n)$. (Simulate exponential random variables for interarrival and service times.)

2 Set $T(n) = T(n-1) + A(n)$. (The arrival time of nth customer

is arrival time of $(n - 1)$th customer plus the interarrival time.)

3 Set $B(n) = \max(D(n - 1), T(n))$. (If arrival time of nth customer occurs before departure time of $n - 1$ customer, the service for nth customer begins when the previous customer departs; otherwise, the nth customer begins service at the time of arrival.)

4 Set $D(n) = B(n) + S(n)$. (Add the time of service to the time service begins to determine departure time.)

5 Set $L(n) = B(n) - T(n)$. (The time spent in line is the difference between the time service begins and the arrival time.)

6 Set $W(n) = D(n) - T(n)$. (The time in the system is the difference between the departure time and arrival time.)

Unlike the frame-by-frame simulation, the arrival–departure method outlined does not directly give the number of customers in the system at any given time. However, for any time t, this can be computed by counting the number of customers for which $T(n) \le t < D(n)$.

The arrivals and departures of 10 customers are shown in Table 7.9-1. The interarrival and service times were generated as exponential random variables with $\lambda_A = .10$ and $\lambda_S = .15$. Outcomes were rounded to the next integer for simplicity of display.

TABLE 7.9-1 Simulation of Arrivals and Departures from an M/M/1 Queuing Process with $\lambda_A = .10$ and $\lambda_S = .15$

Interarrival Time $A(n)$	Service Time $S(n)$	Arrival Time $T(n)$	Time Service Begins $B(n)$	Departure Time $D(n)$	Time in Line $L(n)$	Time in System $W(n)$
1	4	1	1	5	0	4
1	3	2	5	8	3	6
6	13	8	8	21	0	13
14	4	22	22	26	0	4
1	8	23	26	34	3	11
1	9	24	34	43	10	19
30	1	54	54	55	0	1
2	5	56	56	61	0	5
22	3	78	78	81	0	3
1	8	79	81	89	2	10
				Means	1.8	8.6

Now consider simulating the M/M/k queuing process. The key modification to the algorithm for the M/M/1 queue is in the handling of departure times. One customer may come after another yet depart sooner. Thus, we cannot determine when the nth customer begins service based on when the $(n - 1)$th customer completed service, as we did in the M/M/1 process. Instead, we will find the most recent departure times for each of the k servers. If one of the most recent departure times for the

k servers is less than the arrival time of the nth customer, then the nth customer begins service immediately. Otherwise, the nth customer will begin service when a customer has departed from one of the servers.

Simulation of M/M/k Queuing Process

Variables:

n	number of the customer
Nmax	maximum number of customers to go through system
$A(n)$	interarrival time (time between arrival of $(n-1)$th and nth customer
$S(n)$	service time for nth customer
$T(n)$	arrival time of nth customer
j	server number $j = 1, 2, \ldots, k$
$F(j)$	departure time of the customer most recently served by the jth server
Jmin	server for which $F(j)$ is smallest, that is, server with the earliest of the most recent departure times
$B(n)$	time service begins for nth customer
$D(n)$	time nth customer departs from the system
$W(n)$	time nth customer spends in the system
$L(n)$	time nth customer spends waiting in line before beginning service

Initialize:

$n = 0, T(0) = 0, D(0) = 0$

Nmax $=$ predetermined value

$F(j) = 0, j = 1, 2, \ldots, k$

Repeat the following steps for $n = 1$ to Nmax:

1 Generate values for $A(n)$ and $S(n)$.

2 Set $T(n) = T(n-1) + A(n)$.

3a Find Jmin. (Find a server for which $F(J\text{min}) \le F(j)$ for $j = 1, 2, \ldots, k$. In case of a tie, choose the smallest value of Jmin. This is server with the smallest of the most recent departure times.)

3b Set $B(n) = \max(F(J\text{min}), T(n))$. (If $T(n) > F(J\text{min})$ then a server is free and service begins immediately; otherwise, the customer must wait for the earliest departure to occur.)

4a Set $D(n) = B(n) + S(n)$.

4b Set $F(Jmin) = D(n)$. (The departure time for the server which handles the nth customer is updated.)

5 Set $L(n) = B(n) - T(n)$.

6 Set $W(n) = D(n) - T(n)$.

The arrivals and departures of 10 customers from an M/M/2 queue are shown in Table 7.9-2. The interarrival and service times are the same as the ones used in the M/M/1 queue. As expected, the mean time in line and the mean time in the system are smaller in the two-server queue.

TABLE **7.9-2**
Simulation of an M/M/2 Queuing System with $\lambda_A = .10$, $\lambda_S = .15$

Interarrival Time	Service Time	Arrival Time	Server 1	Server 2	Time Service Begins	Departure Time	Time in Line	Time in System
$A(n)$	$S(n)$	$T(n)$	$F(1)$	$F(2)$	$B(n)$	$D(n)$	$L(n)$	$W(n)$
1	4	1	0	0	1	5	0	4
1	3	2	5	0	2	5	0	3
6	13	8	5	5	8	21	0	13
14	4	22	21	5	22	26	0	4
1	8	23	21	26	23	31	0	8
1	9	24	31	26	26	35	2	11
30	1	54	31	35	54	55	0	1
2	5	56	55	35	56	61	0	5
22	3	78	55	61	78	81	0	3
1	8	79	81	61	79	87	0	8
						Mean	0.2	6.0

Exercises 7.9

7.9-1 Listed are the interarrival times and service times for 10 customers in a single-server queuing process. Simulate the amount of time each customer spends in line and the amount of time each spends in the system.

Interarrival times 3 2 1 4 6 8 2 5 9 4
Service times 2 3 1 5 7 7 3 8 1 3

7.9-2 Repeat Exercise 7.9-1 for a two-server queuing process.

7.9-3 Repeat Exercise 7.9-1 for an M/M/3 queuing system. Use the same interarrival and service times as obtained in Exercise 7.9-1.

7.9-4 Simulate the arrivals and departures of 20 customers in an M/M/1 queuing system with an arrival rate of 20 customers per hour and a service rate of 21 customers per hour. Compute the sample means of the amount of time spent in line and the amount of time spent in the system.

7.9-5 Simulate the arrivals and departures of 20 customers in a single-server queuing system in which the interarrival times are uniform [0, 3] random variables and service times are uniform [0, 2] random variables. Compare this system to that of an M/M/1 queue in which the means of the exponential interarrival and service times are the same as those of the uniform random variables.

8

The Distribution of Sums of Random Variables

1-3

SECTION

8.1 Sums of Random Variables

Sums of random variables occur frequently in applications. For instance, the total weight of a shipment of items can be expressed as a sum of the weights of the individual items. Total cost of an appliance can be expressed as the sum of costs of the components that make up the finished product. In this section, we will consider the problem of finding the probability distribution of a sum of independent random variables. First we will consider the case of normal random variables.

THEOREM 8.1-1 Let $X_1, X_2, \ldots, X_n$ be independent, normally distributed random variables with $E(X_i) = \mu_i$ and $\mathrm{STD}(X_i) = \sigma_i$ for $i = 1, 2, \ldots, n$. Then the sum $X = X_1 + X_2 + \cdots + X_n$ has a normal distribution with mean $\mu = \mu_1 + \mu_2 + \cdots + \mu_n$ and standard deviation $\sigma = \sqrt{\sigma_1^2 + \sigma_2^2 + \cdots + \sigma_n^2}$.

Theorems 2.7-3 and 2.7-5 apply in deriving the mean and standard deviation of a sum of independent random variables. The additional information given in Theorem 8.1-1 is the fact that a sum of independent, normally distributed random variables has a normal distribution. The proof of Theorem 8.1-1 is given at the end of this section.

EXAMPLE **8.1-1** Suppose that a car is taken into a station for a tune-up and an oil change. We would like to know the probability that the work will be done in less that 1.5 hours. Assume that the time X_1 to complete a tune-up is a normal random variable with a mean of 60 minutes and a standard deviation of 10 minutes, and the time X_2 to complete an oil change is a normal random variable with a mean of 5 minutes and a standard deviation of 5 minutes. The time for service to be completed is the sum $X = X_1 + X_2$. If the times to complete the two tasks are independent of one another, then the sum has a normal distribution with a mean of $\mu = 60 + 15 = 75$ minutes and a standard deviation of $\sigma = \sqrt{10^2 + 5^2} = 11.2$ minutes. The probability that the job will be done in less than 1.5 hours is

$$\begin{aligned} P(X < 90) &= P\left(Z < \frac{90 - 75}{11.2}\right) \\ &= P(Z < 1.34) \\ &= .9099. \end{aligned}$$

An important special case of Theorem 8.1-1 occurs when the X_i 's are a random sample. If the mean and standard deviation of the X_i 's are denoted by μ and σ, respectively, then the sum of n such independent, normally distributed random variables has a normal distribution with a mean $\mu + \mu + \cdots + \mu = n\mu$ and a standard deviation $\sqrt{\sigma^2 + \sigma^2 + \cdots + \sigma^2} = \sigma\sqrt{n}$. This is illustrated in the following example.

EXAMPLE **8.1-2** Suppose that the weight of college-age males is a normal random variable with a mean of 160 pounds and a standard deviation of 20 pounds. If 10 such randomly selected individuals get on an elevator, then their total weight is a normal random variable with a mean of $(10)(160) = 1600$ pounds and a standard deviation of $20\sqrt{10} = 63.2$ pounds. The total weight has about a 99% chance of falling in the interval $1600 \pm 2.58(63.2)$ pounds, that is from 1437 pounds to 1763 pounds.

In some situations, the distribution of a sum of independent random variables cannot be expressed in a convenient mathematical form. In such cases, the distribution may be obtained by simulation.

EXAMPLE **8.1-3** A salesperson begins at location 1 and visits locations 2, 3, and 4 before returning to

1 as shown in Figure 8.1-1. The time spent at each location is an exponential random variable with a mean of 30 minutes. The time in transition between each location is a normal random variable with a mean of 20 minutes and a standard deviation of 5 minutes. Assume these various times are independent of one another.

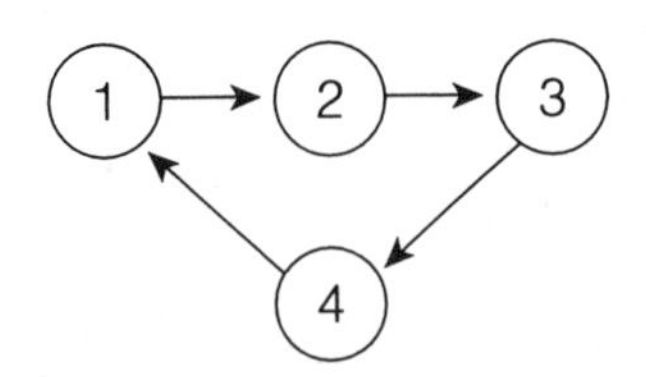

FIGURE **8.1-1**
Salesperson's Route

The random variable of interest is the total time that it takes the salesperson to return to location 1. If we let T_i be the time spent at location i, and T_{ij} be the time in transition from location i to j, then

$$T = T_1 + T_{12} + T_2 + T_{23} + T_3 + T_{34} + T_4 + T_{41}.$$

Finding the mean and variance of T is an application of Theorems 2.7-3 and 2.7-6. Thus,

$$\begin{aligned} E(T) &= E(T_1 + T_2 + T_3 + T_4) + E(T_{12} + T_{23} + T_{34} + T_{14}) \\ &= 4(30) + 4(20) \\ &= 200 \text{ minutes} \\ &= 3.33 \text{ hours}, \\ \text{VAR}(T) &= \text{VAR}(T_1 + T_2 + T_3 + T_4) + \text{VAR}(T_{12} + T_{23} + T_{34} + T_{14}) \\ &= 4(900) + 4(25) \\ &= 3700, \\ \text{STD}(T) &= \sqrt{3700} \\ &= 60.8 \text{ minutes} \\ &= 1.0 \text{ hour}. \end{aligned}$$

The results of 1000 simulated sales trips are summarized by the histogram in Figure 8.1-2. In about 77% of the cases the route will be completed in less than 4 hours. Occasionally there will be rather large times. For example, the maximum time in the 1000 simulations was over 9 hours.

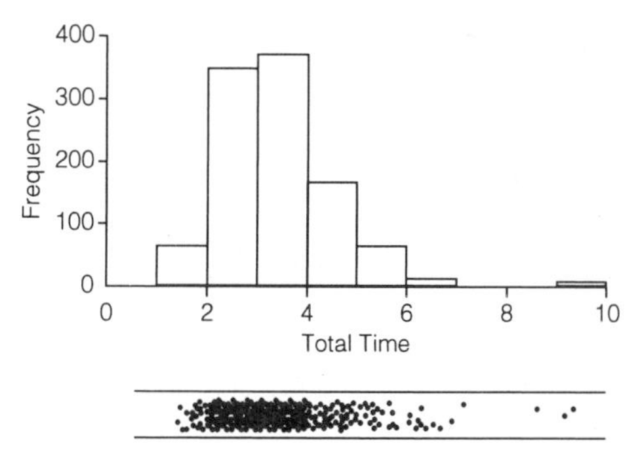

FIGURE **8.1-2** Distribution of Time to Complete Sales Route

Moment-generating functions can be used to derive distributional results about sums of independent random variables.

THEOREM **8.1-2** Let $X_1, X_2, \ldots, X_n$ be independent random variables with moment-generating functions $M_{X_1}(t), M_{X_2}(t), \ldots, M_{X_n}(t)$. If $X = X_1 + X_2 + \cdots + X_n$, then

$$M_X(t) = M_{X_1}(t) M_{X_2}(t) \cdots M_{X_n}(t).$$

Theorem 8.1-2 is verified as follows. First,

$$M_X(t) = E(e^{Xt}) = E\left(e^{(X_1 + X_2 + \cdots + X_n)t}\right) = E\left(e^{X_1 t} e^{X_2 t} \cdots e^{X_n t}\right).$$

Recall that $X_1, X_2, \ldots, X_n$ are independent random variables. Therefore the random variables $e^{X_1 t}, e^{X_2 t}, \ldots, e^{X_n t}$ are also independent. Using the fact that the expected value of a product of independent random variables is the product of the expected values (see Theorem 2.7-4 and Exercise 2.7-9), we have

$$E\left(e^{X_1 t} e^{X_2 t} \cdots e^{X_n t}\right) = E\left(e^{X_1 t}\right) E\left(e^{X_2 t}\right) \cdots E\left(e^{X_n t}\right)$$

and the result follows.

PROOF of Theorem **8.1-1**

Let $X_1, X_2, \ldots, X_n$ be independent, normally distributed random variables with $E(X_i) = \mu_i$ and $\text{STD}(X_i) = \sigma_i$, for $i = 1, 2, \ldots, n$. We will find the moment-generating function of $X = X_1 + X_2 + \cdots + X_n$. From Example 6.5-3, the moment-generating function of a normally distributed random with mean μ and standard deviation σ is $e^{\mu t + \sigma^2 t^2/2}$. Then,

$$\begin{aligned} M_X(t) &= M_{X_1}(t) M_{X_2}(t) \cdots M_{X_n}(t) \\ &= e^{\mu_1 t + \sigma_1^2 t^2/2} e^{\mu_2 t + \sigma_2^2 t^2/2} \cdots e^{\mu_n t + \sigma_n^2 t^2/2} \\ &= e^{(\mu_1 + \mu_2 + \cdots + \mu_n)t + (\sigma_1^2 + \sigma_2^2 + \cdots + \sigma_n^2)t^2/2}. \end{aligned}$$

Therefore, the random variable X has a normal distribution with mean $\mu_1 + \mu_2 + \cdots + \mu_n$ and variance $\sigma_1^2 + \sigma_2^2 + \cdots + \sigma_n^2$.

Exercises 8.1

8.1-1 The means and standard deviations for the time to complete 400 meters in a 1600-meter relay race involving four runners are given in the table. Let T denote the total time to run the race.

Runner	Mean Time to Complete 400 Meters (seconds)	Standard Deviation (seconds)
1	54	1.5
2	56	2.0
3	52	1.5
4	51	1.0

a Find the expected value and standard deviation of T.

b Within what range would you expect the total time to fall?

c Assume that the random variables have a normal distribution. Find the probability that the race is run in less than 204 seconds.

8.1-2 In a certain student population, incomes are normally distributed with a mean of \$1000 per month and a standard deviation of \$200 per month. If we randomly select five students from this population, what is the probability that their total monthly income is greater than \$5500?

8.1-3 Rigging a sail for a sailboard involves six steps. The time (in minutes) it takes to complete each step is a normal random variable with the indicated means and standard deviations.

	Mean	Standard Deviation
1. Unload the equipment	3	2
2. Unroll the sail and slide it on the mast	1.5	1
3. Downhaul the sail	1	.5
4. Adjust the boom length and attach it to mast	3	1.5
5. Outhaul the sail	1	1
6. Adjust the battens	2	.5

If the times are independent random variables, what is the probability it will take more than 15 minutes to rig a sail?

8.1-4 The time (in minutes) that it takes a tactical fighter plane to reach a target may be expressed as $10 + X$, where X has an exponential distribution with a mean of 5 minutes. The time spent over the target is a normal random variable with a mean of 15 minutes and a standard deviation of 4 minutes. The time to return from the target is an exponential random variable with a mean of 5 minutes. The total mission time is the sum of these three times.

a Find the mean and standard deviation of the mission time.

b Simulate the probability distribution of the mission time by repeating the experiment 1000 times, and make a histogram of the results.

c Based on your simulation in part (b), what percent of the time will the mission be completed in less than 40 minutes?

8.1-5 Let $X_1, X_2, \ldots, X_n$ be independent and identically distributed exponential random variables each with mean μ. Find the moment-generating function of $\Sigma_{i=1}^{n} X_i$ and verify that it is the moment-generating function of a gamma random variable with parameters $\alpha = n$ and $\beta = \mu$.

8.1-6 Let X be a standard normal random variable and let Y be a normal random variable with $\mu = 2$ and $\sigma = 3$. Find the moment-generating function for $W = 2X - 3Y$. From this moment-generating function, find $E(W)$ and $\text{VAR}(W)$.

8.1-7 Let $X_1, X_2, \ldots, X_n$ be n independent Poisson random variables with means $\mu_1, \mu_2, \ldots, \mu_n$, respectively. Find the moment-generating function of $X_1 + X_2 + \cdots + X_n$. Show that this sum has a Poisson distribution.

8.1-8 Let $X_1, X_2, \ldots, X_k$ be independent chi-square random variables with degrees of freedom $n_1, n_2, \ldots, n_k$, respectively. Show that the sum has a chi-square distribution with $n_1 + n_2 + \ldots + n_k$ degrees of freedom.

SECTION

8.2 Sums of Random Variables and the Central Limit Theorem

In this section we deal with the sum and the sample mean of a random sample $X_1, X_2, \ldots, X_n$. An important theoretical and practical result is called the *Central Limit Theorem*.

THEOREM 8.2-1 Central Limit Theorem for Sums. Let $X_1, X_2, \ldots, X_n$ be a random sample with $E(X_i) = \mu$ and $\text{STD}(X_i) = \sigma$ for all i. For sufficiently large n, the sum $X_1 + X_2 + \cdots + X_n$ has an approximate normal distribution with mean $n\mu$ and standard deviation $\sigma\sqrt{n}$.

The expected value and standard deviation of the sum of independent and identically distributed random variables can be obtained by applying Theorems 2.7-3 and 2.7-5. The new information that is asserted here is that the sum has an approximate normal distribution provided that the number of random variables making up the sum is "large enough." We can compare this result to that in Theorem 8.1-1. In that theorem, the random variables X_i have normal distributions, and the sum has a normal distribution for any sample size n. In Theorem 8.2-1 we do not require any distributional assumptions about the X_i's except that they all have the same distribution with mean μ and standard deviation σ. In Theorem 8.1-1 the sum is normally distributed for all n, whereas here we can only assert that the distribution is approximately normal and then only for n "large enough." A proof of the Central Limit Theorem using moment-generating functions is given at the end of this section.

EXAMPLE 8.2-1 Consider the distribution of the sum of 16 exponential random variables each of which has a mean of 4. Recall that the standard deviation of an exponential random variable is the same as the mean. The Central Limit Theorem tells us that the sum $X = X_1 + X_2 + \cdots + X_{16}$ will have an approximate normal distribution with a mean of $\mu = (16)(4) = 64$ and a standard deviation of $\sigma = 4\sqrt{16} = 16$. An empirical probability histogram based on a simulation of 1000 sums of 16 exponential random variables is shown in Figure 8.2-1. Notice that this distribution has roughly the "bell-shaped" appearance of the normal distribution, although it is slightly asymmetrical. The frequencies occur as would be predicted from the fact that the sum has an approximate normal distribution. For instance,

$$\begin{aligned} P(X \le 50) &\approx P\left(Z \le \frac{50-64}{16}\right) \\ &= P(Z < -.88) \\ &= .1894. \end{aligned}$$

The corresponding empirical probability is $187/1000 = .187$.

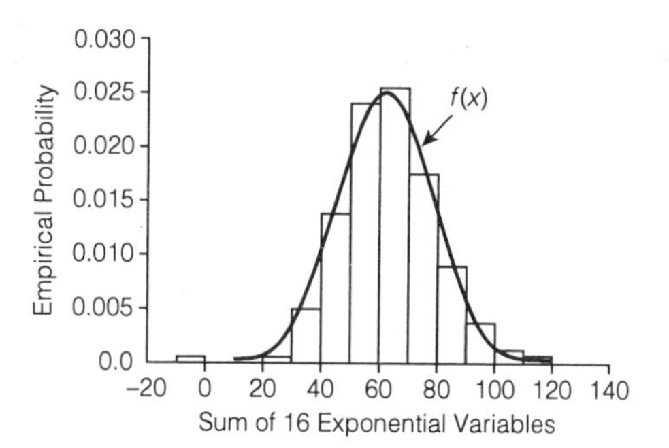

FIGURE **8.2-1** Empirical Probability Histogram of Sum of 16 Exponential Random Variables with Normal Distribution Overlaid

A question that naturally arises in applying the Central Limit Theorem is, how large must n be for the normal approximation to apply? There is no definitive answer to this question. It depends on the probability distribution from which the X_i 's are obtained. For instance, if the X_i 's have a uniform distribution, then a value of n around 6 is adequate for the Central Limit Theorem to be applied (see Exercise 8.2-6). However, a value of n around 16 is needed for the exponential distribution. Of course, the choice of n depends on what we believe is adequate as an approximation. The more accurate we wish the approximations to be, the larger n should be. Generally speaking, the Central Limit Theorem can be applied with relatively small values of n if the original distribution is symmetric or mound shaped and has very little probability in the tails (ends) of the distribution. If the distribution is skewed, as it is with the exponential, or has heavy tails (substantial probability in the tails), then larger values of n are needed for the Central Limit Theorem to apply.

EXAMPLE **8.2-2** The useful life of a cutting blade of a certain machine has a mean of 40 hours and a standard deviation of 8 hours. When one blade fails, it is immediately replaced by an identical new blade. Suppose there are 30 blades on hand. Consider the total number of hours the machine can operate. If we let T_i denote the lifetime of the ith blade, $i = 1, 2, \ldots, 30$, then the total lifetime is $T = T_1 + T_2 + \cdots + T_{30}$. If the T_i 's are assumed to be independent and if $n = 30$ is regarded as sufficiently large for the Central Limit Theorem to apply, then T has an approximate normal distribution with a mean of $(30)(40) = 1200$ hours and a standard deviation of $8\sqrt{30} = 43.8$ hours. The probability that the 30 blades are used up in less than 1250 hours is given by

$$\begin{aligned} P(T \le 1250) &\approx P\left(Z \le \frac{1250 - 1200}{43.8}\right) \\ &= P(Z \le 1.14) \\ &= .8729. \end{aligned}$$

The following corollary to Theorem 8.2-1 is the Central Limit Theorem for sample means.

COROLLARY **8.2-1** Central Limit Theorem for Sample Means. Let $X_1, X_2, \ldots, X_n$ be a random sample with $E(X_i) = \mu$ and $\text{STD}(X_i) = \sigma$. For sufficiently large n the sample mean $\bar{X}$ has an approximate normal distribution with mean μ and standard deviation $\sigma/\sqrt{n}$.

In Theorem 2.7-7 we found that $E(\bar{X}) = \mu$ and $\text{STD}(\bar{X}) = \sigma/\sqrt{n}$. Moreover, from Theorem 6.2-2, any linear function of a normal random variable has a normal distribution. It follows that if the sum has an approximate normal distribution, then the sample mean will as well.

EXAMPLE **8.2-3** The mean age of a population of assistant professors is 30 years with a standard deviation of 6 years. A sample of 100 is selected at random from this population. We will compute the probability that the sample mean is within one year of the mean of the population. We have $E(\bar{X}) = 30$ and $\text{STD}(\bar{X}) = 6/\sqrt{100} = .6$. Thus,

$$\begin{aligned} P(29 < \bar{X} < 31) &= P\left(\frac{29-30}{.6} < Z < \frac{31-30}{.6}\right) \\ &= P(-1.67 < Z < 1.67) \\ &= .9525 - .0475 \\ &= .9050. \end{aligned}$$

If we had not known the true mean of this population, this result shows that we would have about a 90% chance of coming within one year of the true mean simply by averaging the ages of 100 individuals randomly selected from this population. Computations of this sort tell pollsters that relatively accurate information about a population can often be obtained from a relatively small sample of the population.

The normal approximation to the binomial distribution is a special case of the Central Limit Theorem. Suppose that $X_1, X_2, \ldots, X_n$ are independent and identically distributed Bernoulli random variables with probability of success p. Then $X = X_1 + X_2 + \cdots + X_n$ is a binomial random variable. In Section 3.1, we found $E(X_i) = p$ and $\text{VAR}(X_i) = p(1-p)$. Since X is a sum of a random sample, the

Central Limit Theorem asserts that X will have an approximate normal distribution with mean np and standard deviation $\sqrt{np(1-p)}$, for sufficiently large n. This is simply a restatement of the normal approximation to the binomial. The sample fraction $\hat{P} = X/n$ is just the sample mean of a random sample of Bernoulli random variables; hence $\hat{P}$ has an approximate normal distribution with mean p and standard deviation $\sqrt{p(1-p)/n}$.

EXAMPLE 8.2-4 Suppose the probability that a student owns a microcomputer is .7. In a group of 200 randomly selected students, the probability that more than 75% own microcomputers is

$$\begin{aligned} P(\hat{P} > .75) &= P\left(Z > \frac{.75 - .70}{\sqrt{(.70)(.30)/200}}\right) \\ &= P(Z > 1.54) \\ &= .0618. \end{aligned}$$

The following is a proof of the Central Limit Theorem.

PROOF of Theorem 8.2-1 Let $X_1, X_2, \ldots, X_n$ be independent and identically distributed random variables each with mean μ and standard deviation σ. Let

$$Y = \frac{\sum_{i=1}^{n} X_i - n\mu}{\sigma\sqrt{n}} = \sum_{i=1}^{n} \frac{Y_i}{\sqrt{n}},$$

where $Y_i = (X_i - \mu)/\sigma$, for $i = 1, 2, \ldots, n$. Note that the $Y_1, Y_2, \ldots, Y_n$ are also independent and identically distributed each with mean 0 and standard deviation 1. Now,

$$\begin{aligned} M_Y(t) &= M_{Y_1}\left(\frac{t}{\sqrt{n}}\right) M_{Y_2}\left(\frac{t}{\sqrt{n}}\right) \cdots M_{Y_n}\left(\frac{t}{\sqrt{n}}\right) \\ &= \left[M_{Y_1}\left(\frac{t}{\sqrt{n}}\right)\right]^n \end{aligned}$$

since the Y_i have identical moment-generating functions. Using the definition of a moment-generating function, we have

$$M_{Y_1}\left(\frac{t}{\sqrt{n}}\right) = E\left(e^{Y_1 t/\sqrt{n}}\right),$$

$$= E\left(1 + \frac{Y_1 t}{n^{1/2}} + \frac{Y_1^2 t^2}{2n} + \frac{Y_1^3 t^3}{3!n^{3/2}} + \cdots\right)$$

$$= 1 + \frac{t^2}{2n} + \frac{t^3 E(Y_1^3)}{3!n^{3/2}} + \cdots.$$

Next we need to find $\lim_{n\to\infty} M_Y(t)$. To do this we find $\lim_{n\to\infty} \log[M_Y(t)]$ and exponentiate this result to obtain $\lim_{n\to\infty} M_Y(t)$.

$$\ln[M_Y(t)] = n \log\left[1 + \frac{t^2}{2n} + \frac{t^3 E(Y_1^3)}{3!n^{3/2}} + \cdots\right]$$

$$= n \log[1 + w],$$

where $w = t^2/2n + t^3 E(Y_1^3)/3!n^{3/2} + \cdots$. Recall from calculus the series expansion $\log(1 + w) = (w - w^2/2 + \cdots)$, for $|w| < 1$. Clearly, when n is large, $|w| < 1$. Then

$$\ln[M_Y(t)] = n\left[\left(\frac{t^2}{2n} + \frac{t^3 E(Y_1^3)}{3!n^{3/2}} + \cdots\right) - \frac{1}{2}\left(\frac{t^2}{2n} + \frac{t^3 E(Y_1^3)}{3!n^{3/2}} + \cdots\right)^2 + \cdots\right].$$

When we take the limit as $n \to \infty$, all terms but the first go to zero. Therefore,

$$\lim_{n\to\infty} \ln[M_Y(t)] = \frac{t^2}{2},$$

$$\lim_{n\to\infty} M_Y(t) = e^{t^2/2}.$$

This is the moment-generating function of a standard normal random variable. Thus, for large n, $\sum_{i=1}^{n}(X_i - n\mu)/\sigma\sqrt{n}$ is approximately a standard normal random variable. Consequently, $\sum_{i=1}^{n} X_i$ has an approximate normal distribution with mean $n\mu$ and standard deviation $\sigma\sqrt{n}$, for large n.

Exercises 8.2

8.2-1 Assume the distribution of the lifetime of a light bulb is exponential with a mean of 900 hours. Each time a bulb fails it is replaced by a new one identical in characteristics to the old one.

a Find the expected value and standard deviation of the total lifetime of 20 such bulbs.

b Find the probability that the total lifetime of 20 bulbs exceeds 22,000 hours.

c Find the probability that the total lifetime of 20 bulbs is less than 16,000 hours.

8.2-2 Suppose we obtain a random sample of 400 uniform $[0, 4]$ random variables. Let $\bar{T}$ denote the sample mean. Find the probability that $\bar{T}$ is within .06 unit of its expected value.

8.2-3 The profit in a business on any given day has an expected value of \$300 and a standard deviation of \$50.

a Find the probability that the total profit in 30 days of business exceeds \$9500.

b Find the probability that the average profit in 30 days is less than \$290.

8.2-4 Let X denote the sum of the outcomes of 10 tosses of a fair die.

a Use the Central Limit Theorem to find $P(X > 30)$.

b Simulate 100 values of X and determine the fraction of outcomes for which $X > 30$. Compare to the results in part (a).

c Make a histogram of the 100 simulated values to verify that it has the characteristic "bell-shaped" appearance of a normal distribution. Use six intervals in constructing the histogram. Choose the vertical scale so that the area of each rectangle in the histogram equals the fraction of observations in the corresponding interval. Plot the approximate normal distribution along with the histogram. The appearance of the two should be similar.

8.2-5 The weights of frozen pizzas should have a mean weight of 16 ounces and a standard deviation of 1 ounce. Sometimes the mean weight will drift away from the target value of 16. To see whether this is happening, an inspector will select and weigh a random sample of 25 pizzas coming from the production line. A common practice in quality control is to stop the process for repairs when the sample mean is more than $3\sigma/\sqrt{n}$ above or below the target value. Suppose that is done in this case.

a What is the probability that the process will be stopped if the mean of the process is at its target value of 16?

b Suppose that the mean of the process has drifted off target and is now at 16.2 ounces. What is the probability that the process will be stopped?

8.2-6 Let $X = X_1 + X_2 + \cdots + X_n$, where the X_i 's are a random sample from a uniform $[0, 1]$ distribution. Simulate 1000 values of X for $n = 3, 6, 12$. Make histograms of the outcomes. Based on your simulation, what value of n would you recommend in applying the Central Limit Theorem?

8.2-7 Let $\bar{X}$ denote the sample mean of a random sample of size 30 from an exponential distribution with mean $\mu = 1$. Simulate 1000 such sample means.

a Make a histogram and comment on how bell shaped it appears to be.

b Find the fraction of $\bar{X}$'s in the sample of 1000 that are greater than 1.2. Compare with the normal approximation for $P(\bar{X} > 1.2)$.

c Compute the sample mean and sample standard deviation of the 1000 $\overline{X}$'s. Compare these sample values with $E(\overline{X}) = 1$ and $\text{STD}(\overline{X}) = 1/\sqrt{30} = .18$.

8.2-8 Ten percent of the individuals in a certain population are left-handed. In a random sample of 300 find the probability that more than 12% are left-handed.

8.2-9 A baseball player has a probability of .30 of getting on base for each time at bat. Assume that the outcomes are independent from one time at bat to the next. If the player is at bat 400 times, what is the probability he will get on base less than 28% of the time?

SECTION

8.3 Confidence Intervals for Means

Let $X_1, X_2, \ldots, X_n$ be a random sample. Assume that the probability distribution of the X_i's depends on a parameter θ whose value is unknown. In Section 6.6, we discussed how to obtain an estimate of θ by using the method of moments. Such an estimate is called a *point estimate* since it is a single number. Here we are interested in obtaining an *interval estimate* of θ,— that is, a random interval that has a prescribed probability of "capturing" θ.

Definition 8.3-1

A $100p\%$ *confidence interval* for θ is a random interval

$$(L(X_1, X_2, \ldots, X_n), U(X_1, X_2, \ldots, X_n))$$

whose endpoints depend on the random sample such that

$$P(L(X_1, X_2, \ldots, X_n) < \theta < U(X_1, X_2, \ldots, X_n)) = p.$$

The lower and upper endpoints of the confidence interval must be a function of the random sample $X_1, X_2, \ldots, X_n$ and known constants. In particular, the endpoints must not depend on unknown parameters. The value of p is typically .95, so the interval has a large probability of capturing θ.

Suppose that we have a random sample from a normal distribution. We will derive a 95% confidence interval for μ when the numerical value of the standard deviation σ is known. Since the sample mean $\overline{X}$ has a normal distribution with mean μ and standard deviation $\sigma/\sqrt{n}$, it follows that the random variable

$$Z = \frac{\overline{X} - \mu}{\sigma/\sqrt{n}}$$

has a standard normal distribution. Since a standard normal random variable Z has a probability of .95 of falling in the interval (−1.96, 1.96), we have

$$P\left(-1.96 < \frac{\bar{X} - \mu}{\sigma/\sqrt{n}} < 1.96\right) = .95.$$

Rearranging the inequality so that μ is in the middle, we have

$$P\left(\bar{X} - \frac{1.96\sigma}{\sqrt{n}} < \mu < \bar{X} + \frac{1.96\sigma}{\sqrt{n}}\right) = .95.$$

It follows that the random interval $(\bar{X} - 1.96\sigma/\sqrt{n}, \bar{X} + 1.96\sigma/\sqrt{n})$ has a probability of .95 of capturing the mean μ. Since the standard deviation σ is known, numerical values for the lower and upper endpoints of the interval can be obtained upon observing a random sample. Thus, the interval is a 95% confidence interval for μ. For other levels of confidence, we simply replace the 1.96 by the appropriate value from a normal distribution. For instance, replace 1.96 by 2.58 for a 99% confidence interval and by 1.65 for a 90% confidence interval.

EXAMPLE 8.3-1 A certain college gives an entrance examination in mathematics. Experience has shown that the score of a randomly selected student is a normal random variable with a mean of 75 and a standard deviation of 8. This year a sample of 25 students was tutored prior to taking the examination. The average score of this group of students was $\bar{X} = 81$. Suppose that the tutoring may change the mean but not the standard deviation of the distribution. Under this assumption, a 95% confidence interval for the mean examination score after tutoring is

$$\left(81 - 1.96\frac{8}{\sqrt{25}}, 81 + (1.96)\frac{8}{\sqrt{25}}\right) = (78, 84).$$

Thus, we say with 95% confidence, tutoring will increase the mean score from its present value of 75 to somewhere between 78 and 84.

When we say that we have 95% confidence, we are making a statement about the procedure we have used to construct the interval. We are saying that among all possible intervals we could make for a parameter through repeated sampling, on average 95 out of 100 will contain the parameter, and 5 out of 100 will not. In Example 8.3-1, we can imagine taking many samples of size 25, each time making a 95% confidence interval. Of those intervals, an average of 95 out of 100 will capture the mean, and the other 5% will not. Whether the particular interval (78, 84) contains the mean, one never knows; however, one has confidence it does.

How can we construct a confidence interval for a mean when either the standard deviation σ is not known or when the random sample does not have a normal distribution? Fortunately, for large samples there is a way to obtain approximate confidence intervals that has proven to be satisfactory in many practical situations. When the sample size is sufficiently large, the Central Limit Theorem tells us that the sample mean $\bar{X}$ has an approximate normal distribution. Thus, when the standard deviation is known, the derivation for the confidence interval is essentially no different than if the distribution of the random sample were normal. The only caution is rather than being able to assert that the confidence level is $100p\%$, the best we can say is that it is approximately $100p\%$. As far as the standard deviation is concerned, if the sample size is large, the sample standard deviation S may be substituted for σ with no appreciable effect on the level of confidence. Thus, an approximate 95% confidence interval for a mean, when sample sizes are sufficiently large, is $(\bar{X} - 1.96S/\sqrt{n}, \bar{X} + 1.96S/\sqrt{n})$.

EXAMPLE 8.3-2 An experiment was conducted to determine the mean time it would take an automobile oil pump to fail under high-stress racing conditions. A random sample of 30 such pumps gave a sample mean of 205 hours and a sample standard deviation of 41 hours. The approximate 95% confidence interval for the mean time to failure of the automobile pump is

$$\left(205 - 1.96\frac{41}{\sqrt{30}}, 205 + 1.96\frac{41}{\sqrt{30}}\right) = (190, 220).$$

Thus, the average time to failure of a pump will be somewhere between 190 hours and 220 hours with approximately 95% confidence.

The large sample confidence interval can be used to find approximate confidence intervals for the probability of success in a binomial experiment. Let $X_1, X_2, \ldots, X_n$ denote n independent and identically distributed Bernoulli random variables with $E(X_i) = p$, where p is the probability of success. The sample mean of this random sample is just the sample proportion $\hat{P} = \sum_{i=1}^{n} X_i/n$. The sample variance is

$$\begin{aligned} S^2 &= \frac{(\text{number of } X_i\text{'s} = 0)(0 - \hat{P})^2 + (\text{number of } X_i\text{'s} = 1)(1 - \hat{P})^2}{n} \\ &= (1 - \hat{P})(\hat{P})^2 + (\hat{P})(1 - \hat{P})^2 \\ &= \hat{P}(1 - \hat{P}) \end{aligned}$$

Thus, an approximate 95% confidence interval for the probability p is given by

$$\left(\hat{P} - 1.96\sqrt{\frac{\hat{P}(1-\hat{P})}{n}},\ \hat{P} + 1.96\sqrt{\frac{\hat{P}(1-\hat{P})}{n}}\right).$$

EXAMPLE 8.3-3 A sample of 150 hospital records show that 35 are incorrectly filled out. For this sample $\hat{P} = 35/150 = .23$. An approximate 95% confidence interval for the probability that a record is incorrect is

$$\left(.23 - 1.96\sqrt{\frac{.23\,(1-.23)}{150}},\ .23 + 1.96\sqrt{\frac{.23\,(1-.23)}{150}}\right) = (.16, .30).$$

Thus, based on the information in the sample, there is somewhere between a 16% and a 30% chance that a record is incorrect, with approximate 95% confidence.

A confidence interval for a mean when samples are small and the standard deviation is unknown can be obtained if the random variables in the sample have a normal distribution. The form of the confidence interval is $(\bar{X} - c(n)S/\sqrt{n},\ \bar{X} + c(n)S/\sqrt{n})$, where $c(n)$ is an appropriate constant depending on the sample size. For a 95% confidence interval $c(n) > 1.96$. If n is large, say 30 or greater, setting $c(n) = 1.96$ will have negligible effect on the confidence interval. The constant $c(n)$ for a 90%, 95%, and 99% confidence interval and for selected values of n are given in Table 8.3-1. The derivation is based on a distribution called Student's t and may be found in Hogg and Craig (1978), as well as other books on mathematical statistics.

TABLE 8.3-1 Values of $c(n)$

	Sample Size						
Level of Confidence	**5**	**10**	**15**	**20**	**25**	**30**	∞
90	2.38	1.93	1.82	1.77	1.75	1.73	1.65
95	3.10	2.38	2.22	2.15	2.11	2.08	1.96
99	5.15	3.43	3.08	2.94	2.86	2.80	2.58

EXAMPLE 8.3-4 Five samples of a plastic used in packaging were tested for strength. The pounds per square inch required to break each sample are 10.3, 10.6, 10.7, 10.0, 10.8. The sample mean is 10.5, and the sample standard deviation is 0.3. Assuming the strengths

are normally distributed, the 95% confidence interval is

$$\left(10.5 - \frac{2.776(.3)}{\sqrt{5}},\ 10.5 + \frac{2.776(.3)}{\sqrt{5}}\right) = (10.1, 10.9).$$

Exercises 8.3

8.3-1 The time it takes for a certain manufacturing company to assemble a computer is a normal random variable with a mean of 1.2 hours and a standard deviation of 0.2 hour. New procedures are put in place to lower assembly time. A random sample of the 35 items produced under the new procedures gave a mean assembly time of 0.9 hour. Form a 95% confidence interval for the mean time to assemble a computer under the new procedures. Assume the standard deviation remains unchanged.

8.3-2 A random sample of 50 shoppers at a local convenience store showed that they spent an average of \$15.50 with a standard deviation of \$6.25. Form a 95% confidence interval for the average amount spent by a shopper at this store.

8.3-3 Nutritional information on a packaged food product shows the product contains 12 grams of fat. A random sample of five such packages gave readings of grams of fat as $\{12.3, 11.8, 13.2, 12.2, 13.5\}$. Assume that the grams of fat have a normal distribution and form 90%, 95%, and 99% confidence intervals for the mean grams of fat. Based on these data, can we say that labeling on the package is misleading?

8.3-4 Heights of the general male population are normally distributed with a mean of 70 inches and a standard deviation of 3 inches. A certain Texas politician claims that people grow taller in Texas. A random sample of 20 males from Texas gave the following data on heights:

75, 77, 70, 69, 67, 74, 73, 73, 70, 69, 66, 67, 71, 72, 72, 71, 68, 70, 68, 71.

a Form 95% and 99% confidence intervals for the mean of the population, using the known standard deviation of 3.

b Form 95% and 99% confidence intervals for the mean of the population, using the sample standard deviation.

c Comment on the politician's claim.

8.3-5 A random sample of 150 elm trees in a certain area showed that 33 had the Dutch Elm disease. Form an approximate 95% confidence interval for the fraction of trees in this area with the Dutch Elm disease.

8.3-6 A survey of 1500 credit card customers showed that 510 pay their bills in full each month. Form an approximate 90% confidence interval for the fraction of customers in the population who pay their bills in full each month.

8.3-7 Simulate a random sample of size 30 from a uniform $[0, 1]$ distribution.

a Form an approximate 95% confidence interval for the mean. Use the true standard deviation of $1/\sqrt{12}$ in constructing the confidence interval. Does the confidence interval capture the true mean of 0.5?

b Repeat the procedure above 1000 times. What percentage of these confidence intervals contains the true mean of .5? This percentage should be approximately 95%.

8.3-8 This exercise investigates the sensitivity of the confidence interval formula to violation of distributional assumptions. Simulate a random sample of size 5 from an exponential distribution with a mean of 1. Use the 95% confidence interval formula for the normal distribution with known standard deviation to construct a confidence interval for the mean. Repeat this procedure 1000 times, and determine the percentage of the intervals that actually contains the true mean of 1. If this percentage is near 95%, then the fact that the observations were not from a normal distribution would not be of great importance in this case. On the other hand, if the percentage is not near 95%, then using the normal distribution confidence interval for small samples from the exponential distribution would not be recommended.

8.3-9 Suppose $X_1, X_2, \ldots, X_m$ and $Y_1, Y_2, \ldots, Y_n$ are independent random samples from different distributions. Let μ_x and μ_y denote the means of the X_i's and Y_i's, respectively, and let σ_x and σ_y denote their respective standard deviations.

a Show that

$$E(\bar{X} - \bar{Y}) = \mu_x - \mu_y \quad \text{and} \quad \text{VAR}(\bar{X} - \bar{Y}) = \frac{\sigma_x^2}{m} + \frac{\sigma_y^2}{n}.$$

b If the X's and Y's have a normal distribution and if the standard deviations of the X's and Y's are known, show that a 95% confidence interval for $\mu_x - \mu_y$ is

$$\left(\bar{X} - \bar{Y} - 1.96\sqrt{\frac{\sigma_x^2}{m} + \frac{\sigma_y^2}{n}},\ \bar{X} - \bar{Y} + 1.96\sqrt{\frac{\sigma_x^2}{m} + \frac{\sigma_y^2}{n}} \right).$$

c How would the confidence interval be modified for unknown standard deviations and large samples from unknown distributions?

8.3-10 A researcher studied the hourly wage among workers in two different industries. A random sample of 50 from industry A gave a mean hourly wage of $19.50 per hour with a standard deviation of $4.25 per hour. A random sample of size 75 from industry B gave a mean hourly wage of $14.30 per hour and a standard deviation of $3.60 per hour. Use the results of Exercise 8.3-9 to find an approximate 95% confidence interval for the difference between the mean hourly wages of the two industries.

SECTION

8.4 A Random Sum of Random Variables

In this section, we consider sums of random variables in which the number of terms in the sum is also random. Such random processes occur in many areas of application.

Definition 8.4-1

Let $N(t)$ be a Poisson process with rate λ, and let $Y_1, Y_2, \ldots, Y_n$ be independent and identically distributed random variables with mean μ and standard deviation σ. Assume that the Y_i 's are independent of $N(t)$. A process $X(t)$ is said to be a *compound Poisson process* if

$$X(t) = \sum_{i=1}^{N(t)} Y_i.$$

The process $X(t)$ is a random sum of random variables since the number of terms in the sum, $N(t)$, is itself random. Note that $X(t)$ is a continuous-time stochastic process. If the Y_i 's are continuous random variables, then $X(t)$ has continuous states.

EXAMPLE 8.4-1 Let the number of automobile accidents in a city be a Poisson process $N(t)$, and let Y_i be the expense to the city for handling the ith accident. Then the total expense to the city for handling accidents occurring in the interval $[0, t]$ is a compound Poisson process $X(t)$.

EXAMPLE 8.4-2 Suppose the number of pledge calls received by a public television station during a pledge drive is a Poisson process $N(t)$. Let Y_i be the amount pledged by the ith caller. Then the total of the pledges $X(t)$ during the period $[0, t]$ is a compound Poisson process.

THEOREM 8.4-1 Let $X(t)$ be a compound Poisson process. Then

$$E[X(t)] = \lambda t\mu, \qquad \mathrm{VAR}[X(t)] = \lambda t\mu^2 + \lambda t\sigma^2.$$

Verification of this theorem is a direct application of Theorems 2.9-1 and 2.9-2. That is,

$$\begin{aligned} E[X(t)] &= E(E[X(t) \mid N(t)]) \\ &= E(N(t)\mu) \\ &= \lambda t\mu, \end{aligned}$$

and

$$\begin{aligned}\mathrm{VAR}[X(t)] &= E(\mathrm{VAR}[X(t) \mid N(t)]) + \mathrm{VAR}(E[X(t) \mid N(t)]) \\ &= E[N(t)\sigma^2] + \mathrm{VAR}[N(t)\mu] \\ &= \lambda t\sigma^2 + \lambda t\mu^2.\end{aligned}$$

EXAMPLE 8.4-3 The number of students using a check-cashing service at the Student Union is a Poisson process with a rate of $\lambda = 25$ per hour. The amount of each check is a random variable with a mean of \$30 and a standard deviation of \$5. The total dollar amount $X(t)$ of checks cashed in an 8-hour day has a mean of $E[X(t)] = (25)(8)(30) =$ \$6000. The variance is $\mathrm{VAR}[X(t)] = (25)(8)[30^2 + 5^2] = 185{,}000$, and the standard deviation is \$430.12. If the union has \$6860 on hand, the demand should be met most of the time.

EXAMPLE 8.4-4 In Example 8.4-3, there was an average of $(25)(8) = 200$ students per 8-hour period who cashed a check. Suppose that the number of students was not random but was exactly 200 per day. The mean dollar amount of the checks cashed would be \$6000 as before. However, the variance would be $(200)5^2 = 5000$, and the standard deviation would be \$71. Thus, randomness in the number of students has the effect of increasing the standard deviation about six times over what it would be if there were always the same number of students cashing checks.

EXAMPLE 8.4-5 The number of calls for service received by a plumber is a Poisson process with a rate of $\lambda = 4$ per day. The amount of time it takes a plumber to answer a call is an exponential random variable with a mean of 2 hours. The number of service hours in a five-day period has a mean of $(4)(5)(2) = 40$ hours, a variance of $(4)(5)(8) = 160$, and a standard deviation of 12.6 hours.

THEOREM 8.4-2 If $X(t)$ is a compound Poisson process, then for large t, $X(t)$ has an approximate normal distribution with mean and variance as given by Theorem 8.4-1.

This result is intuitively reasonable. For large t, $N(t)$ will be large, and therefore $X(t)$ will be the sum of numerous independent and identically distributed random variables. Thus, the Central Limit Theorem (Theorem 8.2-1) would suggest that $X(t)$ has an approximate normal distribution.

EXAMPLE 8.4-6 The number of defective items produced by a manufacturing process is a Poisson process with a rate of $\lambda = 10$ per hour. The cost of repairing each defective item is a random variable with a mean of \$8 per item and a standard deviation of \$2 per item. The dollar amount of repairing the defective items in a 24-hour period is approximately normally distributed with a mean of $(10)(24)(8) = \$1920$, a variance of $(10)(24)[8^2 + 2^2] = 16{,}320$, and a standard deviation \$127.75. The probability that the dollar amount of repairs in 24 hours exceeds \$2000 is

$$\begin{aligned} P[X(24) > 2000] &\approx P\left[Z > \frac{2000 - 1920}{127.75}\right] \\ &= P[Z > 0.63] \\ &= .2643. \end{aligned}$$

Exercises 8.4

8.4-1 The number of customers buying shoes at a discount shoe store during a weekday is a Poisson process with a rate of nine per hour. The average purchase is \$40 with a standard deviation of \$12. Find the expected value and standard deviation for the total sales over an 8-hour day.

8.4-2 The number of cars coming to a parking garage is a Poisson process with a rate of 30 per hour. The amount charged for parking depends on the time the car spends in the garage. The mean amount is \$6.25, and the standard deviation is \$1.10. Find the expected value and standard deviation for the total amount of money collected over a 12-hour period.

8.4-3 The number of calls for service to the telephone company in a certain area is a Poisson process with a rate of 15 per hour. The time to repair the problem is a random variable with a mean of 90 minutes and a standard deviation of 15 minutes.

- **a** What are the expected value and standard deviation for the total repair time over an 8-hour period?
- **b** The company has enough technicians to cover 200 hours of repair during an 8-hour period. What is the probability that there are enough technicians to cover the required work?

8.4-4 The number of boats for which a certain drawbridge must be raised is a Poisson process with a rate of 0.8 per hour. The average time that the bridge is up is 7 minutes with a standard deviation of 2 minutes. The bridge is raised and lowered individually for each boat. What is the probability that the bridge is up more than 60 minutes in a 12-hour period?

8.4-5 Suppose that it is desired to simulate the traffic flow in a downtown area, and as part of the program it is necessary to construct a module to simulate the raising and lowering of a drawbridge. Suppose that the information in Exercise 8.4-4 applies. Also assume that the amount of time the bridge is up each time is a normal random variable and that the bridge, when it has been lowered, cannot be raised again for 5 minutes. For instance, if the first boat arrives at 9:05 and the bridge is open for 8 minutes, then the earliest a second boat can pass under the bridge is 9:18. Write a program that simulates the raising and lowering of the drawbridge for a 12-hour period. The program should be able to indicate whether the bridge is up or down at any time during the 12 hours, and it should record the total time the bridge is up.

9

Selected Systems Models

1-2.

SECTION

9.1 Distribution of Extremes

The maximum and minimum values of a set of independent random variables are often of interest in modeling various processes. We begin with a theorem for the distribution of the maximum.

THEOREM 9.1-1 Let $X_1, X_2, \ldots, X_n$ be independent random variables with cumulative distribution functions $F_1(x), F_2(x), \ldots, F_n(x)$, respectively. Let $X_{max} = \max(X_1, X_2, \ldots, X_n)$. Then the cumulative distribution function $F_{max}(x)$ of X_{max} is

$$F_{max}(x) = F_1(x) F_2(x) \cdots F_n(x)$$

To see this, we note that $X_{max} \le x$ if and only if $X_i \le x$ for all i. Thus,

$$F_{max}(x) = P(X_1 \le x, X_2 \le x, \ldots, X_n \le x)$$

$$= \prod_{i=1}^{n} P(X_i \le x)$$

$$= \prod_{i=1}^{n} F_i(x).$$

EXAMPLE 9.1-1 Computers with parallel processors allow several sets of instructions to be executed at the same time. Suppose that a certain type of task submitted to a computer with parallel processors can be split into four subtasks each being handled by a different processor. Assume that the time T_i it takes to complete the ith subtask, $i = 1, \ldots, 4$, is an exponential random variable with a mean of 5 seconds. When all subtasks are completed, the task is completed. The time T it takes for this to happen may be expressed as

$$T = \max(T_1, T_2, T_3, T_4).$$

Applying Theorem 9.1-1, we find that the cumulative distribution function of T is

$$F(t) = (1 - e^{-t/5})^4.$$

The probability density function of T is

$$f(t) = \frac{d}{dt}F(t)$$

$$= \frac{4}{5}e^{-t/5}(1 - e^{-t/5})^3, \qquad t \ge 0.$$

A calculus exercise shows that $E(T) = 10.4$ seconds and $\text{STD}(T) = 6.0$ seconds (Exercise 9.1-3).

Suppose instead of being processed in parallel the four subtasks are processed sequentially with the distribution of times to complete the four subtasks being the same as before. In this case, the total time to complete the task is

$$T_S = T_1 + T_2 + T_3 + T_4.$$

In this case, $E(T_S) = 20$ seconds and $\text{STD}(T_S) = 10$ seconds. Thus, on average, a sequential processor would take about twice as long as the parallel processors to complete this task, and the standard deviation would be about 68% greater.

The following theorem gives the distribution of the minimum of a set of independent random variables.

THEOREM 9.1-2 Let X_i, $i = 1, 2, \ldots, n$, satisfy the conditions of Theorem 9.1-1. Let $X_{\min} = \min(X_1, X_2, \ldots, X_n)$. The cumulative distribution function $F_{\min}(x)$ of $X_{\min}$ is

$$F_{\min}(x) = 1 - (1 - F_1(x))(1 - F_2(x)) \cdots (1 - F_n(x)).$$

To see this, notice that $X_{\min} > x$ if and only if $X_i > x$ for all i. Thus,

$$\begin{aligned} P(X_{\min} > x) &= P(X_1 > x, X_2 > x, \ldots, X_n > x) \\ &= \prod_{i=1}^{n} P(X_i > x) \\ &= \prod_{i=1}^{n} [1 - F_i(x)], \end{aligned}$$

and $F_{\min}(x) = 1 - P(X_{\min} > x)$.

EXAMPLE 9.1-2 The amount that a contractor will bid on a certain project is a random variable that has a uniform [10, 30] distribution expressed in thousands of dollars. The cumulative distribution function is

$$F_x(x) = \begin{cases} 0, & x < 10, \\ \dfrac{x - 10}{20}, & 10 \le x \le 30, \\ 1, & x > 30. \end{cases}$$

Suppose that n contractors bid independently on the project. Let X_i denote the bid of the ith contractor. Then the low bid can be expressed as

$$X_{\min} = \min(X_1, X_2, \ldots, X_n).$$

The distribution function of the low bid is

$$F_{\min}(x) = \begin{cases} 0, & x < 10, \\ 1 - \left[1 - \dfrac{x-10}{20}\right]^n, & 10 \le x \le 30, \\ 1, & x > 30. \end{cases}$$

As would be expected, lower bids are more likely to occur when there are more bidders. For instance, if there are $n = 3$ bidders the probability the low bid is below 12,000 is $F_{\min}(12) = .27$. However, if there are $n = 6$ bidders, then $F_{\min}(12) = .47$.

EXAMPLE 9.1-3 The lifetimes of automobile batteries of brand A and brand B are exponentially distributed with means of three and four years, respectively. A delivery company buys one of each brand and installs them in their delivery vans. Let T be the time elapsed before one or the other battery has to be replaced. We will find the probability distribution of T. If T_A and T_B denote the lifetimes of brand A and brand B batteries, then $T = T_{\min} = \min(T_A, T_B)$. Thus,

$$\begin{aligned} F_T(t) &= 1 - e^{-t/3}e^{-t/4} \\ &= 1 - e^{-7t/12}, \quad t > 0. \end{aligned}$$

It follows that T has an exponential distribution with mean $12/7 = 1.7$ years. It may be surprising that the expected time for the first replacement is about half the mean lifetime of the shorter-life battery. On the other hand, the time until the second battery fails will have a mean lifetime longer that the individual lifetimes of the two batteries. The longer of the two lifetimes is $T_{\max} = \max(T_A, T_B)$ and

$$F_{\max}(t) = (1 - e^{-t/3})(1 - e^{-t/4}), \quad t > 0.$$

Unlike the minimum, the maximum does not have an exponential distribution. It can be shown that the mean of this distribution is $37/7 = 5.3$ years. Now since $T_{\min} + T_{\max} = T_A + T_B$, it follows that the total expected lifetime of the two batteries is equal to the sum of the individual expected lifetimes, namely seven years. However, one will fail on average much earlier than the mean of either of the two individual batteries and one will fail on average much later.

Exercises 9.1

9.1-1 Five hikers decide to split up and take separate trails each leading to the same destination. The amount of time it takes to complete each trail is a uniform random variable on the interval 35 to 60 minutes.

a Find the cumulative distribution function of the amount of time it takes all to reach the destination.

b Find the expected amount of time it takes all to reach the destination.

9.1-2 A company dispatches 10 vans at 8 A.M. to pick up packages from customers. When all vans have returned, the packages are immediately placed on a truck for shipment elsewhere. The amount of time each van spends on its route is an exponential random variable with a mean of 2 hours. What is the probability that the truck will be ready to go by noon?

9.1-3 Evaluate the expected value in Example 9.1-1.

9.1-4 The time it takes each of three workers of equal ability to finish a certain task is an exponential random variable with a mean of 2 hours. Suppose management decides to evaluate the workers by having them each perform the task once and ranking them on the basis of the results. Let T denote the best time of the three, and let X denote the worst time.

a Find the probability distribution of T and its expected value.

b Find the probability distribution of X and its expected value.

c What do the computations suggest about one-time, or short-term, evaluation of employees?

9.1-5 Three subcontractors have been designated to build different parts of an experimental aircraft. It has been determined that the number of months it will take each to finish its part is a Weibull random variable. The value of β in the Weibull distribution is .5 for all three, and the values of α are 1, 2, and 4, respectively, for the three subcontractors. All three parts are necessary before final assembly of the aircraft can take place. How many months will the main contractor have to wait to be 90% certain that all three parts are finished?

9.1-6 The initial planning of a shopping complex involves three activities that can take place at the same time: land acquisition, site planning, obtaining commitments from major department stores. The number of months it will take to complete each of these activities can be expressed as $12 + X$, where X is an exponential random variable with a mean of six months. Let T be the amount of time it will take for all activities to be completed.

a Find the cumulative distribution function and probability density function of T.

b Find the expected time to complete the project.

9.1-7 An instructor gives three 1-hour examinations and a final. The grade in the course is determined by dropping the lowest score of the 1-hour examinations and obtaining the sum X of the remaining three scores. The distribution of scores on each 1-hour examination is a normal random variable with a mean of 75 and a standard deviation of 12. The score on the final is a normal random variable with a mean of 80 and a standard deviation of 8. Use simulation to determine the distribution of X. If it takes a score of 270 to get an A, estimate the percentage of the students that will get an A.

9.1-8 Suppose that $X_1, X_2, \ldots, X_n$ are independent and identically distributed random variables each with cumulative distribution function $F(x)$. Let Y_k denote the kth smallest of the X_i's.

Note that $Y_1 = X_{\min}$ and $Y_n = X_{\max}$. The random variable Y_k is called the *kth order statistic*. Find the cumulative distribution function of Y_k. (*Hint*: $Y_k \le x$ if and only if k or more of the X_i's are less than or equal to x.)

SECTION

9.2 Scheduling Problems

In this section, we consider modeling processes that involve the scheduling of activities, some of which can take place at the same time and others of which depend on prior activities. An example is a manufacturing process in which components of a product are made simultaneously at different locations and then are brought together for final assembly. The time it takes to complete a set of scheduled activities is the quantity of interest. As an example of a scheduling problem we will model the establishment of a new office. This example is one for which the critical path method (CPM) of analysis based on fixed times for activities would traditionally be used. However, we include randomness in our model.

The Little Green Computing Machine (LGCM) Company has identified seven major activities that must take place in order for a new office to be established. The activities are

1 Obtain a lease.
2 Draw up remodeling plans.
3 Construction.
4 Order office furniture.
5 Deliver furniture.
6 Advertise for new positions.
7 Make hiring decisions.

Some activities must precede others, and some can proceed in parallel. For instance, the lease must be obtained before any other activity can start, but ordering the office furniture can be done at the same time that hiring takes place. A schematic diagram of the sequence of activities and the interrelationships among them is shown in Figure 9.2-1. All activities must be completed before LGCM Company can move into the new office.

FIGURE 9.2-1
Activities to be Completed to Move into a New Office

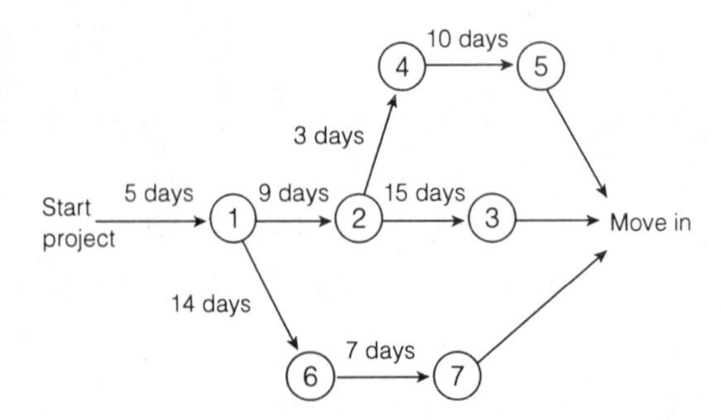

We first consider the nonrandom case. That is, the times to complete each activity are assumed to be fixed and known in advance. In Figure 9.2-1, the time to complete each activity is indicated along the solid arrow leading into the activity. For example, the amount of time to complete the leasing agreement is 5 days, whereas the amount of time to complete construction is 15 days.

Three paths of activities must be completed before LGCM can move in. These paths, along with the amount of time to complete each one, are given in Table 9.2-1.

TABLE **9.2-1**
Fixed-Time Paths for LGCM

Paths	Time (days)
Start → 1 → 2 → 4 → 5 → move in	27
Start → 1 → 2 → 3 → move in	29
Start → 1 → 6 → 7 → move in	26

Since the maximum time for the three paths is 29 days, according to this fixed-time analysis, it will take 29 days to complete the project.

More realistic assumptions would allow the times to complete the activities to be random variables. Let T_i denote the time to complete the ith activity, $i = 1, 2, \ldots, 7$. Assume that all the T_i's except T_6 have exponential distributions with means equal to the completion times in the nonrandom case. For instance, T_1 has an exponential distribution with a mean of five days. The time T_6 is fixed with $T_6 = 14$ days to reflect amount of advertising time. The times to complete the three paths are given in Table 9.2-2.

TABLE **9.2-2**
Random Time Paths for LGCM

Route	Time
Start → 1 → 2 → 4 → 5 → move in	$T_1 + T_2 + T_4 + T_5$
Start → 1 → 2 → 3 → move in	$T_1 + T_2 + T_3$
Start → 1 → 6 → 7 → move in	$T_1 + T_6 + T_7$

Therefore, the time to complete the project is

$$T = \max[(T_1 + T_2 + T_4 + T_5), (T_1 + T_2 + T_3), (T_1 + T_6 + T_7)].$$

The probability distribution of T can be estimated by simulation. The results of one repetition of this experiment are given in Table 9.2-3.

TABLE **9.2-3**
Simulation of Exponential Random Time Paths for LGCM

Unit Random Number	Exponential Completion Times
.91646	$T_1 = -5 \ln(1 - .91646) = 12.4$
.89198	$T_2 = -9 \ln(1 - .89198) = 20.0$
.64809	$T_3 = -15 \ln(1 - .64809) = 15.7$
.16376	$T_4 = -3 \ln(1 - .16376) = 15.7$
.91782	$T_5 = -10 \ln(1 - .91782) = 25.0$
*****	T_6 *************** = 14.0
.53498	$T_7 = -7 \ln(1 - .53498) = 5.4$

The time to complete the project from this simulation is

$$T = \max(57.9, 48.1, 31.8) = 57.9 \approx (58) \text{ days.}$$

The results of 1000 repetitions of this experiment are given in Figure 9.2-2.

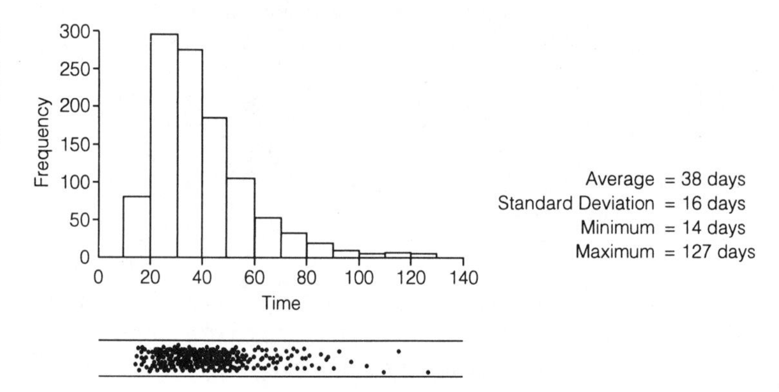

FIGURE 9.2-2 Simulated Distribution of T Assuming Exponential Distributions for the T_i's, $i = 1, 2, 3, 4, 5$, and 7

The amount of variability in the time needed to establish the new office is substantial. The average time is 38 days, and the standard deviation is approximately 16 days. The largest observed time to completion of the project in the 1000 repetitions of the experiment is approximately 127 days, and the smallest observed time is approximately 14 days. If the exponential distributions are the correct probability distributions for the various activities, it becomes apparent how misleading a fixed-time analysis could be. The actual time needed to establish the office may differ substantially from the time of 29 days determined by the fixed-time analysis.

A random-time analysis can give the company an assessment of the effects of random variability on the completion time of the project. One possible consequence of an analysis like this might be a decision by management to exercise tighter controls on the project so that the random variability predicted by this model does not occur. If tighter controls are not possible, the company must plan for the very real possibility that the project will take much longer to complete than the fixed-time analysis would suggest.

EXAMPLE 9.2-1 We reconsider the scheduling problem by considering the effect of changing the probability distributions for the T_i's. We assume that the T_i's have normal distributions except for T_6, which is again fixed at 14 days. The means of the normal distributions will be the same as those of the exponential distributions used before. In addition, we must specify the standard deviations.

Rather than our specifying the standard deviations directly, it is often more convenient to specify intervals within which the completion times for the various activities would likely fall. We choose each likely interval such that it has about a 95% chance of containing the actual completion time for the activity. For the normal distribution, such an interval is $(\mu - 2\sigma, \mu + 2\sigma)$, which has a length of 4σ. Therefore,

$$4\sigma \approx \text{Length of likely interval},$$

$$\sigma \approx \frac{\text{Length of likely interval}}{4}.$$

Suppose that the company comes up with the following likely intervals for the various activities given in Table 9.2-4.

TABLE **9.2-4** Likely Intervals of Activities for LGCM

Time	Likely Interval (days)	Interval Length	σ (days)
T_1	4 to 6	2	0.5
T_2	6 to 12	6	1.5
T_3	9 to 21	12	3.0
T_4	2 to 4	2	0.5
T_5	5 to 15	10	2.5
T_6 (fixed time)	14	—	—
T_7	4 to 10	6	1.5

Again the distribution of T can be estimated by simulation. The results of 1000 repetitions of the experiment are shown in Figure 9.2-3. Here there is a markedly different picture from what was obtained from using the exponential distributions. The average time to establish the office is approximately 30 days with a standard deviation of 3 days. The largest observed time in the 1000 repetitions was approximately 40 days, and the smallest observed time was approximately 22 days. The results are much closer to those obtained in the fixed-time analysis. The fact that these standard deviations are smaller than those of the exponential distributions in Figure 9.2-2 is largely responsible for the differences in the results between the two

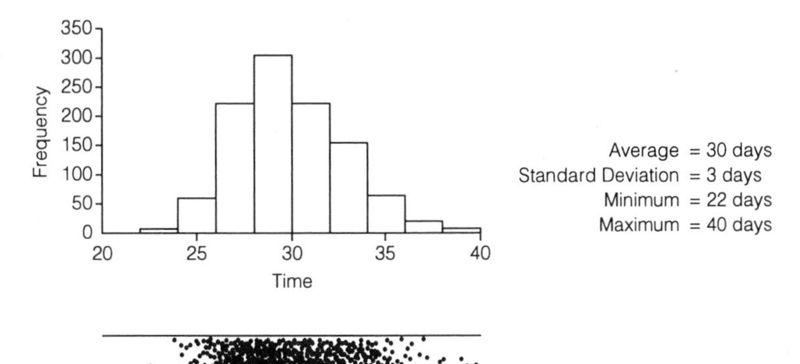

FIGURE **9.2-3** Simulate Distribution of T Assuming Normal Distributions for the Distributions for the T_i's, $i = 1, 2, 3, 4, 5$, and 7

analyses. The symmetrical nature of the normal distribution, with observations tending to cluster around the mean, is also a factor. This case would represent tighter management controls on the variability of the process.

Exercises 9.2

9.2-1 The Statistics Department is planning to participate in the University Open House scheduled for April 5. The display will consist of written material about the department and a computer demonstration. The following diagram shows the activities that must take place. The first organizational meeting is held March 10.

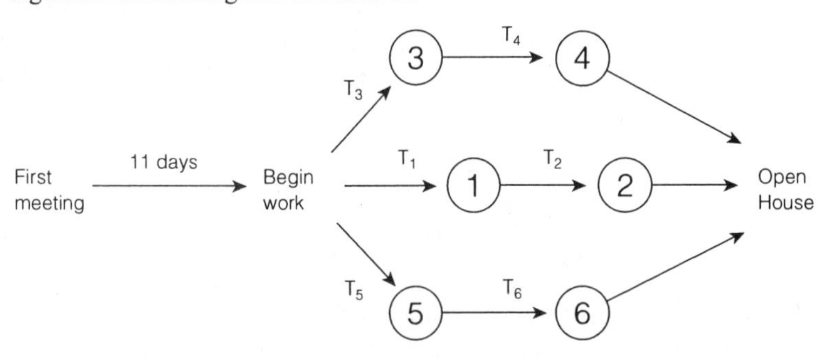

1 First draft of written material is prepared.
2 Final draft of written material is prepared.
3 Computer code for demonstration is written.
4 Computer code is tested and revised.
5 Graduate students are selected to work at Open House.
6 Work times for Open House are arranged.

a Give an equation in terms of the T_i's that represents the time to completion of this project.

b Suppose the T_i's are exponentially distributed random variables with the following means in days:

Time	T_1	T_2	T_3	T_4	T_5	T_6
Means	4	3	4	2	2	1

Based on 1000 simulations, find the sample mean, sample standard deviation, minimum days, and maximum days to complete this project. Do you think it is likely the Statistics Department will have its Open House display ready by April 5?

9.2-2 Modules of a newly developed computer program are related to each other according to the hierarchical diagram.

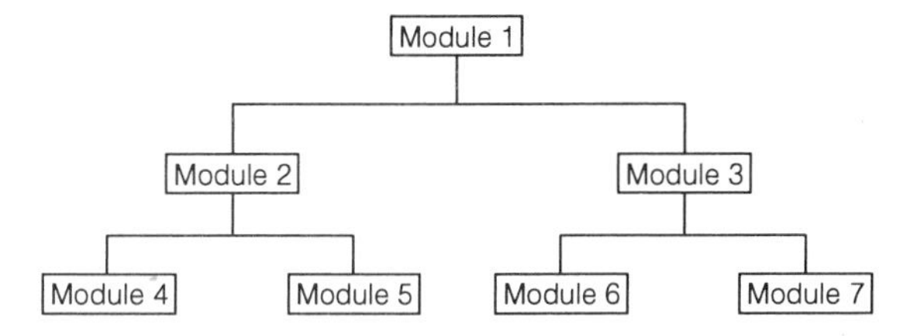

Testing of the program begins with the lowest-level modules (4, 5, 6, 7), which all begin at the same time. Testing on modules that are higher up begins when those below them have been tested. For instance, testing on module 2 begins after modules 4 and 5 have been tested. Let T_i denote the test times for module i, $i = 1, \ldots, 7$. The distributions of the T_i's are independent normal random variables. The likely ranges of these random variables are as follows, where time is expressed in days:

Variable	T_1	T_2	T_3	T_4	T_5	T_6	T_7
Likely range	3 to 6	4 to 8	4 to 8	6 to 8	6 to 8	7 to 9	7 to 9

a Express the time to complete testing on all modules in terms of the T_i's.

b Simulate 1000 values of the time to complete testing. Find the sample mean and sample standard deviation.

c Do a fixed-time analysis by assuming that the time to complete testing of the *i*th module is equal to the mean of T_i, $i = 1, \ldots, 7$. In this case, do you think the fixed-time analysis provides a realistic assessment of the time to complete testing?

9.2-3 A couple has located some land they plan to purchase and then build a house. It is now May, and they wonder if they will be moving into their new home by next May. The following diagram shows the activities that must take place between now and when they could move in:

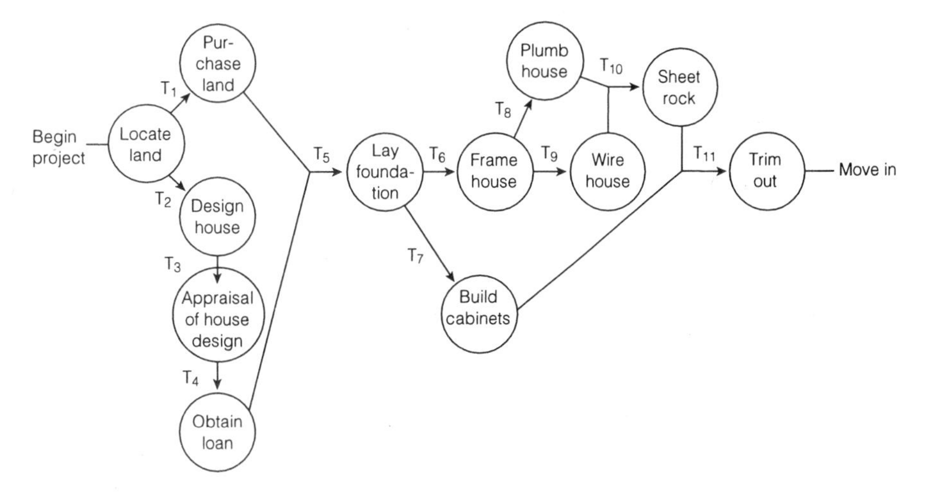

a Give an equation in terms of the T_i's that represents the time to complete this project.

b Suppose the T_i's are exponentially distributed random variables with the following means in weeks:

Time in weeks	T_1	T_2	T_3	T_4	T_5	T_6	T_7	T_8	T_9	T_{10}	T_{11}
Mean	9	6	3	1.5	2	3	8	2	2.5	3	5

Based on 1000 simulations, find the sample mean, sample standard deviation, minimum, and maximum number of weeks to complete the house. Do you think it is likely this couple will move in by the next May?

c Suppose some of the T_i's are normally distributed and some are exponentially distributed as given in the table. Based on these distributions do you think it is likely the couple will move in by May?

Time in weeks	T_1	T_2	T_3	T_4	T_5	T_6	T_7	T_8	T_9	T_{10}	T_{11}
Distribution	Normal	Exponential	Normal	Normal	Exponential	Normal	Normal	Exponential	Normal	Exponential	Normal
Mean or likely interval	(6,12)	6	(1,5)	(0,3)	2	(1,5)	(3,13)	2	(.5,4.5)	3	(2,8)

SECTION

9.3 Sojourns and Transitions for Continuous-Time Markov Processes

The scheduling problems in Section 9.2 studied the total time it took to complete the activities that made up a system. In this section we look more closely at what happens inside the system. We consider systems where states can be revisited and study the time the system spends in the individual states. Also of interest is the time between visits to a state. The types of systems discussed in this section are *continuous-time Markov processes*.

Definition 9.3-1

A process is a *continuous-time Markov process* if the following conditions are satisfied. Let i and j denote states of process.

i There exist nonnegative rates λ_{ij}, $i \neq j$, such that for frames of sufficiently small length Δ

$$P(i \to j) \approx \lambda_{ij}\Delta,$$

$$P(i \to i) \approx 1 - \sum_{i \neq j} \lambda_{ij}\Delta,$$

where the errors of approximation are negligible in comparison to terms involving Δ when Δ is small.

ii Transitions that occur in nonoverlapping frames are independent of one another.

We restrict our attention to continuous-time Markov processes with a finite number of states, denoted $1, 2, \ldots, s$. We may approximate such a process with a discrete-time Markov chain by choosing a frame size to be sufficiently small and using the expressions for $P(i \to j)$ in Definition 9.3-1 as the elements of the one-step transition matrix.

EXAMPLE 9.3-1 A robotic arm in a manufacturing plant moves among three stations during the manufacturing process. Assume the movement of the robot follows a Markov process with transition rates given in Figure 9.3-1, where the rates are expressed in expected number of moves per minute. For instance, the robot moves from station 1 to station 2 at the rate of $\lambda_{12} = 5$ times per minute and to station 3 at the rate of $\lambda_{13} = 3$ times per minute.

FIGURE 9.3-1 Transition Rates of the Robotic Arm

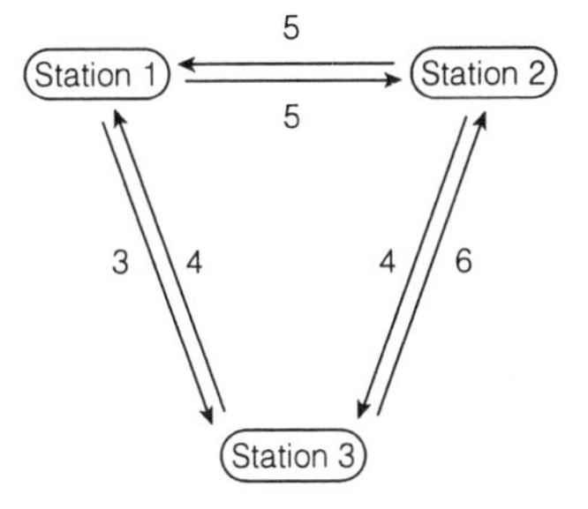

Suppose this process is approximated with Definition 9.3-1 with frame $\Delta = 1$ second. Then $P(1 \to 2) = 5/60$, $P(1 \to 3) = 3/60$, and $P(1 \to 1) = 1 - [(5/60 + (3/60)] = 52/60$. Note that the transition probability $P(1 \to 1)$ represents the probability the robot stays at station 1. Continuing in this way, the continuous-time Markov process is approximated by a Markov chain with the following transition matrix in which each step of the Markov chain denotes 1 second of time:

$$P = \begin{matrix} & \begin{matrix} 1 & 2 & 3 \end{matrix} \\ \begin{matrix} 1 \\ 2 \\ 3 \end{matrix} & \begin{bmatrix} \frac{52}{60} & \frac{5}{60} & \frac{3}{60} \\ \frac{5}{60} & \frac{51}{60} & \frac{4}{60} \\ \frac{4}{60} & \frac{6}{60} & \frac{50}{60} \end{bmatrix} \end{matrix}.$$

Properties of the chain may be used to study the continuous-time process. For example, the steady-state probabilities for states (stations) 1, 2, and 3 are .37, .38, and .26, respectively. These steady-state probabilities imply that the robot spends less time at station 3 than at the other stations.

Note that we would obtain the same steady-state probabilities for Example 9.3-1 regardless of the choice of Δ. The one-step transition matrix used to approximate the continuous-time process has the form

$$P = \begin{array}{c} \\ 1 \\ 2 \\ 3 \end{array} \begin{array}{c} \begin{array}{ccc} 1 & 2 & 3 \end{array} \\ \begin{bmatrix} 1-8\Delta & 5\Delta & 3\Delta \\ 5\Delta & 1-9\Delta & 4\Delta \\ 4\Delta & 6\Delta & 1-10\Delta \end{bmatrix} \end{array}.$$

Denoting the steady-state probability vector by $\pi = (\pi_1, \pi_2, \pi_3)$, we see that the Δ's cancel out of the equation $\pi P = \pi$, which is used to determine steady-state probabilities. Thus, the solution of this equation does not depend on Δ. Moreover, this system of equations may be put in the form

$$(\pi_1, \pi_2, \pi_3) \begin{bmatrix} -8 & 5 & 3 \\ 5 & -9 & 4 \\ 4 & 6 & -10 \end{bmatrix} = (0, 0, 0).$$

This suggests that the steady-state probabilities obtained in Example 9.3-1 are the steady-state probabilities for the continuous-time process and that these probabilities may be expressed as a solution of a system of equations involving the λ_{ij}'s. This is indeed true, and the result is given in Theorem 9.3-1.

THEOREM 9.3-1 Let $P_{ij}(t)$ denote the probability that a Markov process is in state j at time t given that it starts in state i. If it is possible for the chain to visit every state, then there exists a probability vector $\pi = (\pi_1, \pi_2, \ldots, \pi_S)$ such that $\lim_{t\to\infty} P_{ij}(t) = \pi_j$. Let Λ denote the matrix when off-diagonal elements are the rates λ_{ij} and $\lambda_{ii} = -\Sigma_{j\neq i}\lambda_{ij}$. Then π satisfies the matrix equation $\pi\Lambda = 0$ and $\Sigma\,\pi_i = 1$.

As in the case of discrete-time Markov chains, the π_i's can be interpreted as the long-run fraction of time the process is in state i.

EXAMPLE 9.3-2 The long-run fraction of the time the robot in Example 9.3-1 is at each station can be determined by solving the following system of equations:

$$\begin{aligned} -8\pi_1 + 5\pi_2 + 4\pi_3 &= 0, \\ 5\pi_1 - 9\pi_2 + 6\pi_3 &= 0, \\ 3\pi_1 + 4\pi_2 - 10\pi_3 &= 0, \\ \pi_1 + \pi_2 + \pi_3 &= 1. \end{aligned}$$

The solution is $\pi_1 = .37$, $\pi_2 = .38$, $\pi_3 = .26$.

The time spent in a state on one visit is called a *sojourn time*. A *conditional transition probability* is the probability of a transition given that a transition to a new state is made. The following theorem expresses the distribution of sojourn times and the conditional transition probabilities in terms of the rates λ_{ij}.

THEOREM 9.3-2 The sojourn time for state i in a continuous-time Markov process is an exponential random variable with mean

$$\mu_i = \frac{1}{\sum_{j \neq i} \lambda_{ij}}.$$

Given that a transition is made from state i to j, $i \neq j$, the transition occurs according to a Markov chain with one-step transition probabilities:

$$p_{ij} = \frac{\lambda_{ij}}{\sum_{j \neq i} \lambda_{ij}}.$$

To verify this result, first note that the probability that some transition from state i occurs during an infinitesimally small time interval of length Δ is (essentially) $q_i = (\sum_{j \neq i} \lambda_{ij})\Delta$. Thus, the time for a transition to occur has the same distribution as the interarrival time of a Poisson process with rate $\lambda = \sum_{j \neq i} \lambda_{ij}$. Therefore, the sojourn time has an exponential distribution with mean $\mu_i = 1/\lambda$. Since the probability that a transition occurs in an infinitesimally small frame of length Δ is q_i, the conditional probability that this transition is to state j, given that a transition to a new state occurs, is

$$p_{ij} = \frac{\lambda_{ij}\Delta}{q_i} = \frac{\lambda_{ij}}{\sum_{j \neq i} \lambda_{ij}}.$$

EXAMPLE 9.3-3 Consider the robot arm in Example 9.3-1. The amount of time the robot spends at station 1 before moving to the next location is an exponential random variable with mean $\mu_1 = 1/(5+3) = 1/8$ minute. Similarly, the amounts of time spent at stations 2 and 3 are exponential random variables with means $\mu_2 = 1/9$ minute and $\mu_3 = 1/10$ minute. Given that the robot leaves station 1, the robot has probability $p_{12} = 5/(3+5) = 5/8$ of moving to station 2 and $p_{13} = 3/8$ of moving to station 3. Similarly, the probabilities of moving from stations 2 and 3 are $p_{21} = 5/9$, $p_{23} = 4/9$, and $p_{31} = 4/10$, $p_{32} = 6/10$.

The following theorem expresses the steady-state probabilities of a continuous-time Markov chain in terms of the mean sojourn times and transition probabilities in Theorem 9.3-2.

THEOREM 9.3-3 Let π_i, $i = 1, 2, \ldots, s$, denote the steady-state probabilities for a continuous-time Markov process. Suppose that the conditional transition probabilities, p_{ij}, of Theorem 9.3-2 define a regular Markov chain (see Definition 4.6-1). Let p_i denote the steady-state probability for state i from this regular Markov chain, and let μ_i denote the mean sojourn time for state i. Then

$$\pi_i = \frac{p_i \mu_i}{\sum_{i=1}^{s} p_i \mu_i}.$$

To verify this result, note that p_i can be interpreted as the long-run fraction of transitions the process makes to state i from some other state. Each transition to state i is called a *visit* to state i. Thus, in a large number of transitions, say N, the expected number of visits to state i is Np_i, and the expected amount of time spent in state i is $Np_i\mu_i$. Thus, the expected fraction of time spent in state i among many transitions is

$$\pi_i = \frac{Np_i\mu_i}{\sum_{i=1}^{s} Np_i\mu_i} = \frac{p_i\mu_i}{\sum_{i=1}^{s} p_i\mu_i}.$$

EXAMPLE 9.3-4 The matrix of conditional transition probabilities for Example 9.3-3 is

$$P = \begin{array}{c} \\ 1 \\ 2 \\ 3 \end{array}\begin{array}{c} \begin{array}{ccc} 1 & 2 & 3 \end{array} \\ \begin{bmatrix} 0 & \frac{5}{8} & \frac{3}{8} \\ \frac{5}{9} & 0 & \frac{4}{9} \\ \frac{4}{10} & \frac{6}{10} & 0 \end{bmatrix} \end{array}.$$

The steady-state probabilities for the three conditional transitions from stations 1, 2, and 3 are $p_1 = .33$, $p_2 = .37$, and $p_3 = .29$, which are the long-run fraction of visits the robot makes to each station. For instance, 29% of the time the robot moves, it moves to station 3 from one of the other two stations.

The mean sojourn times (minutes) are $\mu_1 = 1/8$, $\mu_2 = 1/9$, and $\mu_3 = 1/10$. This information (i.e., p_1, p_2, p_3 and μ_1, μ_2, μ_3) can be used as described in Theorem 9.3-3 to find the steady-state probabilities for the system. For example, the long-run fraction of the time the robot spends at station 3 is

$$\begin{aligned} \pi_3 &= \frac{(.29)(1/10)}{(.33)(1/8) + (.37)(1/9) + (.29)(1/10)} \\ &= .26. \end{aligned}$$

This is the same value as obtained in Example 9.3-2. Similar computations apply to π_1 and π_2.

One of the advantages of dealing with sojourn times and conditional transition probabilities rather than rates occurs in simulating the behavior of the continuous-time process. We can simply simulate the amount of time spent in each state as an exponential random variable with the appropriate mean, then simulate the transitions when they occur according to the procedures for simulating transitions in Markov chains. For instance, refer to Example 9.3-3; the amount of time the robot spends at station 1 would be simulated as an exponential random variable with mean $1/8$. When the robot moves, the station to which it moves would be simulated as station 2 with probability $5/8$ and station 3 with probability $3/8$. Details of simulating the robot arm example are left as an exercise.

Another advantage of dealing with sojourn times and conditional transition probabilities is that it may be more natural to specify these quantities than rates in modeling some systems. For instance, suppose we are modeling the flow of freight traffic from one division to another on a railroad. The amount of time a freight car spends in each division is something that is recorded routinely and thus can be used to determine the distribution of sojourn times. Records would also be kept of where a freight car goes as it leaves one division for another, so conditional transition probabilities can be estimated by determining the fraction of times a freight car leaves one division for another.

Exercises 9.3

9.3-1 The states of a continuous-time Markov system are 1, 2, 3, 4. When a change is made, the state of the system will either increase by 1 or decrease by 1 provided such a change is possible. The rates at which increases occur are $\lambda_{i,i+1} = 2, i = 1, 2, 3$. The rates at which decreases occur are $\lambda_{i,i-1} = 1, i = 2, 3, 4$. Find the steady-state probabilities.

9.3-2 Five people sit around a table and pass around a bottle of wine. It is always passed to the adjacent person either to the left or to the right. The rate at which it is passed to the left is $c > 0$, and the rate at which it is passed to the right is $d > 0$. Show that each person has the bottle one fifth of the time in the long run.

9.3-3 Items arrive one at a time at a loading dock until 10 have arrived. At that point the items are prepared for shipment and sent away, leaving no items on the dock. The process then starts over. Assume that the number of items at the dock can be modeled as a Markov process with transition rates $\lambda_{i,i+1} = 5$ per hour, $i = 0, 1, \ldots, 9$, and $\lambda_{10,0} = 2$ per hour. Find the steady-state probabilities.

9.3-4 A continuous-time Markov process has states 1, 2, and 3. The mean times spent in these states are 5, 10, and 15 minutes, respectively. The matrix of conditional probabilities of transition from one state to another is

$$P = \begin{array}{c} \\ 1 \\ 2 \\ 3 \end{array} \begin{array}{c} \begin{array}{ccc} 1 & 2 & 3 \end{array} \\ \begin{bmatrix} 0 & .5 & .5 \\ .2 & 0 & .8 \\ .1 & .9 & 0 \end{bmatrix} \end{array}.$$

a Find the rates λ_{ij}.

b Find the steady-state probabilities π_i.

9.3-5 A service truck may be dispatched from the central office to one of five locations. The time spent at the central office during each visit is an exponential random variable with a mean of 20 minutes. The time spent at location 1 during each visit is an exponential random variable with a mean of 45 minutes. The times spent at the other locations during each visit are exponential random variables each with a mean of 75 minutes. Given that the truck is at the central office, the probabilities it is dispatched to locations 1 through 5 are .3, .3, .2, .1, .1, respectively. The truck always returns to the central office before being dispatched again. Find the long-run fraction of time the truck spends at the central office and at each of the five locations.

9.3-6 Consider the rat-in-the-maze example in Sections 4.3 and 4.6. Suppose the amount of time the rat spends in compartment 1 is an exponential random variable with a mean of 60 seconds, and suppose the times it spends in the other compartments are exponential random variables each with a mean of 20 seconds. Assume that when the rat moves from one compartment to another it does so according to the Markov chain defined in Section 4.3.

a Find the steady-state probabilities π_i as given by Theorem 9.3-3.

b Explain why the steady-state probabilities given in Section 4.6 for this example differ from the steady-state probabilities found in part (a). Under what conditions would they be the same?

9.3-7 Write a computer program to simulate the movement of the robot arm in Examples 9.3-1 and 9.3-4. Quantities of interest are the transition number, N, the time at which the Nth transition occurs, $T(N)$, and the state to which the process goes when the Nth transition is made, $S(N)$. Assume the robot starts at station 1. Simulate a sufficient number of transitions so that 10 minutes of movement is completed. At which station is the robot arm located at 1 minute, 2 minutes, . . . , 10 minutes?

SECTION

9.4 Sojourns and Transitions for Continuous-Time Semi-Markov Processes

We now discuss continuous-time processes in which the sojourn times are not exponential random variables. If the distribution of sojourn times is not exponential but the conditional probabilities of transition follow the Markov chain as described in Section 9.3, then the process is called a *semi-Markov* process. Simulating the behavior of a semi-Markov process using sojourn times and conditional transition probabilities is essentially like simulating a Markov process. The only difference is that the procedure for simulating sojourn times must be changed to generate the appropriate nonexponential random variables.

EXAMPLE **9.4-1** Forms that need to be processed begin at station 1 in an office. After station 1, the forms are sent to station 2 or station 3 for additional processing. The forms may then be sent back and forth among these stations until they reach station 4, where the final processing is done. Each time a form is received at station 1, whether for the first time or for a revisit, the number of minutes it spends being processed is uniformly distributed on the interval 3 to 5 minutes. Similarly, the number of minutes a form spends being processed at stations 2 or 3 at each visit is uniformly distributed on the interval 4 to 10 minutes. Finally, the time a form spends at station 4 is exponentially distributed with a mean of 5 minutes. After being processed at stations 1, 2, or 3, the forms are sent to other stations according to a Markov chain with transition matrix

$$P = \begin{array}{c|cccc} & 1 & 2 & 3 & 4 \\ \hline 1 & 0 & .5 & .5 & 0 \\ 2 & .2 & 0 & 0 & .8 \\ 3 & .1 & .6 & 0 & .3 \\ 4 & 0 & 0 & 0 & 1 \end{array}.$$

The state diagram with transitions probabilities are shown in Figure 9.4-1.

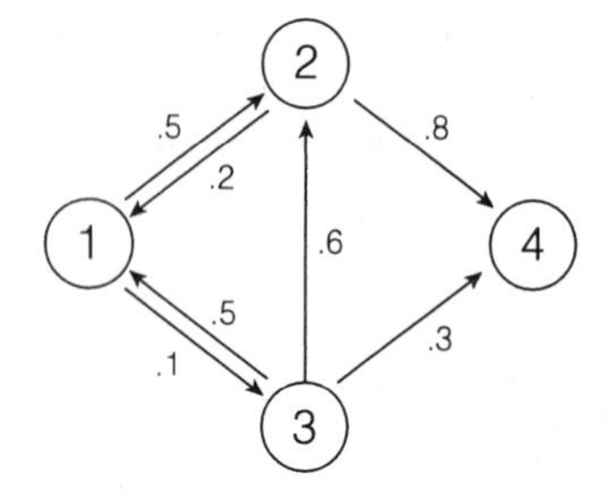

FIGURE 9.4-1
State Diagram and Transition Probabilities for Automated Office System

A form has completed all processing steps when processing at station 4 is finished. The quantity of interest is the time it takes for a form to be completely processed. It is easy to simulate the probability distribution of this quantity. First, the time spent in the current state is simulated according to an exponential or uniform random variable, depending on which is the current state. Next, this time is added to the total time. If the current state is not state 4, a transition is simulated according to the probabilities in the transition matrix and the foregoing procedure is repeated. If the current state is state 4, the simulation ends upon obtaining the time spent in state 4. The results of 1000 repetitions of the simulation are summarized in Figure 9.4-2. The sample mean is approximately 21.1 minutes, but in about 5% of the cases the times are more than double this amount.

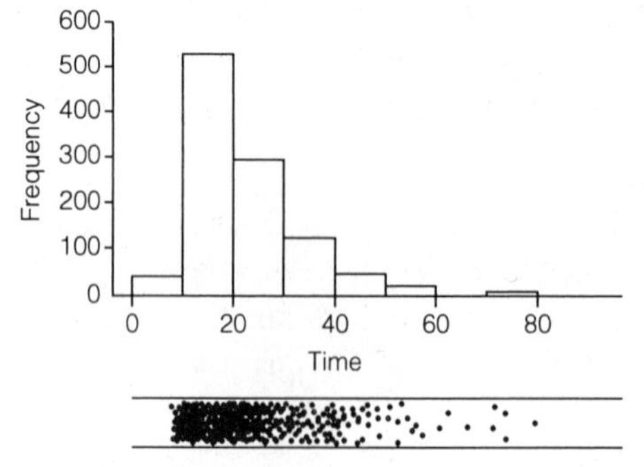

Average = 21.1 minutes
Standard Deviation = 10.0 minutes
Minimum = 8.1 minutes
Maximum = 80.0 minutes

FIGURE 9.4-2
Empirical Distribution of Processing Time for Automated Office System

The previous example used sojourn times and conditional transition probabilities to study the total time it takes to get through the system. It was found that extremely large processing times occasionally occur. It is interesting to examine a system such as in Example 9.4-1 to find out what changes could be made to reduce such extremes. Suggestions for modifications of the system in Example 9.4-1 are given in Exercise 9.4-4.

The mean time and variance it takes to move completely through a semi-Markov process can sometimes be obtained without simulation from the sojourn times and conditional transition probabilities. This is illustrated in the next example.

EXAMPLE 9.4-2 The procedure for diagnosing the fault in an electrical system involves up to three diagnostic tests. Test 1 is performed for the most likely causes of the problem, and this test leads to a diagnosis of the problem 75% of the time. If test 1 does not reveal the problem, test 2 is performed. Given test 1 failed to reveal the problem, a diagnosis is reached 60% of the time with test 2. If tests 1 and 2 fail, test 3 is performed. Having eliminated the possible causes associated with tests 1 and 2, a technician can make a diagnosis with test 3. The times to perform tests 1 and 2 are normally distributed random variables with means of 50 minutes and standard deviations of 10 minutes, and the time for test 3 is a normal random variable with a mean of 30 minutes and a standard deviation of 5 minutes. The times to complete the tests are independent of one another, and the chance of a diagnosis being reached at each test is independent of the amount of time it takes to perform the test. A diagram of the testing procedure is shown in Figure 9.4-3. The random variable of interest is the total time T that it takes to make a diagnosis.

FIGURE 9.4-3 Diagram of Diagnostic Procedure

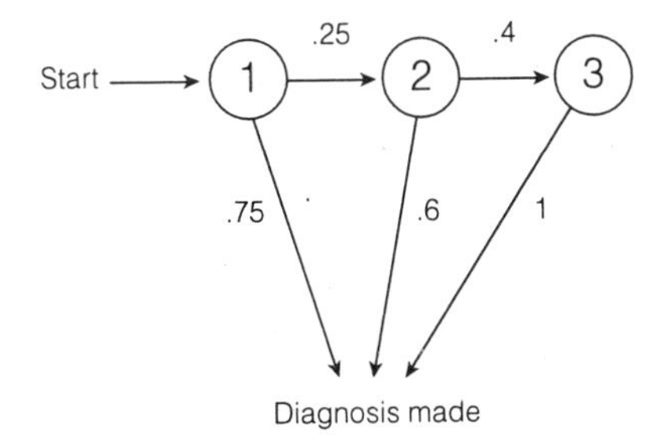

To compute $E(T)$ and $\mathrm{VAR}(T)$, we may use Theorems 2.9-1 and 2.9-2. Let T_i denote the time to complete the ith test. Let X be a random variable that denotes the number of tests performed. The total diagnosis time T may be expressed as

$$\begin{aligned} T &= T_1, && \text{if } X = 1; \\ T &= T_1 + T_2, && \text{if } X = 2; \\ T &= T_1 + T_2 + T_3, && \text{if } X = 3. \end{aligned}$$

Now, $E(T \mid X)$ can have three possible values, depending on whether $X = 1$, $X = 2$, or $X = 3$, where $P(X = 1) = .75$, $P(X = 2) = (1 - .75)(.60)$, and $P(X = 3) = (1 - .75)(1 - .60)(1)$. The probability distribution of $E(T \mid X)$ is

$E(T \mid X)$	50	100	130
$p(x)$	.75	.15	.10

Therefore,

$$E(T) = E[E(T \mid X)] = 50(.75) + 100(.15) + 130(.10)$$
$$= 65.5 \text{ minutes,}$$

$$\text{VAR}(E(T \mid X)) = [50^2(.75) + 100^2(.15) + 130^2(.10)] - (65.5)^2$$
$$= 774.75.$$

The variance $\text{VAR}(T \mid X)$ is a random variable that can have three possible values depending on whether $X = 1$, $X = 2$, or $X = 3$. Its probability distribution is

$\text{VAR}(T \mid X)$	100	200	225
$p(x)$	.75	.15	.10

Thus,

$$E[\text{VAR}(T \mid X)] = 100(.75) + 200(.15) + 225(.10)$$
$$= 127.5.$$

Therefore,

$$\text{VAR}(T) = \text{VAR}[E(T \mid X)] + E[\text{VAR}(T \mid X)]$$
$$= 774.75 + 127.5$$
$$= 902.25.$$

The standard deviation of T, 30.04 minutes, is about 47% of the mean of 65.5 minutes. This is typical of distributions of times in which there is a considerable skew to the distribution.

The values of $E(T)$ and $\text{VAR}(T)$ do not necessarily give us all the information we would like about the system. In this example, some notion about the distribution of T would be useful. This information is easily obtained through simulation. The estimated distribution of T was obtained by simulating 1000 diagnoses, and the histogram is shown in Figure 9.4-4. Note that the distribution has a large peak around 50 minutes, which corresponds to those times involving just test 1, and there is a smaller peak around 100 corresponding to those tests involving test 1 and test 2.

FIGURE 9.4-4
Time to Reach Diagnosis of Electrical Fault

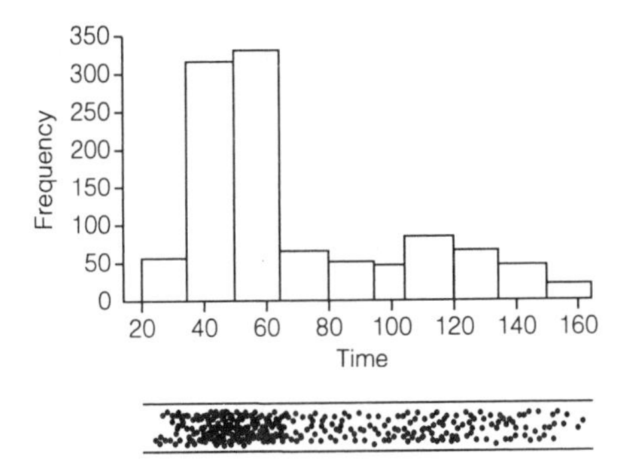

Exercises 9.4

9.4-1 A robot is used to deliver mail to five stations. It traverses a fixed route as shown in the figure. The robot takes 50 seconds to go from the start to station 1, 50 seconds to go from station 5 to the start, and 70 seconds to go from station to station. The probability that the robot stops at each station is .8. Given that the robot stops, the time that it spends there is an exponential random variable with a mean of 90 seconds. All random variables are independent. The quantity of interest is the time T it takes the robot to deliver the mail and return to the starting point.

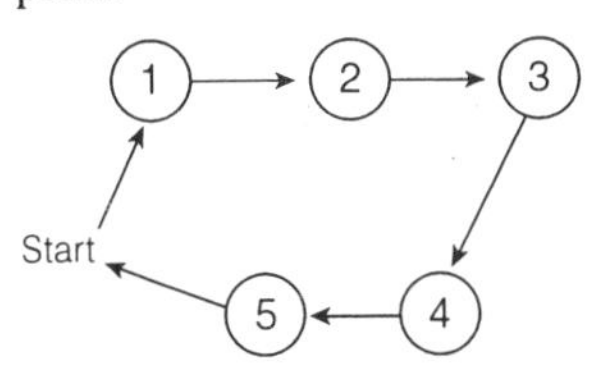

a Obtain 1000 values of T by simulation. Find the sample mean, sample standard deviation, and empirical probability distribution for the random variable T.

b Find $E(T)$ and $\text{STD}(T)$, using techniques similar to those in Example 9.4-2.

9.4-2 The time it takes to assemble an item at a factory is a normal random variable with a mean of 30 minutes and a standard deviation of 5 minutes. After the item is assembled it is inspected. Inspection time is a uniform random variable on the interval 4 to 6 minutes. The probability that the item passes the inspection is .75. If it fails, the item is reworked and the fault is corrected. Rework time is an exponential random variable with a mean of 20 minutes. All random variables of this system are independent. The process is diagrammed. Let T be the total time an item spends in the system.

a Simulate 1000 values of T. Obtain the sample mean, sample standard deviation, and empirical probability distribution of the simulated values.

b Explain how costs to the company could be affected by both the mean and the standard deviation of the time the item spends in the system.

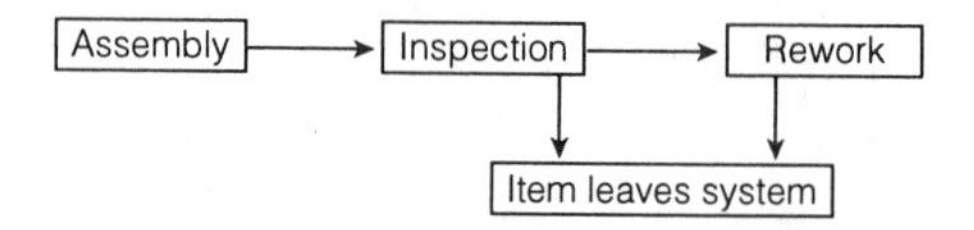

9.4-3 Referring to Exercise 9.4-2, vary the probability p that an item passes inspection. Theoretically obtain $E(T)$ and $STD(T)$ in terms of p and plot $E(T)$ versus p and $STD(T)$ versus p. Discuss the implications of these results on the cost to the company of having to inspect and rework items.

9.4-4 Modify Example 9.4-1 in two ways to see which factors most affect the variance of the time for a form to be processed. In each instance, simulate 1000 values of the time to complete the processing, and obtain the sample mean, sample standard deviation, and empirical probability distribution of the data.

a Change the time to process a form at station 4 so that it is uniformly distributed on the interval 4 to 6 minutes.

b Change the following transition probabilities for station 3: $P(3 \to 2) = .1$, and $P(3 \to 4) = .8$.

c Compare your results in parts (a) and (b) to those obtained in Example 9.4-1.

9.4-5 Use the following procedure to derive $E(T)$ for Example 9.4-1. Let m_s denote the expected number of visits to state s, starting with $s = 1$, and let μ_s denote the expected amount of time spent in state s each visit. Show that

$$E(T) = \sum_{s=1}^{4} \mu_s m_s.$$

Note that the values of m_s can be found directly from the distribution of sojourn times as specified in the assumptions of Example 9.4-1. The values of μ_s for $s = 1, 2$, and 3 can be found from the results of Theorem 4.7-4 as follows. Let $\mathbf{Q}$ be the matrix consisting of the transitions among the nonabsorbing states of matrix $\mathbf{P}$. It follows from Theorem 4.7-4 that the elements of the first row of $(\mathbf{I} - \mathbf{Q})^{-1}$ are the desired values of μ_s for $s = 1, 2, 3$. Since there is only one visit to state 4, we have $\mu_4 = 1$. Use this to show $E(T) = 21.4$ minutes.

9.4-6 For the purpose of studying the flow of customers through an enclosed shopping mall, the mall is divided into four areas: parking lot, anchor store A, anchor store B, other interior locations. The lengths of the first visits the customer makes to store A, store B, and the other interior locations are exponential random variables with means 20, 30, and 40 minutes, respectively. For each subsequent visit to these locations, the means are halved. For instance, the mean time spent in store A on the second visit is 10 minutes and on the third visit is 5 minutes. When a transition is made among locations, it is made according to a Markov chain whose one-step transition matrix is

$$P = \begin{array}{r c} & \begin{array}{cccc} 1 & 2 & 3 & 4 \end{array} \\ \begin{array}{r} \text{Parking lot } 1 \\ \text{Store A } 2 \\ \text{Store B } 3 \\ \text{Other interior } 4 \end{array} & \left[\begin{array}{cccc} 0 & .3 & .2 & .5 \\ .5 & 0 & 0 & .5 \\ .6 & 0 & 0 & .4 \\ .5 & .3 & .2 & 0 \end{array}\right] \end{array}.$$

The customer begins in the parking lot. The random variable of interest is the amount of time a customer spends inside the mall before returning to the parking lot. Simulate 1000 values of this random variable, and find the sample mean, sample standard deviation, and empirical probability distribution based on your data.

SECTION

9.5 Sojourns and Transitions for Queuing Processes

The M/M/k queuing process and the variations discussed in Chapter 7 are continuous-time Markov processes in which the states are the numbers of customers in the system. For such processes, the probability that the number of customers increases by 1 in a small frame of length Δ can be approximated by

$$P(i \to i+1) \approx a_i \Delta,$$

where a_i is the system arrival rate with i customers in the system, and the probability of a decrease of 1 can be approximated by

$$P(i \to i-1) \approx s_i \Delta,$$

where s_i is the system service rate when there are i customers in the system. Since the process can essentially have only one increase or one decrease in a small time frame, we have

$$P(i \to i) = 1 - a_i \Delta - s_i \Delta.$$

The derivations of the distributions of sojourn times and conditional transition probabilities are the same as those in Theorem 9.3-2 with $\lambda_{i,i+1} = a_i$, $\lambda_{i,i-1} = s_i$, $\lambda_{ii} = -a_i - s_i$, and $\lambda_{ij} = 0$, otherwise, for $i = 0, 1, 2, \ldots$. These results are summarized in Theorem 9.5-1.

THEOREM 9.5-1 Let a_i denote the system arrival rate, and let s_i denote the system service rate of a continuous-time Markov queuing process when there are i customers in the system. The sojourn time for state i is an exponential random variable with mean $\mu_i = 1/(a_i + s_i)$. When a transition is made to a new state, it occurs according to a Markov chain with the following transition probabilities:

$$P(i \to i+1) = \frac{a_i}{a_i + s_i},$$

$$P(i \to i-1) = \frac{s_i}{a_i + s_i}.$$

EXAMPLE 9.5-1 Consider an M/M/3 queuing process in which customers arrive at the rate of five customers per minute and service takes place at the rate of four services per minute when the servers are busy. For this process, $a_i = 5,\ i = 0, 1, 2, \ldots;\ s_0 = 0,\ s_1 = 4,\ s_2 = 8$, and $s_i = 12$ for $i \geq 3$. Table 9.5-1 gives the mean time the system spends in each state during each visit to that state and the conditional transition probabilities from the various states given that a transition to a new state is made.

TABLE 9.5-1 Mean Time Spent in State i and Conditional Transition Probabilities for M/M/3 Queue of Example 9.5-1

		Conditional Transition Probabilities	
State i	**Mean time in State (minutes)**	$i \to i + 1$	$i \to i - 1$
0	$\frac{1}{5}$	1	0
1	$\frac{1}{9}$	$\frac{5}{9}$	$\frac{4}{9}$
2	$\frac{1}{13}$	$\frac{5}{13}$	$\frac{8}{13}$
$i \geq 3$	$\frac{1}{17}$	$\frac{5}{17}$	$\frac{12}{17}$

The following algorithm, based on Theorem 9.5-1, describes a way to simulate an M/M/k queuing process. The algorithm terminates the simulation after a predetermined number of transitions have been made. It is easy to modify the algorithm to terminate the simulation after a predetermined length of time.

Simulation of the M/M/k Queue Using Sojourn Times and Conditional Transition Probabilities

Variables:

I	present state of the system
$A(I)$	arrival rate for state I
$S(I)$	service rate for state I
X	sojourn time for the present state
N	transition number
NMAX	number of transitions to be observed
$T(N)$	time at which Nth transition is made
STATE(N)	state to which the Nth transition has been made

Initialize:

Set NMAX, $A(I)$, $S(I)$ at predetermined values.

Set $S(0) = 0, I = 0, T(0) = 0$, STATE(0) $= 0$.

Repeat for $N = 1, 2, \ldots, \text{NMAX}$:

1 Generate a unit random number u_1.

Set $X = \dfrac{-\ln(u_1)}{A(I) + S(I)}$, exponential sojourn time in state I.

2 Set $T(N) = T(N-1) + X$, time of Nth transition.

3 Generate a unit random number u_2.

If $u_2 \le \dfrac{A(I)}{A(I) + S(I)}$, then increment $I = I + 1$; else decrement $I = I - 1$.

4 Set $\text{STATE}(N) = I$.

EXAMPLE **9.5-2** Table 9.5-2 shows 10 transitions of an M/M/3 process in which the arrival rate is two customers per unit of time and the service rate is one customer per unit of time. Here we have $a_i = 1,\ i = 0, 1, 2, \ldots;\ s_0 = 0,\ s_1 = 1,\ s_2 = 2,\ s_i = 3,\ i \ge 3$. The system starts in state 0.

TABLE **9.5-2** Simulation of M/M/3 Queuing Process

I	N	u_1	X	$T(N)$	u_2	$\frac{A(I)}{A(I)+S(I)}$	a or s	STATE(N)
0	1	.75563	0.28	0.28	.31261	1	a	1
1	2	.74082	0.15	0.43	.18332	1/2	a	2
2	3	.91334	0.03	0.46	.71699	1/3	s	1
1	4	.49659	0.35	0.81	.43296	1/2	a	2
2	5	.74113	0.10	0.91	.01203	1/3	a	3
3	6	.61873	0.12	1.03	.53694	1/4	s	2
2	7	.30120	0.40	1.43	.83006	1/3	s	1
1	8	.61878	0.24	1.67	.95122	1/2	s	0
0	9	.36055	1.02	2.69	.36448	1	a	1
1	10	.54781	0.30	2.99	.71603	1/2	s	0

Note: a = Nth transition is an arrival.
s = Nth transition is a service.

Instead of an exponential distribution for the amount of time spent in each state, we may use an arbitrary distribution as we did with semi-Markov processes in Section 9.4. An example is given in Exercise 9.5-5.

Exercise 9.5

9.5-1 Using the methods of this section, simulate 50 transitions in an M/M/1 queue with arrival rate of 2 and service rate of 2.5 customers per minute. Suppose we are interested not in the steady-state behavior of the queue but in its behavior in the early stages, say at time 1 minute. Use simulation to find an empirical distribution of the number in the queue at this time.

9.5-2 Suppose the rate of arrival in a single-server queue depends on the number in the system. Specifically, suppose that the arrival rate when i customers are in the system is $a_i = 2/(i+1)$ customers per minute. Suppose that the service rate is 2.5 customers per minute. Simulate this process for 50 transitions. Compare the behavior of this system to that in Exercise 9.5-1.

9.5-3 The Ajax Rental Center rents canoes at the rate of one a day. Customers use the canoes for an average of three days. Suppose the rental center has five canoes for rent, and at the start all five are available. The quantity of interest is the expected amount of time it will take for all canoes to be rented out. Estimate this quantity by using the simulation techniques of this section.

9.5-4 Refer to Exercise 9.5-3. Suppose the rental center begins with all five canoes rented out. Write a program to simulate the amount of time it takes for all five to be back at the rental center.

9.5-5 Modify Example 9.5-2 to allow for the following nonexponential sojourn time. If there are i customers in the system, $i = 0, 1, 2, 3$, assume that the time the system spends in state i has a uniform $[0, 2/(i+1)]$ distribution. For $i > 3$, the distribution is the same as $i = 3$, that is, uniform $[0, 1/2]$. Simulate 10 transitions.

10

Reliability Models

SECTION

10.1 The Reliability Function

Suppose that a company guarantees its product for one year. For the product to be judged reliable, it should have a high probability of functioning as expected throughout the guarantee period. A measure of the reliability is the probability that the product does not fail during this period. If T represents the time to failure, this measure of reliability can be expressed as

$$R = P(T > 1 \text{ year}).$$

This is a useful index of how well the product will perform. For instance, if R = .999, then only one item in a thousand on average will fail in the one-year period.

Our concern will be in obtaining values of R for simple systems. Instead of considering R for just a single time, we obtain R as a function of time.

Definition 10.1-1

If T denotes the time to failure of a system, then the *reliability at time t* is denoted $R(t)$ and is defined to be

$$R(t) = P(T > t).$$

The reliability at time t, $R(t)$, is called the *reliability function* and is the probability that the time to failure will exceed t. The reliability function can be expressed in terms of the cumulative distribution function of T. That is,

$$R(t) = 1 - P(T \le t) = 1 - F(t).$$

EXAMPLE 10.1-1 Suppose that the time to failure has an exponential distribution with mean $\mu = 1/\lambda$. Since the cumulative distribution function is $F(t) = 1 - e^{-\lambda t}$, the reliability function is

$$R(t) = e^{-\lambda t}.$$

EXAMPLE 10.1-2 If the time to failure is a normal random variable, the reliability function does not have a simple form. However, reliabilities may be obtained from Table 4 of Appendix A. For example, if the time to failure is a normal random variable with mean 1000 hours and standard deviation 100 hours, then the reliability at time $t = 850$ hours is

$$\begin{aligned} R(850) &= P(T > 850) \\ &= P\left(Z > \frac{850 - 1000}{100}\right) \\ &= P(Z > -1.5) \\ &= .9332. \end{aligned}$$

An important problem in many applications is to determine the reliability of a system in terms of the reliabilities of its components. We first consider *series systems* and *parallel systems*, as shown schematically in Figure 10.1-1. Think of these systems as representing electrical devices in which current is supposed to flow from A to B. The system will fail if all paths of current from A to B are blocked. In the series system, it takes only one failed component to block the path, whereas with the parallel system it takes all components to fail to keep the current from going from A to B.

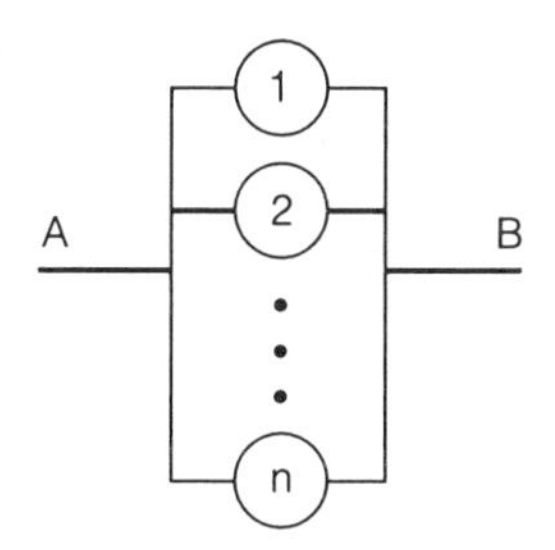

FIGURE 10.1-1
Series and Parallel Systems

THEOREM 10.1-1 Let a system consist of n components with reliability functions $R_1(t), R_2(t), \ldots, R_n(t)$. Assume that the components fail independently of one another. The reliability function of the series system is

$$R_S(t) = \prod_{i=1}^{n} R_i(t).$$

The reliability function of the parallel system is

$$R_P(t) = 1 - \prod_{i=1}^{n} (1 - R_i(t)).$$

These results are applications of Theorems 9.1-1 and 9.1-2 on the distribution of the maximum and minimum of a sequence of independent random variables. To see this, let T denote the time to failure of the system in question and let T_i, $i = 1, 2, \ldots, n$, denote the time to failure of the ith component. The series system will have a time to failure greater than t if and only if all the components have time to failure greater than t, which in turn occurs if and only if the minimum of $T_1, T_2, \ldots, T_n$ is greater than t. Letting $F_i(t)$ denote the cumulative distribution function of T_i, we have

$$\begin{aligned} R_S(t) &= P(T > t) \\ &= P(\min(T_1, T_2, \ldots, T_n) > t) \\ &= \prod_{i=1}^{n} [1 - F_i(t)] \\ &= \prod_{i=1}^{n} R_i(t). \end{aligned}$$

For the parallel system, it is easier to compute $P(T \leq t) = 1 - R_P(t)$ than it is to compute $R_P(t)$ directly. Since a parallel system fails if and only if all components fail, it follows that $T \leq t$ if and only if $\max(T_1, T_2, \ldots, T_n) \leq t$. Thus,

$$\begin{aligned} 1 - R_P(t) &= P(T \leq t) \\ &= P(\max(T_1, T_2, \ldots, T_n) \leq t) \\ &= \prod_{i=1}^{n} F_i(t) \\ &= \prod_{i=1}^{n} (1 - R_i(t)). \end{aligned}$$

Again, note that the independence of component failures is essential to the validity of the expressions in Theorem 10.1-1. Not all systems have this characteristic. For instance, if a common source of failure, such as an electrical shock, causes components to fail simultaneously, then the foregoing expressions would not be correct.

EXAMPLE 10.1-3 The times to failure of three components are exponentially distributed with means in thousands of hours given by $\mu_1 = 2.0$, $\mu_2 = 2.5$, and $\mu_3 = 4.0$, respectively. The reliability functions of the components are $R_1(t) = e^{-.5t}$, $R_2(t) = e^{-.4t}$, and $R_3(t) = e^{-.25t}$. If the components fail independently of one another, the reliability functions for the series and parallel systems are

$$\begin{aligned} R_S(t) &= e^{-1.15t}, \\ R_P(t) &= 1 - (1 - e^{-.5t})(1 - e^{-.4t})(1 - e^{-.25t}). \end{aligned}$$

The probability that the series system lasts more than 1000 hours is $R_S(1) = .32$, and the probability that the parallel system lasts more than 1000 hours is $R_P(1) = .97$.

Many systems are composed of series and parallel subsystems. The reliability of these systems may be computed by repeated application of the results of Theorem 10.1-1.

EXAMPLE 10.1-4 Suppose the three components in Example 10.1-3 are arranged as shown in Figure 10.1-2 and fail independently of one another. Components 1 and 2 constitute a parallel subsystem with reliability function

$$R_{12}(t) = 1 - (1 - e^{-.5t})(1 - e^{-.4t}).$$

This subsystem is in series with component 3. Thus, the system reliability is

$$\begin{aligned} R(t) &= R_{12}(t)\, R_3(t) \\ &= [1 - (1 - e^{-.5t})(1 - e^{-.4t})]e^{-.25t}. \end{aligned}$$

The reliability at 1000 hours is R(1) = .68.

FIGURE 10.1-2
A System Consisting of Series and Parallel Subsystems

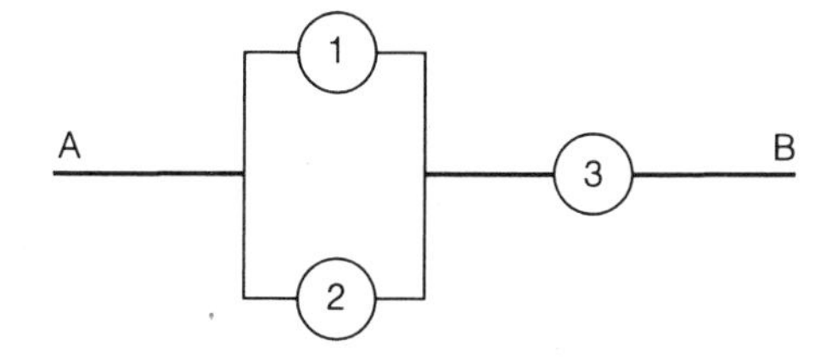

EXAMPLE 10.1-5 Independent components are configured as in Figure 10.1-3. The times to failure are measured in thousands of hours and have exponential distributions with means μ_i as indicated in Figure 10.1-3.

FIGURE 10.1-3
System for Example 10.1-5 with Exponential Times to Failure.

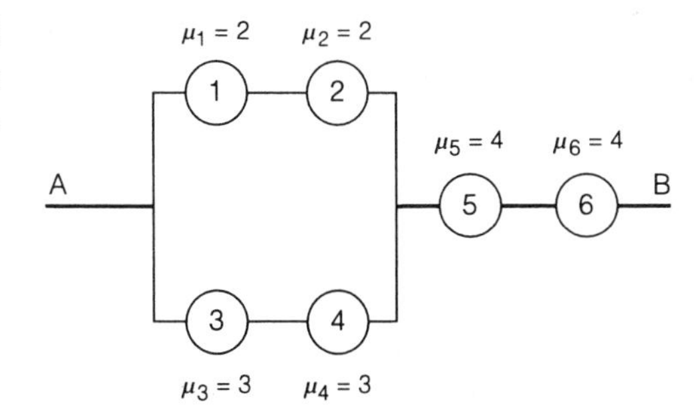

We note that components 1 and 2 form a series subsystem with reliability function $R_{12}(t) = R_1(t)R_2(t) = e^{-t}$, and similarly components 3 and 4 have reliability function $R_{34}(t) = R_3(t)R_4(t) = e^{-2t/3}$. Since the combination of 1 and 2 is in parallel with the combination of 3 and 4, components 1, 2, 3, and 4 have reliability function

$$R_{1234}(t) = 1 - (1 - R_{12}(t))(1 - R_{34}(t)).$$

Finally the combination of these four components is in series with 5 and 6. Thus,

$$\begin{aligned} R(t) &= R_{1234}(t)R_5(t)R_6(t) \\ &= [1 - (1 - e^{-t})(1 - e^{-2t/3})]e^{-t/2}. \end{aligned}$$

If $t = 2.0$ thousand hours, then $R(2.0) = .1336$.

In simulating the times to failure of a system, it is often useful to express these times in terms of the times to failure of the components. For instance, if we have simulated times to failure $T_1, T_2, \ldots, T_n$ of n components, then the times to failure T_S and T_P of series and parallel systems may be expressed as

$$T_S = \min(T_1, T_2, \ldots, T_n) \quad \text{and} \quad T_P = \max(T_1, T_2, \ldots, T_n).$$

For systems containing series and parallel subsystems, the time to failure of the system can be expressed as a combination of minima and maxima of the times to failure of the components. This is illustrated in the following example.

EXAMPLE 10.1-6 Consider the system shown in Figure 10.1-3. Let $T_1, T_2, \ldots, T_6$ denote the failure times of the components 1 through 6, respectively. The time to failure of the series subsystem consisting of components 1 and 2 is $\min(T_1, T_2)$, and for 3 and 4 it is $\min(T_3, T_4)$. It follows that the parallel subsystem that results from combining 1 and 2 with 3 and 4 has time to failure $\max[\min(T_1, T_2), \min(T_3, T_4)]$. Since this subsystem is in series with components 5 and 6, the system time to failure is

$$T = \min(\max[\min(T_1, T_2), \min(T_3, T_4)], T_5, T_6).$$

For instance, if we observe $T_1 = 3.4$, $T_2 = 2.6$, $T_3 = 1.5$, $T_4 = 4.0$, $T_5 = 3.8$, and $T_6 = 5.1$, then

$$\begin{aligned} T &= \min(\max[\min(3.4, 2.6), \min(1.5, 4.0)], 3.8, 5.1) \\ &= \min(\max[2.6, 1.5], 3.8, 5.1) \\ &= \min(2.6, 3.8, 5.1) \\ &= 2.6. \end{aligned}$$

Exercises 10.1

10.1-1 Suppose that the time to failure is a uniform [0, 5] random variable. Find and graph the reliability function.

10.1-2 Suppose the hours to failure of an active flashlight battery have probability density function

$$f(x) = \begin{cases} \dfrac{12}{x^3}, & x > 2, \\ 0, & \text{otherwise.} \end{cases}$$

a Find and graph the reliability function.

b For a certain flashlight to operate, two such batteries must be placed in series. Find the reliability function of the series system.

c Define the half-life of a system to be the time at which reliability reaches 1/2. Compute the half-lives for the single battery and the two-battery series system.

10.1-3 A computer has two disk drives. The time to failure of each, in thousands of hours, is a Weibull random variable with $\alpha = 8$ and $\beta = 2$. Suppose we define a system failure to occur when one or the other of the two disks fails. What is the reliability function of the system?

10.1-4 Consider four components with the following reliability functions:

$$R_1(t) = e^{-t}, \ R_2(t) = e^{-2t}, \ R_3(t) = e^{-3t}, \ R_4(t) = e^{-4t}.$$

a Find the reliability function for these four components in series.

b Find the reliability function for the components in parallel.

c Find the reliability function for the following system.

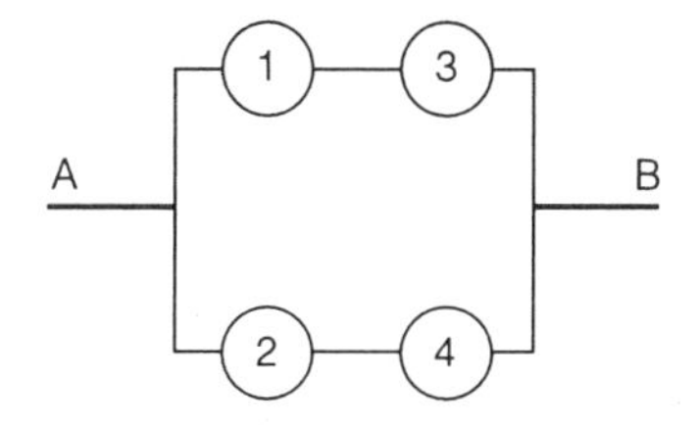

10.1-5 Consider the diagrammed system where the times to failure are in thousands of hours and have exponential distributions with means as indicated. Find and graph the reliability function.

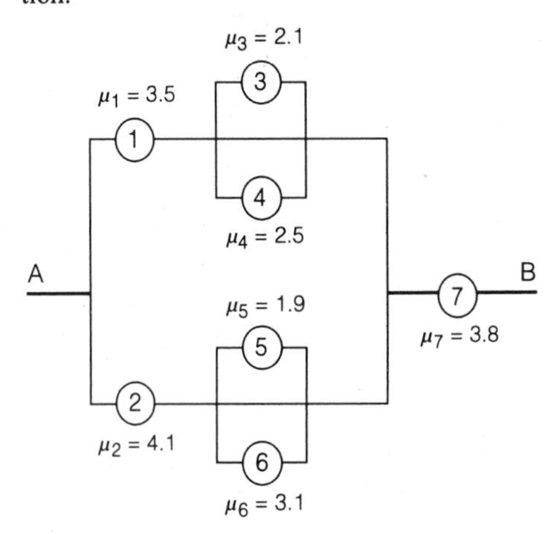

10.1-6 Consider the system in Exercise 10.1-5. Let $T_1, T_2, \ldots, T_7$ denote the independent failure times for the seven components.

a Express the time to failure algebraically as a combination of minima and maxima.

b Simulate 100 times to failure for this system. Place the observed times to failure in order from smallest to largest. At each observed time to failure, estimate $R(t)$ as the fraction of observed times to failure greater than t. Plot the estimated $R(t)$ and the actual $R(t)$ on the same graph. Comment on how well the estimated reliability function approximates the true reliability function.

10.1-7 A spacecraft has two computers. The time to failure of each is an exponential random variable with a mean of 2000 hours. In order for a certain experiment to be completed, at least one of the two computers must function throughout the mission. How long can the mission last so that the probability that the experiment is successful is .995.

10.1-8 For certain systems, reliability may involve units other than time to failure. For instance, for automobile tires one may wish to consider miles to failure, where a failure is defined to be a worn-out tread. Suppose that miles to failure of a certain brand of tire, in thousands of miles, is a Weibull random variable with $\alpha = 45$ and $\beta = 3$.

a If four tires on an automobile fail independently of one another (a questionable assumption perhaps), what is the reliability function of the set of four tires, assuming that failure of the set occurs when the first tire fails?

b Suppose we define the minimum life expectancy of the set of four tires to be the mileage at which the reliability is .95 and the maximum life expectancy to be the mileage at which the reliability is .05. Find the minimum and maximum life expectancies for the set of four tires in part (a).

SECTION

10.2 Hazard Rate

In Section 10.1 we defined the reliability function and showed how it could be computed for several simple systems. Another useful function for describing the reliability of a system is the *hazard rate.* It also goes by the name *failure-rate function, intensity function,* or *force of mortality.*

Definition 10.2-1

Let T be the time to failure of a system. Let $f(t)$ be the probability density function of T, and let $R(t)$ be the reliability function. Then, *hazard rate* $h(t)$ is defined by

$$h(t) = \frac{f(t)}{R(t)}.$$

Consider the interpretation of $h(t)$. The probability of failure in a small time frame t to $t + \Delta$ given that the component has survived to time t is computed as

$$\begin{aligned} P(t < T < t + \Delta \mid T > t) &= \frac{P(T < t + \Delta, T > t)}{P(T > t)} \\ &= \frac{\int_t^{t+\Delta} f(s)\, ds}{R(t)} \\ &\approx h(t)\Delta. \end{aligned}$$

Thus $h(t)\Delta$ is approximately equal to the probability of failure in time t to $t + \Delta$ given that survival has occurred at least to time t. The function $h(t)$ is a measure of the risk of failure at time t given survival to time t.

If $h(t)$ is an increasing function of t, then the system wears out with time. For instance, the probability of failure just after 1000 hours given that the system has lasted 1000 hours would be greater than the probability of failure just after 500 hours given that the system has lasted 500 hours. It should come as no surprise that many systems exhibit this wear-out property. On the other hand, if $h(t)$ is a decreasing function of t, then the system reliability improves with time. Some electronic items exhibit a decreasing hazard rate in the early stages of operation. The period of time in which the reliability improves with age is sometimes called the "burn-in" period.

EXAMPLE 10.2-1 If the time to failure is an exponential random variable with mean $\mu = 1/\lambda$, then the hazard rate is a constant. That is,

$$h(t) = \frac{\lambda e^{-\lambda t}}{e^{-\lambda t}} = \lambda.$$

There is neither wear-out nor improvement in reliability for an exponential system. The no-wear-out property of the exponential distribution makes it a plausible model for systems whose failures are caused by random factors that originate outside the system itself. For instance, the time to failure of an electronic component could reasonably be modeled by the exponential distribution if the failures were due to random electrical surges originating from lightning or other such sources. The no-wear-out property of the exponential distribution is the same as the lack-of-memory property discussed in Section 6.1. Systems that have the exponential distribution for time to failure do not "remember" how long they have been operating.

EXAMPLE 10.2-2 The hazard rate of a Weibull random variable can be increasing or decreasing, depending on whether $\beta > 1$ or $\beta < 1$. Specifically,

$$\begin{aligned} h(t) &= \frac{(\beta/\alpha)(t/\alpha)^{\beta-1} e^{-(t/\alpha)^{\beta}}}{e^{-(t/\alpha)^{\beta}}} \\ &= \left(\frac{\beta}{\alpha}\right)\left(\frac{t}{\alpha}\right)^{\beta-1}. \end{aligned}$$

The flexibility of the Weibull distribution to include both increasing or decreasing hazard rates, as well as the constant hazard rate when $\beta = 1$, is one of the features that make it attractive as a model for time to failure.

EXAMPLE 10.2-3 Consider a system of two identical components that fail independently of one another and are configured in parallel. Assume that distributions of the components have identical exponential distributions each with mean $\mu = 1/2$. The reliability functions of the components are $R_1(t) = R_2(t) = e^{-2t}$. From Theorem 10.1-1, the reliability function of the parallel system is

$$\begin{aligned} R_P(t) &= 1 - (1 - e^{-2t})(1 - e^{-2t}) \\ &= 2e^{-2t} - e^{-4t}. \end{aligned}$$

The cumulative distribution function is

$$\begin{aligned} F(t) &= 1 - R_P(t) \\ &= 1 - 2e^{-2t} + e^{-4t}, \qquad t > 0, \end{aligned}$$

and the probability density function is

$$f(t) = 4e^{-2t} - 4e^{-4t}, \qquad t > 0.$$

This gives a hazard rate of

$$h(t) = \frac{4e^{-2t} - 4e^{-4t}}{2e^{-2t} - e^{-4t}}.$$

A plot of the hazard rate is shown in Figure 10.2-1. It begins at zero and increases to a limiting value of 2. Thus, the system exhibits the wear-out property even though the individual components do not. Moreover, the limiting hazard rate of the system as $t \to \infty$ is the hazard rate of an individual component, $\lambda = 2$. This is a reasonable result. The system has a lower hazard rate than an individual component for all times, but as the system ages and one of the components fails, the system will behave as a single-component system.

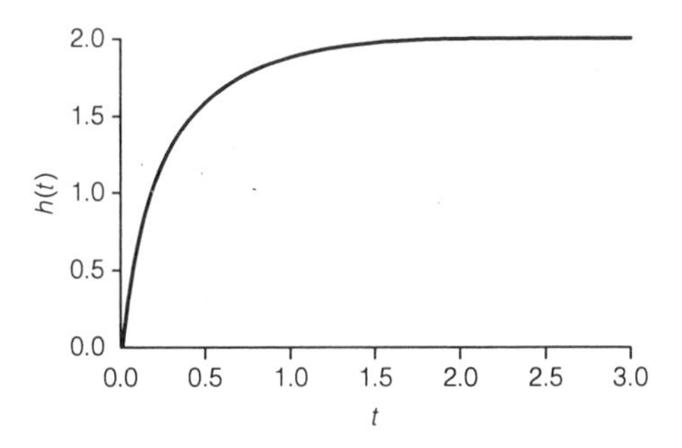

FIGURE **10.2-1** Hazard Rate of the Two-component Parallel System of Example 10.2-3

Some systems have hazard rates that are initially decreasing, then relatively constant, then increasing. An automobile, for instance, may initially have factory defects that must be corrected, but reliability improves as corrections are made. Then for a time the hazard rate will be more or less constant. Finally, the hazard rate will increase as the automobile reaches an age where it starts to wear out. Such hazard rates have been described as "bathtub" shaped as shown in Figure 10.2-2. The hazard rate for the lifetime of a human being is like this. The decreasing phase is the period from infancy to early childhood. The constant phase is from childhood through middle age, and the increasing phase is old age.

FIGURE 10.2-2 A "Bathtub"-shaped Hazard Rate

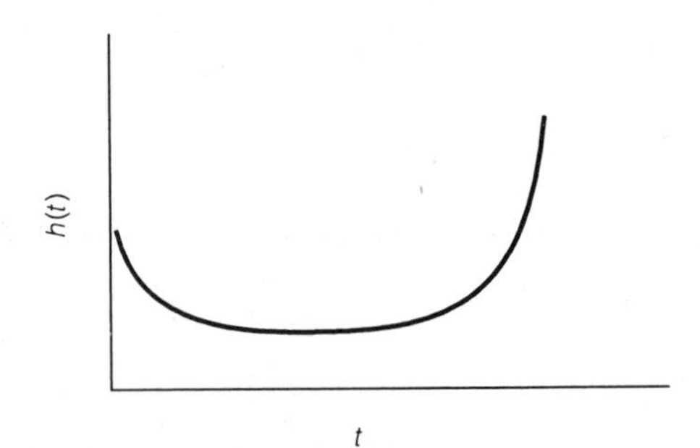

There is a useful relationship between the hazard rate and the reliability function given by the following theorem.

THEOREM 10.2-1 Let a random variable T have hazard rate $h(t)$ and reliability function $R(t)$. Then

$$R(t) = e^{-\int_0^t h(s)\, ds}.$$

To verify this result, we first note that $\frac{d}{dt}R(t) = -f(t)$. Thus,

$$\begin{aligned}\frac{d}{dt}\ln(R(t)) &= \frac{1}{R(t)}\frac{d}{dt}R(t)\\ &= \frac{-f(t)}{R(t)}\\ &= -h(t).\end{aligned}$$

Integrating both sides, we find

$$\ln(R(t)) = -\int_0^t h(s)\, ds$$

from which Theorem 10.2-1 follows.

Having obtained $R(t)$ from $h(t)$, we can obtain the cumulative distribution function and density function as

$$F(t) = 1 - e^{-\int_0^t h(s)\,ds}$$

$$f(t) = -\frac{d}{dt}R(t) = h(t)e^{-\int_0^t h(s)\,ds}.$$

EXAMPLE 10.2-4 Suppose that the hazard rate is

$$h(t) = e^t, \qquad 0 < t.$$

This represents a system that wears out at an exponential rate. Now

$$\begin{aligned} R(t) &= e^{-\int_0^t e^s\,ds} \\ &= e^{-(e^t - 1)}, \qquad t > 0, \end{aligned}$$

and

$$F(t) = 1 - e^{-(e^t - 1)}, \qquad t > 0,$$

$$f(t) = e^t e^{-(e^t - 1)}, \qquad t > 0.$$

A graph of $f(t)$ is shown in Figure 10.2-3.

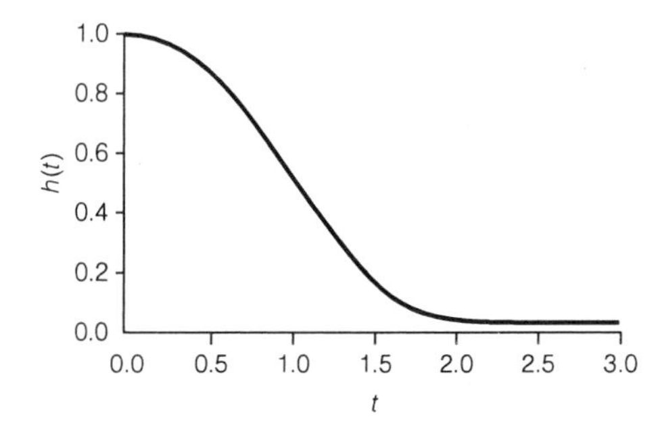

FIGURE 10.2-3 Graph of Density Function with Exponential Hazard Rate

Exercises 10.2

10.2-1 A system has three independent components arranged in parallel. The components have exponential distributions with means 1, 1/2, and 1/3, respectively. Find and plot the hazard rate of the system.

10.2-2 A system has n identical components arranged in series. The times to failure of the components are independent, Weibull random variables with $\alpha = 1$ and $\beta = 2$. Compare the hazard rate of the system with $n = 3$ components to the one with $n = 2$ components. Which wears out faster in the sense measured by the hazard rate?

10.2-3 The lifetime, in years, of an automobile battery is uniformly distributed on the interval $[1, 5]$. Find the hazard rate.

10.2-4 A useful estimate of the hazard rate can be obtained directly from observed time to failure data when the number of observations is sufficiently large. Divide the range of the data into intervals $t_0 < t_1 < t_2 < \cdots < t_n$. Then

$$h(t_i) = \frac{\text{Number of failure times in the interval } (t_i < t \le t_{i+1})}{(t_{i+1} - t_i)(\text{Number of failure times} > t_i)}.$$

Simulate 200 times to failure of a Weibull random variable with $\alpha = 1$ and $\beta = 2$. Graph the estimated hazard rate and corresponding values of the actual hazard rate, using 10 equally spaced intervals to estimate $h(t)$.

10.2-5 A piece of fabric is placed on a machine and subjected to stress until it tears. The hazard rate is $h(t) = 2e^{.5t}$, $t > 0$. Find the probability that the fabric lasts more than $t = 1$ hour before tearing.

10.2-6 A random variable has hazard rate $h(t) = t^2 + t$, $t > 0$. Find the reliability function $R(t)$ and the density function $f(t)$.

10.2-7 A machine operates for 200 hours with a constant hazard rate $h(t) = .002$, $t \le 200$. At 200 hours of service, parts better than the original replace existing parts, giving a hazard rate of $h(t) = .001$, $t > 200$.

a Find the reliability function $R(t)$ and the density function $f(t)$.

b Find the expected time to failure.

SECTION

10.3 Renewal Processes

The Poisson process is a counting process in which the times between counts are independent and identically distributed exponential random variables. In this section we generalize the Poisson process by allowing times between counts to have distributions other than the exponential.

Definition 10.3-1

A *renewal process* is a counting process in which the times between the counted outcomes are independent and identically distributed nonnegative random variables.

The name *renewal process* comes from typical applications of such processes. For example, consider replacing a light bulb in a lamp when it burns out with one of the same type as the original, and suppose this process is repeated again and again. It is reasonable to assume that the times between replacements are independent and identically distributed random variables. The distribution of these random variables might be exponential, Weibull, or some other type. Since the lamp is "renewed" each time a light bulb is replaced, the process is called a renewal process.

In the reliability context, the outcomes being counted are called renewals, and the times between counts are called renewal times. These times are denoted as T_1, $T_2, \ldots$, with $E(T_i) = \mu$ and $\text{STD}(T_i) = \sigma$. The number of renewals that occur in the interval $[0, t]$ is denoted $N(t)$.

EXAMPLE **10.3-1** Certain mechanical and electrical systems experience wear-out because of the cumulative effect of shocks that occur to the systems over time. Automobile shock absorbers are like this. We consider systems satisfying the following three assumptions: the shocks to the system occur according to a Poisson process with rate λ; the system fails as soon as k shocks have occurred; and the system is fixed up (renewed) to its original condition as soon as it fails, at which time the cycle of shocks and failures begins anew. What we wish to determine is the distribution of $N(t)$, the number of renewals in an interval of length t. Now, zero renewals occur if there have been from zero to $k - 1$ shocks, one renewal occurs if there have been from k to $2k - 1$ shocks, and so on. Thus,

$$\begin{aligned} P(N(t) = n) &= P(nk \leq \text{number of shocks} \leq (n+1)k - 1) \\ &= \sum_{j=nk}^{(n+1)k-1} e^{-\lambda t}\frac{(\lambda t)^j}{j!}. \end{aligned}$$

If $\lambda = .5$, and the system fails when it has received a total of three shocks, then the probability of one renewal in $t = 4$ units of time is

$$\begin{aligned} P(N(4) = 1) &= \sum_{j=3}^{5} e^{-2}\frac{2^j}{j!} \\ &= .3067. \end{aligned}$$

There is an important relationship between the distribution of the renewal times and the number of renewals $N(t)$.

THEOREM 10.3-1 Let $S_n = T_1 + T_2 + \cdots + T_n$, where T_i denotes the time between the $(i-1)$th and ith renewal, $i = 1, 2, \ldots, n$. Let $N(t)$ denote the number of renewals in an interval of length t. Then

$$P(S_n \leq t) = P(N(t) \geq n).$$

To verify Theorem 10.3-1 first note that the random variable S_n is the time that it takes for n renewals to occur. We note that $S_n \leq t$ if and only if there are at least n renewals in the interval $[0, t]$. That is, the events $(S_n \leq t)$ and $(N(t) \geq n)$ are equivalent and, hence, have the same probability.

We will use Theorem 10.3-1 to give an approximation for $P(N(t) \geq n)$ when n is large. When n is large, we may apply the Central Limit Theorem with $E(S_n) = n\mu$ and $\text{VAR}(S_n) = n\sigma^2$. Therefore,

$$\begin{aligned} P(N(t) \geq n) &= P(T_1 + T_2 + \cdots + T_n \leq t) \\ &\approx P\left(Z \leq \frac{t - n\mu}{\sigma\sqrt{n}}\right). \end{aligned}$$

EXAMPLE 10.3-2 The ribbon on a certain printer has a lifetime that is uniformly distributed on the interval one to three months. We will find the probability of 15 or more replacements in 24 months. The mean and standard deviation of the uniform $[1, 3]$ random variable are $\mu = 2$ and $\sigma = .58$. Thus,

$$\begin{aligned} P(N(24) \geq 15) &= P\left(Z \leq \frac{24 - (15)(2)}{.58\sqrt{15}}\right) \\ &= P(Z \leq -2.67) \\ &= 0.0038. \end{aligned}$$

Thus, it is virtually certain that there will be fewer than 15 ribbons needed for the printer in 24 months.

Besides looking at a renewal process when the number of renewals, n, becomes large, it is also of interest to consider the distribution of $N(t)$ when t becomes large.

THEOREM 10.3-2 For large t, $N(t)$ has an approximate normal distribution with mean $\mu_{N(t)} = t/\mu$ and variance $\sigma^2_{N(t)} = t\sigma^2/\mu^3$.

An intuitive justification of the theorem is based on the observation that the random variable $(T_1 + T_2 + \cdots + T_n)/n$ behaves as the random variable $t/N(t)$ for large n and t, since in both cases time is in the numerator and the number of renewals that occur in that time is in the denominator. Thus, the random variable

$$Z = \frac{(T_1 + T_2 + \cdots + T_n) - n\mu}{\sigma\sqrt{n}},$$

which has an approximate normal distribution for large n, behaves for large t as the random variable

$$\frac{t - N(t)\mu}{\sigma\sqrt{N(t)}} = \frac{t/\mu - N(t)}{(\sigma/\mu)\sqrt{N(t)}}.$$

Since $N(t)/t$ behaves as $n/(T_1 + T_2 + \cdots + T_n)$, which converges to $1/\mu$, we may replace $\sqrt{N(t)}$ in the denominator with $(t/\mu)^{1/2}$. Thus, the random variable

$$Z = \frac{t/\mu - N(t)}{\sigma t^{1/2}/\mu^{3/2}}$$

behaves for large t as a standard normal random variable. As a result $N(t)$ has an approximate normal distribution with mean t/μ and variance $t\sigma^2/\mu^3$.

EXAMPLE 10.3-3 In Example 10.3-2, the mean and variance of the number of ribbon replacements in 24 months are approximately $24/2 = 12$ and $(24)(.58)^2/2^3 = 1.0$, respectively. Thus, we are 95% confident that the number of replacements will fall in the interval 12 ± 2, or 10 to 14.

Many systems (air traffic control systems, weapons systems, etc.) require the interaction of humans and machines. An interesting problem in the area of reliability is to model the performance of such individuals as they carry out their responsibilities. We will describe a model for one such system.

We assume that an individual is confronted with a stream of tasks each of which requires action to be taken. An air traffic controller, for instance, must take action as airplanes move into an assigned area of responsibility. Each time an action is taken, there is a chance that a mistake is made. The quantity of interest is the number of mistakes, $M(t)$, that occur in the interval $[0, t]$. The following assumptions define a *simple task-stream process*.

Definition 10.3-2

A process $M(t)$ that counts the number of mistakes made by an individual in a system is said to be a *simple task-stream process* if

i The number of tasks that confront an individual in the interval $[0, t]$ is a Poisson process $N(t)$ with rate λ.

ii The individual can deal with each task instantly.

iii Each time the task is dealt with, there is a probability p that a mistake is made.

iv Mistakes are independent from one task to the next.

Since the process begins anew each time a task arrives, it is a renewal process. The following theorem gives the distribution of $M(t)$.

THEOREM 10.3-3 In a simple task-stream process, the number of mistakes made by an individual in the interval $[0, t]$ is a Poisson process with rate λp.

To verify Theorem 10.3-3, first note that when there are $N(t) = n$ tasks to be done, the probability that k mistakes are made is a binomial probability; that is,

$$P(k \text{ mistakes} \mid N(t) = n) = \binom{n}{k} p^k (1-p)^{n-k}.$$

Since we must have $N(t) \geq k$ tasks to have the possibility of k mistakes, the unconditional probability is

$$P(k \text{ mistakes}) = \sum_{n=k}^{\infty} P(k \text{ mistakes} \mid N(t) = n) P(N(t) = n)$$

$$= \sum_{n=k}^{\infty} \binom{n}{k} p^k (1-p)^{n-k} e^{-\lambda t} \frac{(\lambda t)^n}{n!}$$

$$= e^{-\lambda t} \frac{(p\lambda t)^k}{k!} \sum_{n=k}^{\infty} \frac{[(1-p)\lambda t]^{n-k}}{(n-k)!}$$

$$= e^{-p\lambda t} \frac{(p\lambda t)^k}{k!}.$$

It follows that $M(t)$ is a Poisson process with rate λp.

The *human performance reliability* of such a system is defined to be the probability that there are no mistakes in the time interval $[0, t]$:

$$\text{HR}(t) = P(M(t) = 0) = e^{-\lambda p t}.$$

EXAMPLE 10.3-4 In a certain manufacturing plant, warning signals are monitored by an operator who must take corrective action to prevent the system from being shut down. The number of signals per hour is a Poisson process with rate $\lambda = .5$ per hour. The probability that an incorrect action is taken is $p = .10$. The probability that the operator makes no mistakes in 8 hours is $\text{HR}(8) = e^{-8(.5)(.1)} = .67$. The probability that an operator makes more than one mistake in a 40-hour week is $P(M(40) \geq 2) = 1 - P(M(40) \leq 1)$, which can be found in Table 3 of Appendix A for a mean of $40(.5)(.1) = 2$. We find $P(M(40) \geq 2) = .594$. Improvements in the system can take place by reducing the causes of warning signals or by reducing an individual's probability of making a mistake. Having a model such as this one can enable the plant manager to determine which is the more cost-effective way to make improvements.

Exercises 10.3

10.3-1 A messenger service receives requests for service according to a Poisson process at a rate of one every 6 minutes. The service dispatches a messenger whenever four requests have arrived.

a Find the probability that three or more messengers must be dispatched in an hour.

b Find the probability that the time between dispatches is less than 1/2 hour.

10.3-2 Suppose that the messenger service in Exercise 10.3-1 has only three messengers on duty and that it takes an hour for each messenger to complete delivery of the four messages. Simulate 8 hours of activity of the service. Determine whether there is a messenger always available when time for a dispatch occurs.

10.3-3 Suppose the time between requests to connect to a computer network is a uniform $[0, 2]$ random variable where time is in minutes. Assume that the number of requests is a renewal process. Use the normal approximation to find a likely range for the number of requests in an hour.

10.3-4 A Poisson process $N(t)$ has rate $\lambda = 1$. Find $P(N(10) > 12)$, using Table 3 of Appendix A, and use the normal approximation to find this probability. Comment on the accuracy of the normal approximation in this case.

10.3-5 In a telephone survey, the number of calls completed by a caller is a Poisson process with a rate of eight per hour. The probability that the caller makes a mistake in recording information is $p = .15$. Find the human performance reliability function $\mathrm{HR}(t)$. At what time t is the human performance reliability less than .20?

10.3-6 The lifetime of a car battery is an exponential random variable with a mean of one year. The battery is replaced when it fails. Suppose we record the lifetime of the battery that is in the car at $t = 4$ years. For instance, if the lifetimes of the first three batteries are 2.5 years, .6 years, and 1.4 years, then the third battery is in the car at $t = 4$ and its lifetime is 1.4 years. Simulate this process 1000 times and determine the sample mean of the lifetimes of the batteries that are in place at time $t = 4$. You should observe that this average is greater than 1, which is the lifetime of a battery selected at random. This phenomenon is called the renewal paradox. Intuitively explain why this happens.

SECTION 10.4 Maintained Systems

Systems that can fail and be repaired are considered in this section. We first consider a system that can be characterized as operable (up) or inoperable (down) at any instant in time. The system is assumed to go through continual up–down cycles. When the system goes down, it is repaired to its original condition, and the cycle is repeated. Let $T_U(i)$ and $T_D(i)$ be independent continuous random variables that denote the uptime and downtime, respectively, for the system during the ith cycle. Assume that across the cycles the $T_U(i)$'s are independent and identically distributed and that the $T_D(i)$'s are independent and identically distributed.

THEOREM 10.4-1 Let $E(T_U(i)) = \mu_U$ and $E(T_D(i)) = \mu_D$. Then

$$\lim_{t\to\infty} P(\text{system is up at time } t) = \frac{\mu_U}{\mu_U + \mu_D}.$$

This result is intuitively reasonable. The expected time for the system to go through an up–down cycle is $\mu_U + \mu_D$, and the expected time for the system to be up during a cycle is μ_U. Thus, in the long run, the fraction of time the system is up is $\mu_U/(\mu_U + \mu_D)$. Likewise, the long-run fraction of time the system is down is $\mu_D/(\mu_U + \mu_D)$.

EXAMPLE **10.4-1** The time in hours that a soft drink bottling machine will operate is a Weibull random variable with $\alpha = 150$ and $\beta = 1.5$. The repair time is a Weibull random variable with $\alpha = 8$ and $\beta = 1.2$. From the results in Section 6.4, we find $E(T_U) = (150)(.896) = 134.4$ and $E(T_D) = (8)(.940) = 7.52$. Thus, the long-run fraction of time the system is up is $134.4/(7.52 + 134.4) = .95$.

Next, we consider a system that has components that can fail independently of one another. As a component fails, it is repaired by a technician who handles the components on a first-come, first-served basis. Our interest is in the number of components that are operable at any point in time. A network of computers, in which the computers are regarded as the components, is an example of the type of system we have in mind. In Section 4.1 we discussed a discrete-time Markov model for a repairable system. Here we introduce a continuous-time Markov process as a model for such a system.

Let the number of components in the system be denoted by k. Let i be the number of components that are up, and assume that the system has the following transition rates:

$$\begin{aligned} &\text{Transition rate from } i \text{ to } i-1 = i\mu, \quad i = 1, 2, \ldots, k; \\ &\text{Transition rate from } i \text{ to } i+1 = \lambda, \quad i = 0, 1, \ldots, k-1. \end{aligned} \qquad (10.4\text{-}1)$$

If there are i components operating at time t, the probability that there is a breakdown in the system in a small time interval of length Δ is approximately $i\mu\Delta$. The quantity μ can be interpreted as the failure rate of an individual component. The probability that the repair on a broken component is finished in a small time interval of length Δ is approximately $\lambda\Delta$. The quantity λ can be interpreted as the repair rate of the technician.

The important thing to note about this system is that it has the same structure as the M/M/k queue with finite capacity as studied in Section 7.8. The repair rate λ is analogous to the arrival rate in an M/M/k queue. Think of the components "arriving" on-line after they have been repaired. The failure rate μ of an individual component is analogous to the service rate of an individual server. Think of a failure of a component as a completion of service in the M/M/k queue. Since there can be at most k

components in operation, the maintained system is like a finite-capacity queue that can have at most k customers in the system at any time.

We may apply the balance equations in Section 7.7 to determine the steady-state probabilities for the number of components that are in operation at any time. Let i denote the number of components in operation at any given time. Using the notation of Section 7.7, we have

$$a_i = \lambda, \qquad i = 0, 1, \ldots, k-1,$$

$$s_i = i\mu, \qquad i = 1, 2, \ldots, k,$$

$$\frac{a_0 a_1 \cdots a_{i-1}}{s_1 s_2 \cdots s_i} = \frac{\lambda^i}{\mu^i \cdot i!}, \qquad i = 1, 2, \ldots, k.$$

Application of Equations (7.7-6) and (7.7-7) gives the steady-state probabilities for the number of operational components in a Markov system maintained by a single technician as

$$\pi_i = \pi_0 \frac{\lambda^i}{\mu^i i!}, \qquad i = 1, 2, \ldots, k,$$

$$\pi_0 = \left(1 + \sum_{i=1}^{k} \frac{\lambda^i}{\mu^i i!}\right)^{-1}. \tag{10.4-2}$$

EXAMPLE **10.4-2** A building has three air-conditioning units that operate independently of one another. The failure rate of each is one breakdown every 120 days. The technician can perform repairs at the rate of one unit every five days. If the Markov model for the system applies, the steady-state probabilities are

$$\pi_0 = \left(1 + \sum_{i=1}^{3} \frac{(1/5)^i}{(1/120)^i i!}\right)^{-1} = .00038,$$

$$\pi_1 = \pi_0 \frac{1/5}{1/120} = .00918,$$

$$\pi_2 = \pi_0 \frac{(1/5)^2}{(1/120)^2 2!} = .11005,$$

$$\pi_3 = \pi_0 \frac{(1/5)^3}{(1/120)^3 3!} = .88040.$$

The expected number of days per year that all units are down is (.00038)(365) = .14, and the expected number of days per year that all units are up is (.88040)(365) = 321.

Now consider repairable systems with more than one technician. Let c denote the number of technicians where $c \le k$. The technicians are assumed to work independently of one another, each with the same repair rate λ. If c or more components are down, then all c technicians will work on the repairs. If fewer than c components are down, then the number of technicians that will work on repairs will equal the number of down components. Again using the notation for the balance equations and letting i denote the number of components in operation, we have

$$\begin{aligned} a_i &= c\lambda, & i &= 0, 1, \ldots, k-c \\ a_i &= (k-i)\lambda, & i &= k-c+1, \ldots, k, \\ s_i &= i\mu, & i &= 1, 2, \ldots, k, \end{aligned}$$

as parameters for a Markov system. Thus,

$$\frac{a_0 a_1 \cdots a_{i-1}}{s_1 s_2 \cdots s_i} = \frac{(c\lambda)^i}{\mu^i i!}, \qquad i = 1, 2, \ldots, k-c+1,$$

$$\frac{a_0 a_1 \cdots a_{i-1}}{s_1 s_2 \cdots s_i} = \frac{c^{k-c+1}\lambda^i}{\mu^i i!}(c-1)\cdots(k-i+1), \qquad i = k-c+2, \ldots, k.$$

In the following example, we apply these results to a two-technician system.

EXAMPLE 10.4-3 Suppose that a second technician is assigned to the system described in Example 10.4-2. The appropriate terms for determining steady-state probabilities for the number of operational air conditioners are

$$a_0 = \frac{2}{5}, \qquad a_1 = \frac{2}{5}, \qquad a_2 = \frac{1}{5},$$

$$s_1 = \frac{1}{120}, \qquad s_2 = \frac{2}{120}, \qquad s_3 = \frac{3}{120},$$

$$\frac{a_0}{s_1} = 48, \qquad \frac{a_0 a_1}{s_1 s_2} = 1152, \qquad \frac{a_0 a_1 a_2}{s_1 s_2 s_3} = 9216.$$

Therefore,

$$\pi_0 = (1 + 48 + 1152 + 9216)^{-1} = \frac{1}{10{,}417} = (9.6)10^{-5},$$

$$\pi_1 = \frac{48}{10{,}417} = 00461,$$

$$\pi_2 = \frac{1152}{10{,}417} = .11059,$$

$$\pi_3 = \frac{9216}{10{,}417} = .88471.$$

The addition of the second technician has not appreciably changed the probabilities of having two or three operational air conditioners. The chance of a total failure (i.e., all three air conditioners being down) has changed from 3.8×10^{-4} to 9.6×10^{-5}. This represents a change from one total failure about every 2600 days to one total failure about every 10,500 days. Although in either the one- or two-technician system, the chance of a total failure is small, a total failure occurs much less frequently with the two-technician system.

Exercises 10.4

10.4-1 An automobile assembly line must occasionally be stopped for "glitches." The time in hours that the line is up is a gamma random variable with $\alpha = 8$ and $\beta = 2$, and the time in hours it takes to fix a glitch is a gamma random variable with $\alpha = .25$ and $\beta = 2$. Find the long-run probability that the line is up.

10.4-2 A grinding tool must be repaired on average every 20 hours. The repair takes an average of 1 hour. Suppose we have a choice of either increasing the average life of the tool by 10% or decreasing the average repair time by 10%. Both choices cost the same. In terms of the probability that the grinding tool is up, which would be the better approach?

10.4-3 A repairable system has two machines and one technician. The ratio of the rate of repair λ to the rate of failure μ is 10. Find the long-run probability that both machines are up.

10.4-4 Compare the steady-state probabilities for the two-machine system with one technician to the two-machine system with two technicians when $\lambda/\mu = 1, 5, 10, 50$. What do these results say about the desirability of adding a second technician?

10.4-5 In a system with k machines and k technicians, find the probability that at least one machine is up.

10.4-6 Presently a system has one machine and one technician with $\lambda/\mu = 10$. The probability that the system is up is $10/11 = .91$. It is desired to have a system in which the probability that at least one machine is up is .995. This can be done by adding more machines, adding more technicians along with the additional machines, decreasing the rate of failure μ, or increasing the rate of repair λ. Consider various such alternatives and speculate on which might be the most cost-effective approach under various circumstances.

SECTION

10.5 Nonhomogeneous Poisson Process and Reliability Growth

Recall from Section 7.2 that the Poisson process is obtained as a limit of the Bernoulli counting process as the frame size approaches zero. The Bernoulli counting process has constant probability of a success each frame. We can modify the Bernoulli counting process so that the probability of a success is not constant from frame to frame. The limit of such a process as frame size approaches zero is a *nonhomogeneous Poisson process.*

Definition 10.5-1

Let $N(t)$ denote a process that counts the number of times some outcome of interest occurs in the interval $[0, t]$. The process is said to be a nonhomogeneous Poisson process if

i $N(0) = 0$.

ii The counts in nonoverlapping intervals are independent.

iii There exists a differentiable, increasing function $m(t)$, called the *mean function,* such that $m(0) = 0$ and

$$P(N(t) - N(s) = k) = e^{-(m(t) - m(s))} \frac{(m(t) - m(s))^k}{k!}, \quad \text{for } s < t \text{ and } k = 0, 1, 2, \ldots.$$

The *intensity function* $\lambda(t)$ is defined as

$$\lambda(t) = \frac{d}{dt} m(t).$$

The difference $N(t) - N(s)$ is the number of times the event of interest occurs in the interval $(s, t]$. This has a Poisson distribution with mean $m(t) - m(s)$. In particular, $N(t)$ has a Poisson distribution with mean $m(t)$. For a small frame of length Δ the probability that an outcome of interest occurs in the interval $[t, t + \Delta]$ is approximately $\lambda(t)\Delta$. The Poisson process defined in Section 7.2 has mean function $m(t) = \lambda t$, and $\lambda(t) = \lambda$.

We will apply the nonhomogeneous Poisson process as a model for the number of failures of a system or the number of defects found in a system. In this context, an increasing intensity function implies that failures or defects occur with increasing frequency as time increases, possibly due to wear-out. On the other hand, a decreasing intensity function implies that the system is getting more reliable with time, possibly due to "bugs" in the system being removed.

If $\lambda(t)$ is a decreasing function of time, then we say that the system exhibits *reliability growth.* Systems undergoing development often exhibit reliability growth as mistakes are corrected and designs are improved. For instance, the rate at which mistakes occur in filling out forms in a new office may decrease as clerks become more familiar with procedures. Software usually exhibits reliability growth as it undergoes development from its inception until the finished product.

Of special interest is the *Weibull process* where the intensity function is

$$\lambda(t) = \lambda\beta t^{\beta-1}$$

and the mean function is

$$m(t) = \lambda t^{\beta}.$$

If $\beta > 0$, the intensity function increases with t, and if $\beta < 0$ it decreases. If we let T_1 denote the time to the first failure, then

$$P(T_1 > t) = P(N(t) = 0) = e^{-\lambda t^{\beta}}.$$

Thus, the cumulative distribution function of the time to first failure is

$$F_1(t) = 1 - e^{-\lambda t^{\beta}}.$$

It follows that T_1 has a Weibull distribution. Unlike the renewal process, the time between the $(n-1)$th and nth failure depends on when the $(n-1)$th failure occurred. That is,

$$\begin{aligned} P(T_n > t+s \mid T_{(n-1)} = s) &= P(N(t+s) - N(s) = 0) \\ &= e^{-(\lambda(t+s)^{\beta} - \lambda s^{\beta})}. \end{aligned}$$

EXAMPLE 10.5-1 As new computer equipment is installed, the daily failures can be modeled as a Weibull process with $\lambda = 2$ and $\beta = .5$. The probability of more than five failures in the first 10 days can be computed by noting that $m(10) = (2)(10)^{.5} = 6.3$. Thus,

$$\begin{aligned} P(N(10) > 5) &= \sum_{k=6}^{\infty} e^{-6.3}\frac{(6.3)^k}{k!} \\ &= .60. \end{aligned}$$

To find the probability of more than five failures in the 10th through the 20th day, we note that $m(20) - m(10) = (2)(20)^{.5} - (2)(10)^{.5} = 2.6$, so

$$\begin{aligned} P(N(20) - N(10) > 5) &= \sum_{k=6}^{\infty} e^{-2.6}\frac{2.6^k}{k!} \\ &= .05. \end{aligned}$$

Thus, the reliability has substantially improved from the first to the second 10-day period, and it will continue to improve thereafter.

EXAMPLE 10.5-2 In Example 10.5-1 suppose we wish to know how long development will have to continue until the probability of one or more failures in a one-day period is .10. This probability is

$$\begin{aligned} P(N(t+1) - N(t) \geq 1) &= 1 - e^{-(2(t+1)^{1/2} - 2t^{1/2})} \\ &= .1. \end{aligned}$$

The equation that needs to be solved for t is

$$2(t+1)^{.5} - 2t^{.5} = -\ln(.90).$$

By graphing

$$y(t) = 2(t+1)^{.5} - 2t^{.5}$$

as in Figure 10.5-1 and noting where the line $y = -\ln(.90)$ crosses the graph, we find the solution $t = 89$ days.

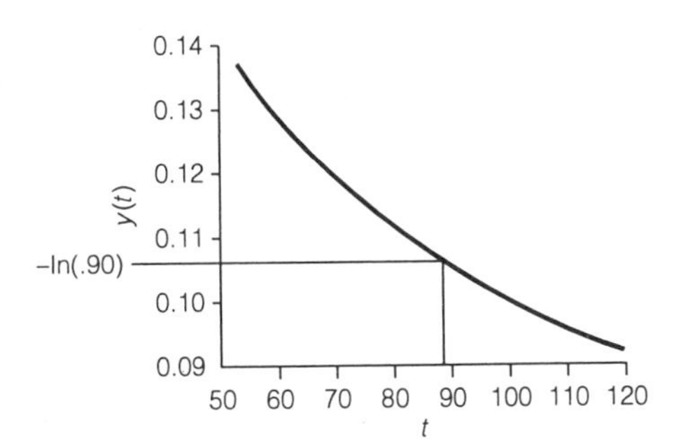

FIGURE 10.5-1 Graph of $y(t) = 2(t+1)^{.5} - 2t$ for Example 10.5-2

Exercises 10.5

10.5-1 The number of mistakes made per day by a technician in a training program is a Weibull process with $\lambda = 1$ and $\beta = .2$.

a What is the probability that at least one mistake occurs in the first half-day?

b What is the probability of three or fewer mistakes in the first two days?

c Suppose training has taken place for three days. What is the probability that at least one mistake occurs within the next day?

d How long must training take place so that the probability that at least one mistake occurs in the next day is .05?

10.5-2 In the beta test phase of software development, potential customers try the software as they would use it in practice. Suppose the number of changes (corrections or modifications) per day to the software undergoing beta testing is a Weibull process with $\lambda = 5$ and $\beta = .1$.

a Suppose beta testing has taken place for 15 days. What is the probability that at least two changes in the software will be made in the next five days?

b Suppose that beta testing has taken place for 300 days, and then the product is released. Assuming that changes take place in practice as they would under beta testing, what is the probability that there will be no changes in the next 150 days?

10.5-3 As a new engine undergoes testing and development, the number of design changes per month has a nonhomogeneous Poisson process rate function

$$\lambda(t) = \frac{1}{1+t}.$$

a What is the mean function?

b Find the probability that there are at least three or more design changes in two months.

c Testing will stop at s months when the probability is .90 that there is no design change within the following month. How long will testing be done?

Appendix

TABLE 1 Unit Random Numbers

Line/Col.	(1)	(2)	(3)	(4)	(5)	(6)	(7)	(8)	(9)	(10)	(11)	(12)	(13)	(14)
1	10480	15011	01536	02011	81647	91646	69179	14194	62590	36207	20969	99570	91291	90700
2	22368	46573	25595	85393	30995	89198	27982	53402	93965	34095	52666	19174	39615	99505
3	24130	48360	22527	97265	76393	64809	15179	24830	49340	32081	30680	19655	63348	58629
4	42167	93093	06243	61680	07856	16376	39440	53537	71341	57004	00849	74917	97758	16379
5	37570	39975	81837	16656	06121	91782	60468	81305	49684	60672	14110	06927	01263	54613
6	77921	06907	11008	42751	27756	53498	18602	70659	90655	15053	21916	81825	44394	42880
7	99562	72905	56420	69994	98872	31016	71194	18738	44013	48840	63213	21069	10634	12952
8	96301	91977	05463	07972	18876	20922	94595	56869	69014	60045	18425	84903	42508	32307
9	89579	14342	63661	10281	17453	18103	57740	84378	25331	12566	58678	44947	05585	56941
10	85475	36857	43342	53988	53060	59533	38867	62300	08158	17983	16439	11458	18593	64952
11	28918	69578	88231	33276	70997	79936	56865	05859	90106	31595	01547	85590	91610	78188
12	63553	40961	48235	03427	49626	69445	18663	72695	52180	20847	12234	90511	33703	90322
13	09429	93969	52636	92737	88974	33488	36320	17617	30015	08272	84115	27156	30613	74952
14	10365	61129	87529	85689	48237	52267	67689	93394	01511	26358	85104	20285	29975	89868
15	07119	97336	71048	08178	77233	13916	47564	81056	97735	85977	29372	74461	28551	90707
16	51085	12765	51821	51259	77452	16308	60756	92144	49442	53900	70960	63990	75601	40719
17	02368	21382	52404	60268	89368	19885	55322	44819	01188	65255	64835	44919	05944	55157
18	01011	54092	33362	94904	31273	04146	18594	29852	71585	85030	51132	01915	92747	64951
19	52162	53916	46369	58586	23216	14513	83149	98736	23495	64350	94738	17752	35156	35749
20	07056	97628	33787	09998	42698	06691	76988	13602	51851	46104	88916	19509	25625	58104
21	48663	91245	85828	14346	09172	30168	90229	04734	59193	22178	30421	61666	99904	32812
22	54164	58492	22421	74103	47070	25306	76468	26384	58151	06646	21524	15227	96909	44592
23	32639	32363	05597	24200	13363	38005	94342	28728	35806	06912	17012	64161	18296	22851
24	29334	27001	87637	87308	58731	00256	45834	15398	46557	41135	10367	07684	36188	18510
25	02488	33062	28834	07351	19731	92420	60952	61280	50001	67658	32586	86679	50720	94953

SOURCE: CRC (1966). *Handbook of Tables for Probability and Statistics*, 2nd ed. Cleveland, OH: CRC Press.

TABLE 1 Unit Random Numbers (continued)

Line/Col.	(1)	(2)	(3)	(4)	(5)	(6)	(7)	(8)	(9)	(10)	(11)	(12)	(13)	(14)
26	81525	72295	04839	96423	24878	82651	66566	14778	76797	14780	13300	87074	79666	95725
27	29676	20591	68086	26432	46901	20849	89768	81536	86645	12659	92259	57102	80428	25280
28	00742	57392	39064	66432	84673	40027	32832	61362	98947	96067	64760	64584	96096	98253
29	05366	04213	25669	26422	44407	44048	37937	63904	45766	66134	75470	66520	34693	90449
30	91921	26418	64117	94305	26766	25940	39972	22209	71500	64568	91402	42416	07844	69618
31	00582	04711	87917	77341	42206	35126	74087	99547	81817	42607	43808	76655	62028	76630
32	00725	69884	62797	56170	86324	88072	76222	36086	84637	93161	76038	65855	77919	88006
33	69011	65797	95876	55293	18988	27354	26575	08625	40801	59920	29841	80150	12777	48501
34	25976	57948	29888	88604	67917	48708	18912	82271	65424	69774	33611	54262	85963	03547
35	09763	83473	73577	12908	30883	18317	28290	35797	05998	41688	34952	37888	38917	88050
36	91567	42595	27958	30134	04024	86385	29880	99730	55536	84855	29080	09250	79656	73211
37	17955	56349	90999	49127	20044	59931	06115	20542	18059	02008	73708	83517	36103	42791
38	46503	18584	18845	49618	02304	51038	20655	58727	28168	15475	56942	53389	20562	87338
39	92157	89634	94824	78171	84610	82834	09922	25417	44137	48413	25555	21246	35509	20468
40	14577	62765	35605	81263	39667	47358	56873	56307	61607	49518	89656	20103	77490	18062
41	98427	07523	33362	64270	01638	92477	66969	98420	04880	45585	46565	04102	46880	45709
42	34914	63976	88720	82765	34476	17032	87589	40836	32427	70002	70663	88863	77775	69348
43	70060	28277	39475	46473	23219	53416	94970	25832	69975	94884	19661	72828	00102	66794
44	53976	54914	06990	67245	68350	82948	11398	42878	80287	88267	47363	46634	06541	97809
45	76072	29515	40980	07391	58745	25774	22987	80059	39911	96189	41151	14222	60697	59583
46	90725	52210	83974	29992	65831	38857	50490	83765	55657	14361	31720	57375	56228	41546
47	64364	67412	33339	31926	14883	24413	59744	92351	97473	89286	35931	04110	23726	51900
48	08962	00358	31662	25388	61642	34072	81249	35648	56891	69352	48373	45578	78547	81788
49	95012	68379	93526	70765	10593	04542	76463	54328	02349	17247	28865	14777	62730	92277
50	15664	10493	20492	38391	91132	21999	59516	81652	27195	48223	46751	22923	32261	85653

TABLE 2 Binomial Cumulative Distribution Function

$$B(x;n,p) = \sum_{k=0}^{n} \binom{n}{k} p^k (1-p)^{n-k}$$

n	x	0.05	0.10	0.15	0.20	0.25	0.30	0.35	0.40	0.45	0.50	0.55	0.60	0.65	0.70	0.75	0.80	0.85	0.90	0.95
4	0	0.8145	0.6561	0.5220	0.4096	0.3164	0.2401	0.1785	0.1296	0.0915	0.0625	0.0410	0.0256	0.0150	0.0081	0.0039	0.0016	0.0005	0.0001	0.0000
	1	0.9860	0.9477	0.8905	0.8192	0.7383	0.6517	0.5630	0.4752	0.3910	0.3125	0.2415	0.1792	0.1265	0.0837	0.0508	0.0272	0.0120	0.0037	0.0005
	2	0.9995	0.9963	0.9880	0.9728	0.9492	0.9163	0.8735	0.8208	0.7585	0.6875	0.6090	0.5248	0.4370	0.3483	0.2617	0.1808	0.1095	0.0523	0.0140
	3	1.0000	0.9999	0.9995	0.9984	0.9961	0.9919	0.9850	0.9744	0.9590	0.9375	0.9085	0.8704	0.8215	0.7599	0.6836	0.5904	0.4780	0.3439	0.1855
5	0	0.7738	0.5905	0.4437	0.3277	0.2373	0.1681	0.1160	0.0778	0.0503	0.0312	0.0185	0.0102	0.0053	0.0024	0.0010	0.0003	0.0001	0.0000	0.0000
	1	0.9774	0.9185	0.8352	0.7373	0.6328	0.5282	0.4284	0.3370	0.2562	0.1875	0.1312	0.0870	0.0540	0.0308	0.0156	0.0067	0.0022	0.0005	0.0000
	2	0.9988	0.9914	0.9734	0.9421	0.8965	0.8369	0.7648	0.6826	0.5931	0.5000	0.4069	0.3174	0.2352	0.1631	0.1035	0.0579	0.0266	0.0086	0.0012
	3	1.0000	0.9995	0.9978	0.9933	0.9844	0.9692	0.9460	0.9130	0.8688	0.8125	0.7438	0.6630	0.5716	0.4718	0.3672	0.2627	0.1648	0.0815	0.0226
	4	1.0000	1.0000	0.9999	0.9997	0.9990	0.9976	0.9947	0.9898	0.9815	0.9688	0.9497	0.9222	0.8840	0.8319	0.7627	0.6723	0.5563	0.4095	0.2262
6	0	0.7351	0.5314	0.3771	0.2621	0.1780	0.1176	0.0754	0.0467	0.0277	0.0156	0.0083	0.0041	0.0018	0.0007	0.0002	0.0001	0.0000	0.0000	0.0000
	1	0.9672	0.8857	0.7765	0.6554	0.5339	0.4202	0.3191	0.2333	0.1636	0.1094	0.0692	0.0410	0.0223	0.0109	0.0046	0.0016	0.0004	0.0001	0.0000
	2	0.9978	0.9842	0.9527	0.0911	0.8306	0.7443	0.6471	0.5443	0.4415	0.3438	0.2553	0.1792	0.1174	0.0705	0.0376	0.0170	0.0059	0.0013	0.0001
	3	0.9999	0.9987	0.9941	0.9830	0.9624	0.9295	0.8826	0.8208	0.7447	0.6562	0.5585	0.4557	0.3529	0.2557	0.1694	0.0989	0.0473	0.0158	0.0022
	4	1.0000	0.9999	0.9996	0.9984	0.9954	0.9891	0.9777	0.9590	0.9308	0.8906	0.8364	0.7667	0.6809	0.5798	0.4661	0.3446	0.2235	0.1143	0.0328
	5	1.0000	1.0000	1.0000	0.9999	0.9998	0.9993	0.9982	0.9959	0.9917	0.9844	0.9723	0.9533	0.9246	0.8824	0.8220	0.7379	0.6229	0.4686	0.2649
7	0	0.6983	0.4783	0.3206	0.2097	0.1335	0.0824	0.0490	0.0280	0.0152	0.0078	0.0037	0.0016	0.0006	0.0002	0.0001	0.0000	0.0000	0.0000	0.0000
	1	0.9556	0.8503	0.7166	0.5767	0.4449	0.3294	0.2338	0.1586	0.1024	0.0625	0.0357	0.0188	0.0090	0.0038	0.0013	0.0004	0.0001	0.0000	0.0000
	2	0.9962	0.9743	0.9262	0.8520	0.7564	0.6471	0.5323	0.4199	0.3164	0.2266	0.1529	0.0963	0.0556	0.0288	0.0129	0.0047	0.0012	0.0002	0.0000
	3	0.9998	0.9973	0.9879	0.9667	0.9294	0.8740	0.8002	0.7102	0.6083	0.5000	0.3917	0.2898	0.1998	0.1260	0.0706	0.0333	0.0121	0.0027	0.0002
	4	1.0000	0.9998	0.9988	0.9953	0.9871	0.9712	0.9444	0.9037	0.8471	0.7734	0.6836	0.5801	0.4677	0.3529	0.2436	0.1480	0.0738	0.0257	0.0038
	5	1.0000	1.0000	0.9999	0.9996	0.9987	0.9962	0.9910	0.9812	0.9643	0.9375	0.8976	0.8414	0.7662	0.6706	0.5551	0.4233	0.2834	0.1497	0.0444
	6	1.0000	1.0000	1.0000	1.0000	0.9999	0.9998	0.9994	0.9984	0.9963	0.9922	0.9848	0.9720	0.9510	0.9176	0.8665	0.7903	0.6794	0.5217	0.3017

SOURCE: Bain, L.J. and Englehardt, M. (1992). *Introduction to Probability and Mathematical Statistics*, 2nd ed. Belmont, CA: Duxbury Press.

TABLE 2 Binomial Cumulative Distribution Function (continued)

n	x	p = 0.05	0.10	0.15	0.20	0.25	0.30	0.35	0.40	0.45	0.50	0.55	0.60	0.65	0.70	0.75	0.80	0.85	0.90	0.95
8	0	0.6634	0.4305	0.2725	0.1678	0.1001	0.0576	0.0319	0.0168	0.0084	0.0039	0.0017	0.0007	0.0002	0.0001	0.0000	0.0000	0.0000	0.0000	0.0000
	1	0.9428	0.8131	0.6572	0.5033	0.3671	0.2553	0.1691	0.1064	0.0632	0.0352	0.0181	0.0085	0.0036	0.0013	0.0004	0.0001	0.0000	0.0000	0.0000
	2	0.9942	0.9619	0.8948	0.7969	0.6785	0.5518	0.4278	0.3154	0.2201	0.1445	0.0885	0.0498	0.0253	0.0113	0.0042	0.0012	0.0002	0.0000	0.0000
	3	0.9996	0.9950	0.9786	0.9437	0.8862	0.8059	0.7064	0.5941	0.4770	0.3633	0.2604	0.1737	0.1061	0.0580	0.0273	0.0104	0.0029	0.0004	0.0000
	4	1.0000	0.9996	0.9971	0.9896	0.9727	0.9420	0.8939	0.8263	0.7396	0.6367	0.5230	0.4059	0.2936	0.1941	0.1138	0.0563	0.0214	0.0050	0.0004
	5	1.0000	1.0000	0.9998	0.9988	0.9958	0.9887	0.9747	0.9502	0.9115	0.8555	0.7799	0.6846	0.5722	0.4482	0.3215	0.2031	0.1052	0.0381	0.0058
	6	1.0000	1.0000	1.0000	0.9999	0.9996	0.9987	0.9964	0.9915	0.9819	0.9648	0.9368	0.8936	0.8309	0.7447	0.6329	0.4967	0.3428	0.1869	0.0572
	7	1.0000	1.0000	1.0000	1.0000	1.0000	0.9999	0.9998	0.9993	0.9983	0.9961	0.9916	0.9832	0.9681	0.9424	0.8999	0.8322	0.7275	0.5695	0.3366
9	0	0.6302	0.3874	0.2316	0.1342	0.0751	0.0404	0.0207	0.0101	0.0046	0.0020	0.0008	0.0003	0.0001	0.0000	0.0000	0.0000	0.0000	0.0000	0.0000
	1	0.9288	0.7748	0.5995	0.4362	0.3003	0.1960	0.1211	0.0705	0.0385	0.0195	0.0091	0.0038	0.0014	0.0004	0.0001	0.0000	0.0000	0.0000	0.0000
	2	0.9916	0.9470	0.8591	0.7382	0.6007	0.4628	0.3373	0.2318	0.1495	0.0898	0.0498	0.0250	0.0112	0.0043	0.0013	0.0003	0.0000	0.0000	0.0000
	3	0.9994	0.9917	0.9661	0.9144	0.8343	0.7297	0.6089	0.4826	0.3614	0.2539	0.1658	0.0994	0.0536	0.0253	0.0100	0.0031	0.0006	0.0001	0.0000
	4	1.0000	0.9991	0.9944	0.9804	0.9511	0.9012	0.8283	0.7334	0.6214	0.5000	0.3786	0.2666	0.1717	0.0988	0.0489	0.0196	0.0056	0.0009	0.0000
	5	1.0000	0.9999	0.9994	0.9969	0.9900	0.9747	0.9464	0.9006	0.8342	0.7461	0.6386	0.5174	0.3911	0.2703	0.1657	0.0856	0.0339	0.0083	0.0006
	6	1.0000	1.0000	1.0000	0.9997	0.9987	0.9957	0.9888	0.9750	0.9502	0.9102	0.8505	0.7682	0.6627	0.5372	0.3993	0.2618	0.1409	0.0530	0.0084
	7	1.0000	1.0000	1.0000	1.0000	0.9999	0.9996	0.9986	0.9962	0.9909	0.9805	0.9615	0.9295	0.8789	0.8040	0.6997	0.5638	0.4005	0.2252	0.0712
	8	1.0000	1.0000	1.0000	1.0000	1.0000	1.0000	0.9999	0.9997	0.9992	0.9980	0.9954	0.9899	0.9793	0.9596	0.9249	0.8658	0.7684	0.6126	0.3698
10	0	0.5987	0.3487	0.1969	0.1074	0.0563	0.0282	0.0135	0.0060	0.0025	0.0010	0.0003	0.0001	0.0000	0.0000	0.0000	0.0000	0.0000	0.0000	0.0000
	1	0.9139	0.7361	0.5443	0.3758	0.2440	0.1493	0.0860	0.0464	0.0233	0.0107	0.0045	0.0017	0.0005	0.0001	0.0000	0.0000	0.0000	0.0000	0.0000
	2	0.9885	0.9298	0.8202	0.6778	0.5256	0.3828	0.2616	0.1673	0.0996	0.0547	0.0274	0.0123	0.0048	0.0016	0.0004	0.0001	0.0000	0.0000	0.0000
	3	0.9990	0.9872	0.9500	0.8791	0.7759	0.6496	0.5138	0.3823	0.2660	0.1719	0.1020	0.0548	0.0260	0.0106	0.0035	0.0009	0.0001	0.0000	0.0000
	4	0.9999	0.9984	0.9901	0.9672	0.9219	0.8497	0.7515	0.6331	0.5044	0.3770	0.2616	0.1662	0.0949	0.0473	0.0197	0.0064	0.0014	0.0001	0.0000
	5	1.0000	0.9999	0.9986	0.9936	0.9803	0.9527	0.9051	0.8338	0.7384	0.6230	0.4956	0.3669	0.2485	0.1503	0.0781	0.0328	0.0099	0.0016	0.0001
	6	1.0000	1.0000	0.9999	0.9991	0.9965	0.9894	0.9740	0.9452	0.8980	0.8281	0.7340	0.6177	0.4862	0.3504	0.2241	0.1209	0.0500	0.0128	0.0010
	7	1.0000	1.0000	1.0000	0.9999	0.9996	0.9984	0.9952	0.9877	0.9726	0.9453	0.9004	0.8327	0.7384	0.6172	0.4474	0.3222	0.1798	0.0702	0.0115
	8	1.0000	1.0000	1.0000	1.0000	1.0000	0.9999	0.9995	0.9983	0.9955	0.9893	0.9767	0.9536	0.9140	0.8507	0.7560	0.6242	0.4557	0.2639	0.0861
	9	1.0000	1.0000	1.0000	1.0000	1.0000	1.0000	1.0000	0.9999	0.9997	0.9990	0.9975	0.9940	0.9865	0.9718	0.9437	0.8926	0.8031	0.6513	0.4013

TABLE 2 Binomial Cumulative Distribution Function (continued)

		p																		
n	x	0.05	0.10	0.15	0.20	0.25	0.30	0.35	0.40	0.45	0.50	0.55	0.60	0.65	0.70	0.75	0.80	0.85	0.90	0.95
15	0	0.4633	0.2059	0.0874	0.0352	0.0134	0.0047	0.0016	0.0005	0.0001	0.0000	0.0000	0.0000	0.0000	0.0000	0.0000	0.0000	0.0000	0.0000	0.0000
	1	0.8290	0.5490	0.3186	0.1671	0.0802	0.0353	0.0142	0.0052	0.0017	0.0005	0.0001	0.0000	0.0000	0.0000	0.0000	0.0000	0.0000	0.0000	0.0000
	2	0.9638	0.8159	0.6042	0.3980	0.2361	0.1268	0.0617	0.0271	0.0107	0.0037	0.0011	0.0003	0.0001	0.0000	0.0000	0.0000	0.0000	0.0000	0.0000
	3	0.9945	0.9444	0.8227	0.6482	0.4613	0.2969	0.1727	0.0905	0.0424	0.0176	0.0063	0.0019	0.0005	0.0001	0.0000	0.0000	0.0000	0.0000	0.0000
	4	0.9994	0.9873	0.9383	0.8358	0.6865	0.5155	0.3519	0.2173	0.1204	0.0592	0.0255	0.0093	0.0028	0.0007	0.0001	0.0000	0.0000	0.0000	0.0000
	5	0.9999	0.9978	0.9832	0.9389	0.8516	0.7216	0.5643	0.4032	0.2608	0.1509	0.0769	0.0338	0.0124	0.0037	0.0008	0.0001	0.0000	0.0000	0.0000
	6	1.0000	0.9997	0.9964	0.9819	0.9434	0.8689	0.7548	0.6098	0.4522	0.3036	0.1818	0.0950	0.0422	0.0152	0.0042	0.0008	0.0001	0.0000	0.0000
	7	1.0000	1.0000	0.9994	0.9958	0.9827	0.9500	0.8868	0.7869	0.6535	0.5000	0.3465	0.2131	0.1132	0.0500	0.0173	0.0042	0.0006	0.0000	0.0000
	8	1.0000	1.0000	0.9999	0.9992	0.9958	0.9848	0.9578	0.9050	0.8182	0.6964	0.5478	0.3902	0.2452	0.1311	0.0566	0.0181	0.0036	0.0003	0.0000
	9	1.0000	1.0000	1.0000	0.9999	0.9992	0.9963	0.9876	0.9662	0.9231	0.8491	0.7392	0.5968	0.4357	0.2784	0.1484	0.0611	0.0168	0.0022	0.0001
	10	1.0000	1.0000	1.0000	1.0000	0.9999	0.9993	0.9972	0.9907	0.9745	0.9408	0.8796	0.7827	0.6481	0.4845	0.3135	0.1642	0.0617	0.0127	0.0006
	11	1.0000	1.0000	1.0000	1.0000	1.0000	0.9999	0.9995	0.9981	0.9937	0.9824	0.9576	0.9095	0.8273	0.7031	0.5387	0.3518	0.1773	0.0556	0.0055
	12	1.0000	1.0000	1.0000	1.0000	1.0000	1.0000	0.9999	0.9997	0.9989	0.9963	0.9893	0.9729	0.9383	0.8732	0.7639	0.6020	0.3958	0.1841	0.0362
	13	1.0000	1.0000	1.0000	1.0000	1.0000	1.0000	1.0000	1.0000	0.9999	0.9995	0.9983	0.9948	0.9858	0.9647	0.9198	0.8329	0.6814	0.4510	0.1710
	14	1.0000	1.0000	1.0000	1.0000	1.0000	1.0000	1.0000	1.0000	1.0000	1.0000	0.9999	0.9995	0.9984	0.9953	0.9866	0.9648	0.9126	0.7941	0.5367
20	0	0.3585	0.1216	0.0388	0.0115	0.0032	0.0008	0.0002	0.0000	0.0000	0.0000	0.0000	0.0000	0.0000	0.0000	0.0000	0.0000	0.0000	0.0000	0.0000
	1	0.7358	0.3917	0.1756	0.0692	0.0243	0.0076	0.0021	0.0005	0.0001	0.0000	0.0000	0.0000	0.0000	0.0000	0.0000	0.0000	0.0000	0.0000	0.0000
	2	0.9245	0.6769	0.4049	0.2061	0.0913	0.0355	0.0121	0.0036	0.0009	0.0002	0.0000	0.0000	0.0000	0.0000	0.0000	0.0000	0.0000	0.0000	0.0000
	3	0.9841	0.8670	0.6477	0.4114	0.2252	0.1071	0.0444	0.0160	0.0049	0.0013	0.0003	0.0000	0.0000	0.0000	0.0000	0.0000	0.0000	0.0000	0.0000
	4	0.9974	0.9568	0.8298	0.6296	0.4148	0.2375	0.1182	0.0510	0.0189	0.0059	0.0015	0.0003	0.0000	0.0000	0.0000	0.0000	0.0000	0.0000	0.0000
	5	0.9997	0.9887	0.9327	0.8042	0.6172	0.4164	0.2454	0.1256	0.0553	0.0207	0.0064	0.0016	0.0003	0.0000	0.0000	0.0000	0.0000	0.0000	0.0000
	6	1.0000	0.9976	0.9781	0.9133	0.7858	0.6080	0.4166	0.2500	0.1299	0.0577	0.0214	0.0065	0.0015	0.0003	0.0000	0.0000	0.0000	0.0000	0.0000
	7	1.0000	0.9996	0.9941	0.9679	0.8982	0.7723	0.6010	0.4159	0.2520	0.1316	0.0580	0.0210	0.0060	0.0013	0.0002	0.0000	0.0000	0.0000	0.0000
	8	1.0000	0.9999	0.9987	0.9900	0.9591	0.8867	0.7624	0.5956	0.4143	0.2517	0.1308	0.0565	0.0196	0.0051	0.0009	0.0001	0.0000	0.0000	0.0000
	9	1.0000	1.0000	0.9998	0.9974	0.9861	0.9520	0.8782	0.7553	0.5914	0.4119	0.2493	0.1275	0.0532	0.0171	0.0039	0.0006	0.0000	0.0000	0.0000
	10	1.0000	1.0000	1.0000	0.9994	0.9961	0.9829	0.9468	0.8725	0.7507	0.5881	0.4086	0.2447	0.1218	0.0480	0.0139	0.0026	0.0002	0.0000	0.0000
	11	1.0000	1.0000	1.0000	0.9999	0.9991	0.9949	0.9804	0.9435	0.8692	0.7483	0.5857	0.4044	0.2376	0.1133	0.0409	0.0100	0.0013	0.0001	0.0000
	12	1.0000	1.0000	1.0000	1.0000	0.9998	0.9987	0.9940	0.9790	0.9420	0.8684	0.7480	0.5841	0.3990	0.2277	0.1018	0.0321	0.0059	0.0004	0.0000
	13	1.0000	1.0000	1.0000	1.0000	1.0000	0.9997	0.9985	0.9935	0.9786	0.9423	0.8701	0.7500	0.5834	0.3920	0.2142	0.0867	0.0219	0.0024	0.0000
	14	1.0000	1.0000	1.0000	1.0000	1.0000	1.0000	0.9997	0.9984	0.9936	0.9793	0.9447	0.8744	0.7546	0.5836	0.3828	0.1958	0.0673	0.0113	0.0003
	15	1.0000	1.0000	1.0000	1.0000	1.0000	1.0000	1.0000	0.9997	0.9985	0.9941	0.9811	0.9490	0.8818	0.7625	0.5852	0.3704	0.1702	0.0432	0.0026
	16	1.0000	1.0000	1.0000	1.0000	1.0000	1.0000	1.0000	1.0000	0.9997	0.9987	0.9951	0.9840	0.9556	0.8929	0.7748	0.5886	0.3523	0.1330	0.0159
	17	1.0000	1.0000	1.0000	1.0000	1.0000	1.0000	1.0000	1.0000	1.0000	0.9998	0.9991	0.9964	0.9879	0.9645	0.9087	0.7939	0.5951	0.3231	0.0755
	18	1.0000	1.0000	1.0000	1.0000	1.0000	1.0000	1.0000	1.0000	1.0000	1.0000	0.9999	0.9995	0.9979	0.9924	0.9757	0.9308	0.8244	0.6083	0.2642
	19	1.0000	1.0000	1.0000	1.0000	1.0000	1.0000	1.0000	1.0000	1.0000	1.0000	1.0000	1.0000	0.9998	0.9992	0.9968	0.9885	0.9612	0.8784	0.6415
	20	1.0000	1.0000	1.0000	1.0000	1.0000	1.0000	1.0000	1.0000	1.0000	1.0000	1.0000	1.0000	1.0000	1.0000	1.0000	1.0000	1.0000	1.0000	1.0000

TABLE 3 Poisson Cumulative Distribution Function

$$F(x;\mu) = \sum_{k=0}^{x} e^{-\mu}\mu^k/k!$$

	μ									
x	0.1	0.2	0.3	0.4	0.5	0.6	0.7	0.8	0.9	1.0
0	0.9048	0.8187	0.7408	0.6730	0.6065	0.5488	0.4966	0.4493	0.4066	0.3679
1	0.9953	0.9825	0.9631	0.9384	0.9098	0.8781	0.8442	0.8088	0.7725	0.7358
2	0.9998	0.9989	0.9964	0.9921	0.9856	0.9769	0.9659	0.9526	0.9371	0.9197
3	1.0000	0.9999	0.9997	0.9992	0.9982	0.9966	0.9942	0.9909	0.9865	0.9810
4		1.0000	1.0000	0.9999	0.9998	0.9996	0.9992	0.9986	0.9977	0.9963
5				1.0000	1.0000	1.0000	0.9999	0.9998	0.9997	0.9994
6							1.0000	1.0000	1.0000	0.9999

	μ									
x	2.0	3.0	4.0	5.0	6.0	7.0	8.0	9.0	10.0	15.0
0	0.1353	0.0498	0.0183	0.0067	0.0025	0.0009	0.0003	0.0001	0.0000	
1	0.4060	0.1991	0.0916	0.0404	0.0174	0.0073	0.0030	0.0012	0.0005	
2	0.6767	0.4232	0.2381	0.1247	0.0620	0.0296	0.0138	0.0062	0.0028	0.0000
3	0.8571	0.6472	0.4335	0.2650	0.1512	0.0818	0.0424	0.0212	0.0103	0.0002
4	0.9473	0.8153	0.6288	0.4405	0.2851	0.1730	0.0996	0.0550	0.0293	0.0009
5	0.9834	0.9161	0.7851	0.6160	0.4457	0.3007	0.1912	0.1157	0.0671	0.0028
6	0.9955	0.9665	0.8893	0.7622	0.6063	0.4497	0.3134	0.2068	0.1301	0.0076
7	0.9989	0.9881	0.9489	0.8666	0.7440	0.5987	0.4530	0.3239	0.2202	0.0180
8	0.9998	0.9962	0.9786	0.9319	0.8472	0.7291	0.5925	0.4557	0.3328	0.0374
9	1.0000	0.9989	0.9919	0.9682	0.9161	0.8305	0.7166	0.5874	0.4579	0.0699
10		0.9997	0.9972	0.9863	0.9574	0.9015	0.8159	0.7060	0.5830	0.1185
11		0.9999	0.9991	0.9945	0.9799	0.9466	0.8881	0.8030	0.6968	0.1848
12		1.0000	0.9997	0.9980	0.9912	0.9730	0.9362	0.8758	0.7916	0.2676
13			0.9999	0.9993	0.9964	0.9872	0.9658	0.9261	0.8645	0.3632
14			1.0000	0.9998	0.9986	0.9943	0.9827	0.9585	0.9165	0.4657
15				0.9999	0.9995	0.9976	0.9918	0.9780	0.9513	0.5681
16				1.0000	0.9998	0.9990	0.9963	0.9889	0.9730	0.6641
17					0.9999	0.9996	0.9984	0.9947	0.9857	0.7489
18					1.0000	0.9999	0.9994	0.9976	0.9928	0.8195
19						1.0000	0.9997	0.9989	0.9965	0.8752
20							0.9999	0.9996	0.9984	0.9170
21							1.0000	0.9998	0.9993	0.9469
22								0.9999	0.9997	0.9673
23								1.0000	0.9999	0.9805
24									1.0000	0.9888
25										0.9938
26										0.9967
27										0.9983
28										0.9991
29										0.9996
30										0.9998
31										0.9999
32										1.0000

SOURCE: Bain, L.J. and Englehardt, M. (1992). *Introduction to Probability and Mathematical Statistics*, 2nd ed. Belmont, CA: Duxbury Press.

TABLE 4 Standard Normal Cumulative Distribution Function

$$F(z) = \int_{-\infty}^{z} \frac{1}{\sqrt{2\pi}} e^{-t^2/2}\, dt$$

z	0.00	0.01	0.02	0.03	0.04	0.05	0.06	0.07	0.08	0.09
0.0	0.5000	0.5040	0.5080	0.5120	0.5160	0.5199	0.5239	0.5279	0.5319	0.5359
0.1	0.5398	0.5438	0.5478	0.5517	0.5557	0.5596	0.5636	0.5675	0.5714	0.5753
0.2	0.5793	0.5832	0.5871	0.5910	0.5948	0.5987	0.6026	0.6064	0.6103	0.6141
0.3	0.6179	0.6217	0.6255	0.6293	0.6331	0.6368	0.6406	0.6443	0.6480	0.6517
0.4	0.6554	0.6591	0.6628	0.6664	0.6700	0.6736	0.6772	0.6808	0.6844	0.6879
0.5	0.6915	0.6950	0.6985	0.7019	0.7054	0.7088	0.7123	0.7157	0.7190	0.7224
0.6	0.7257	0.7291	0.7324	0.7357	0.7389	0.7422	0.7454	0.7486	0.7517	0.7549
0.7	0.7580	0.7611	0.7642	0.7673	0.7704	0.7734	0.7764	0.7794	0.7823	0.7852
0.8	0.7881	0.7910	0.7939	0.7967	0.7995	0.8023	0.8051	0.8078	0.8106	0.8133
0.9	0.8159	0.8186	0.8212	0.8238	0.8264	0.8289	0.8314	0.8340	0.8365	0.8389
1.0	0.8413	0.8438	0.8461	0.8485	0.8508	0.8531	0.8554	0.8577	0.8599	0.8621
1.1	0.8643	0.8665	0.8686	0.8708	0.8729	0.8749	0.8770	0.8790	0.8810	0.8830
1.2	0.8849	0.8869	0.8888	0.8907	0.8925	0.8944	0.8962	0.8980	0.8997	0.9015
1.3	0.9032	0.9049	0.9066	0.9082	0.9099	0.9115	0.9131	0.9147	0.9162	0.9177
1.4	0.9192	0.9207	0.9222	0.9236	0.9251	0.9265	0.9279	0.9292	0.9306	0.9319
1.5	0.9332	0.9345	0.9357	0.9370	0.9382	0.9394	0.9406	0.9418	0.9429	0.9441
1.6	0.9452	0.9463	0.9474	0.9484	0.9495	0.9505	0.9515	0.9525	0.9535	0.9545
1.7	0.9554	0.9564	0.9573	0.9582	0.9591	0.9599	0.9608	0.9616	0.9625	0.9633
1.8	0.9641	0.9649	0.9656	0.9664	0.9671	0.9678	0.9686	0.9693	0.9699	0.9706
1.9	0.9713	0.9719	0.9726	0.9732	0.9738	0.9744	0.9750	0.9756	0.9761	0.9767
2.0	0.9772	0.9778	0.9783	0.9788	0.9793	0.9798	0.9803	0.9808	0.9812	0.9817
2.1	0.9821	0.9826	0.9830	0.9834	0.9838	0.9842	0.9846	0.9850	0.9854	0.9857
2.2	0.9861	0.9864	0.9868	0.9871	0.9875	0.9878	0.9881	0.9884	0.9887	0.9890
2.3	0.9893	0.9896	0.9898	0.9901	0.9904	0.9906	0.9909	0.9911	0.9913	0.9916
2.4	0.9918	0.9920	0.9922	0.9925	0.9927	0.9929	0.9931	0.9932	0.9934	0.9936
2.5	0.9938	0.9940	0.9941	0.9943	0.9945	0.9946	0.9948	0.9949	0.9951	0.9952
2.6	0.9953	0.9955	0.9956	0.9957	0.9959	0.9960	0.9961	0.9962	0.9963	0.9964
2.7	0.9965	0.9966	0.9967	0.9968	0.9969	0.9970	0.9971	0.9972	0.9973	0.9974
2.8	0.9974	0.9975	0.9976	0.9977	0.9977	0.9978	0.9979	0.9979	0.9980	0.9981
2.9	0.9981	0.9982	0.9982	0.9983	0.9984	0.9984	0.9985	0.9985	0.9986	0.9986
3.0	0.9987	0.9987	0.9987	0.9988	0.9988	0.9989	0.9989	0.9989	0.9990	0.9990
3.1	0.9990	0.9991	0.9991	0.9991	0.9992	0.9992	0.9992	0.9992	0.9993	0.9993
3.2	0.9993	0.9993	0.9994	0.9994	0.9994	0.9994	0.9994	0.9995	0.9995	0.9995
3.3	0.9995	0.9995	0.9995	0.9996	0.9996	0.9996	0.9996	0.9996	0.9996	0.9997
3.4	0.9997	0.9997	0.9997	0.9997	0.9997	0.9997	0.9997	0.9997	0.9997	0.9998

γ	0.90	0.95	0.975	0.99	0.995	0.999	0.9995	0.99995	0.999995
z_γ	1.282	1.645	1.960	2.326	2.576	3.090	3.291	3.891	4.417

SOURCE: Bain, L.J. and Englehardt, M. (1992). *Introduction to Probability and Mathematical Statistics*, 2nd ed. Belmont, CA: Duxbury Press.

TABLE 5 Incomplete Gamma Function

$$F(y;\alpha) = \int_0^y \frac{1}{\Gamma(\alpha)} t^{\alpha-1} e^{-t}\, dt$$

	α									
y	**1**	**2**	**3**	**4**	**5**	**6**	**7**	**8**	**9**	**10**
1	.632	.264	.080	.019	.004	.001	.000	.000	.000	.000
2	.865	.594	.323	.143	.053	.017	.005	.001	.000	.000
3	.950	.801	.577	.353	.185	.084	.034	.012	.004	.001
4	.982	.908	.762	.567	.371	.215	.111	.051	.021	.008
5	.993	.960	.875	.735	.560	.384	.238	.133	.068	.032
6	.998	.983	.938	.849	.715	.554	.398	.256	.153	.084
7	.999	.993	.970	.918	.827	.699	.550	.401	.271	.170
8	1.000	.997	.986	.958	.900	.809	.687	.547	.407	.283
9		.999	.994	.979	.945	.884	.793	.676	.544	.413
10		1.000	.997	.990	.971	.933	.870	.780	.667	.542
11			.999	.995	.985	.962	.921	.857	.768	.659
12			1.000	.998	.992	.980	.954	.911	.845	.758
13				.999	.996	.989	.974	.946	.900	.834
14				1.000	.998	.994	.986	.968	.938	.891
15					.999	.997	.992	.982	.963	.930

References

Bain, L. J., and Englehardt, M. (1992). *Introduction to Probability and Mathematical Statistics*, 2nd ed. Belmont, CA: Duxbury Press.

Bhat, U. N. (1984). *Elements of Applied Stochastic Processes*, 2nd ed. New York: Wiley.

CRC (1966). *Handbook of Tables for Probability and Statistics*, 2nd ed. Cleveland, OH: CRC Press.

Hogg, R. V., and Craig, A. T. (1978). *Introduction to Mathematical Statistics*, 4th ed. New York: Macmillan.

Kennedy, W. J., and Gentle, J. E. (1980). *Statistical Computing*. New York: Dekker.

Knuth, D. E. (1973). *The Art of Computer Programming*, Vol 2. Reading, MA: Addison-Wesley.

Answers to Selected Odd-Numbered Exercises

Chapter 1

1.1-1 **a** $S = \{$(red, blue), (red, yellow), (blue,red), (blue, yellow), (yellow, red), (yellow, blue)$\}$
b 4

1.1-3 **a** $S = \{(1,2,3), (1,3,2), (2,1,3), (2,3,1), (3,1,2), (3,2,1)\}$
b 2

1.1-5 **a** $S = \{0,1,2\}$
b $S = \{$(defective, defective), (defective, good) (good, defective), (good, good)$\}$
c $S = \{(1,2), (1,3), (1,4), (2,1), (2,3), (2,4), (3,1), (3,2), (3,4), (4,1) (4,2), (4,3)\}$
d 1, 1, 2 **e** 1, 1, 2

1.2-1 .7, .2, .8

1.2-3 **b** 6/56 **c** 20/56 **d** 30/56
e 21/56

1.2-5 20/36

1.2-7 **a** $S = \{$(pass 1, pass 2), (pass 1, fail 2) (fail 1, pass 2), (fail 1 fail 2)$\}$
b 1/3, 1/6, 1/3, 1/6 **c** 1/2

1.2-9 **a** 1/6 **b** 1/3

1.3-1 **b** 1/6

1.4-1 144

1.4-3 17, 1820

1.4-5 **a** .4196 **b** .4616 **c** 1

1.4-7 **a** 1440 **b** 2000 **c** 9000

1.4-13 .01865

1.4-15 34,650

1.5-1 1/3

1.5-5 **a** .25 **b** .75 **c** .4375

1.5-9 **a** $.591\overline{6}$ **b** $.3\overline{3}$

1.5-11 **a** .2655 **b** .3139

1.6-3 **a** .125, .625, $.6\overline{6}$ **b** no

1.6-5 .504

1.6-7 .818

1.6-9 $1-(1-p_1)(1-p_2)(1-p_3)$

1.6-11 .4219

Chapter 2

2.1-1 **a**

x	2	3	4	5	6	7	8	9	10	11	12
$p(x)$	1/36	2/36	3/36	4/36	5/36	6/36	5/36	4/36	3/36	2/36	1/36

b

x	1	2	3	4	5	6
$p(x)$	1/36	3/36	5/36	7/36	9/36	11/36

2.1-3

y	0	20	40	60
$p(y)$	4/35	18/35	12/35	1/35

2.1-5 **a**

x	0	1	2	3	4
$p(x)$	.20	.30	.30	.10	.10

b .20

2.1-7 $F(x) = \sum_{i=0}^{x} .1(.9)^i$

2.2-1 **a** .1, .7 **b**

y	0	1	2
$p(y)$	.3	.4	.3

c 1/3

2.2-3 **b** 1/3

2.2-5 **a** .42 **b** .45

2.3-1 3.5

2.3-3 $75.00

2.3-7 38278

2.4-1 **a** 3.5, 1.71 **b** (.08, 6.92)

2.4-3 **a** 1.85, 1.62 **b** $747.50, $567.33

2.6-1 6.9, 3.56

2.6-3 3.6, 2.703

2.6-7 **a** 3.3, 1.269

2.7-1 **a** 2.93, −.22, 3.15, 6.52 **b** no

2.7-3 **a** 12.25 **b** 12.25 **c** 1.429

2.7-5 7, 2.415

2.7-7 262.22, 133.10, 262.5, 138.9

2.8-1 **a** 1.41, .8729, 1.1, .9434
b .159, .1931 **c** 7.805, 4.032

2.9-1 −.27, 2.77

2.9-3 7.4025, 4.3025

Chapter 3

3.1-1 .1001, .2670, .6056, .3214

3.1-3 .015625

3.1-5 100, 9.1287

3.1-7 .3621

3.1-9 **a** .55, .05 **b** (.45, .65)
c (.50, .60), (.53, .57)

3.2-1 **a** $\frac{1}{3}\left(\frac{2}{3}\right)^{x-1}$ **b** .0658 **c** .2963
d 3

3.2-3 **a** .125, .1875, .1875 **b** (1, 11)

3.2-7 7.5, 3.354

3.3-1 **a** (0, 1, 2, . . . , 9) **b** 5.4, 1.25
c (0, 1, 2, . . . , 5), 3, .9474

3.3-3 **a** .85968

3.3-5 .00988

3.4-1 .006567

3.4-3 $305, $71.34

3.5-1 .1353, .2707, .5413, .594

3.5-3 **a** .6703 **b** .0616

3.5-5 (19, 41)

3.6-2 .3, .7, .3

Chapter 4

4.2-1

$$\begin{array}{c} \\ 1 \\ 2 \\ 3 \\ 4 \\ 5 \\ 6 \end{array} \begin{array}{c} \begin{array}{cccccc} 1 & 2 & 3 & 4 & 5 & 6 \end{array} \\ \left[\begin{array}{cccccc} 0 & .5 & 0 & 0 & 0 & .5 \\ .5 & 0 & .5 & 0 & 0 & 0 \\ 0 & .5 & 0 & .5 & 0 & 0 \\ 0 & 0 & .5 & 0 & .5 & 0 \\ 0 & 0 & 0 & .5 & 0 & 0 \\ .5 & 0 & 0 & 0 & .5 & .5 \end{array}\right] \end{array}$$

4.2-3

$$\begin{array}{c|cccccc} & 0 & 1 & 2 & 3 & 4 & \cdots \\ \hline 0 & 0 & 1 & 0 & 0 & 0 & \cdots \\ 1 & 1/k & 0 & (k-1)/k & 0 & 0 & \cdots \\ 2 & 0 & 2/k & 0 & (k-2)/k & 0 & \cdots \\ 3 & 0 & 0 & 3/k & 0 & (k-3)/k & \cdots \\ \vdots & \vdots & \vdots & \vdots & \vdots & \vdots & \vdots \end{array}$$

4.2-5 **a** 0, 2, 4; .36, .48, .16

4.2-7

$$\begin{array}{c|cccccc} & 1 & 2 & 3 & 4 & 5 & 6 \\ \hline 1 & P(A) & P(C) & P(B) & 0 & 0 & 0 \\ 2 & P(B) & P(A) & 0 & 0 & P(C) & 0 \\ 3 & P(A) & 0 & P(B) & P(C) & 0 & 0 \\ 4 & 0 & 0 & P(A) & P(B) & 0 & P(C) \\ 5 & 0 & P(A) & 0 & 0 & P(C) & P(B) \\ 6 & 0 & 0 & 0 & P(B) & P(A) & P(C) \end{array}$$

4.3-1 **a** .0032 **b** .56, .35

4.3-3 **a** $\left(\frac{1}{3}, \frac{1}{6}, \frac{1}{6}, \frac{1}{3}\right)$
b $(.27\bar{7}, .22\bar{2}, .22\bar{2}, .27\bar{7})$

4.3-5 .1875

4.3-7 **a** .378 **b** (.394, .2902, .3158)

4.4-1 **a**

$$\begin{array}{c} \\ 0 \\ 1 \end{array}\begin{array}{c} \begin{array}{cc} 0 & 1 \end{array} \\ \begin{bmatrix} .96 & .04 \\ .80 & .20 \end{bmatrix} \end{array}$$

b

Backlog	0	1
Monday	.96	.04
Tuesday	.9536	.0464
Wednesday	.9226	.0474
Thursday	.9524	.0476
Friday	.9524	.0476

4.4-3

$$P^6 = \begin{array}{c} \\ 0 \\ 1 \\ 2 \\ 3 \\ 4 \end{array}\begin{array}{c} \begin{array}{ccccc} 0 & 1 & 2 & 3 & 4 \end{array} \\ \begin{bmatrix} .5025 & .3024 & .1516 & .0396 & .0040 \\ .4820 & .3075 & .1619 & .0440 & .0046 \\ .4415 & .3173 & .1823 & .0531 & .0058 \\ .3823 & .3303 & .2127 & .0670 & .0076 \\ .3071 & .3439 & .2526 & .0862 & .0103 \end{bmatrix} \end{array}$$

4.6-1 (.625, .375)

4.6-3 **a** (.4656, .38, .1544) **b** 2.148

4.6-5 $\left(\frac{b}{a+b}, \frac{a}{a+b}\right)$

4.6-9 $\left(\frac{1}{6}, \frac{1}{6}, \frac{1}{6}, \frac{1}{6}, \frac{1}{6}, \frac{1}{6}\right)$

4.7-1 **a** .72 **b** .45

4.7-3 **a**

t	1	2	3	4	5	6	7	8	9	10
$F(t)$	.4	.68	.841	.922	.962	.981	.991	.996	.998	.999

b 2.2189

4.7-5 **b** 1.608

Chapter 5

5.1-1 **a** 1/8 **b** 3/16

5.1-3 **a** .064 **b** .2599

5.1-5 **a** .2325 **b** .3679

c $f(x) = \begin{cases} .2e^{-.2x}, & x > 0 \\ 0, & \text{otherwise} \end{cases}$

5.2-1 **a** 2/3, 2/3 **b** 1

5.2-3 $23.3\overline{3}$, 4.714

5.2-5

$$G(y) = \begin{cases} 1, & y > 1 \\ \frac{1}{16}[15 + 2y - y^2], & -3 \le y \le 1, \\ 0, & y < -3 \end{cases}$$

$$g(y) = \begin{cases} \frac{1}{8}[1 - y], & -3 \le y \le 1 \\ 0, & \text{otherwise} \end{cases}$$

5.2-7 $g(y) = e^{-y}, \; y > 0$

5.2-9 **a**

$$G(y) = \begin{cases} 1, & y > 4 \\ 1 - \frac{\sqrt{4-y}}{2}, & 0 \le y \le 4 \\ 0, & y < 0 \end{cases}$$

b

$$g(y) = \begin{cases} \frac{1}{4\sqrt{4-y}}, & 0 \le y \le 4, \\ 0, & \text{otherwise} \end{cases}$$

5.2-11 $\frac{1}{2\sqrt{y}}e^{-\sqrt{y}}, \; y > 0$

5.3-1 **b**

$$F(x) = \begin{cases} 1, & x > 1 \\ x^2, & 0 \le x \le 1 \\ 0, & x < 0 \end{cases}$$

5.3-3 **b**

$$F(y) = \begin{cases} 1, & y > b \\ \frac{(y-a)^2}{(b-a)^2}, & a \le y \le b \\ 0, & y < a \end{cases}$$

5.3-5 **b** $2.58\overline{3}$, 1.0769

5.4-1 **b** 1 **c** 1/4

d
$$f_X(x) = \begin{cases} 1, & 0 \le x \le 1 \\ 0, & \text{otherwise} \end{cases},$$
$$f_Y(y) = \begin{cases} 1, & 0 \le y \le 1 \\ 0, & \text{otherwise} \end{cases}$$
e yes

5.4-3 **a** .0625 **b** .015625 **c** .015625

5.4-5 **a** $.2916\overline{6}$ **b** $.83\overline{3}$

5.4-7 **a** $.5,\ .75,\ .6\overline{6}$ **b** .125, .125
c $.6\overline{6}$

5.4-9 **a** $\frac{2l}{9}$ for $0 \le l \le 3$ **b** (.58, 3)

5.4-11 **a** $\frac{1}{2\sqrt{y_1}} e^{-\sqrt{y_1}},\quad \frac{1}{2\sqrt{y_2}} e^{-\sqrt{y_2}}$

5.4-13 **c**

$$f(y) = \begin{cases} \frac{1}{2(10)^4}\left[e^{-y/(10)^4}\left(2e^{-y/(10)^4} - e^{-2}\right) - 1\right], & -2(10)^4 \le y \le 0 \\ \frac{1}{2(10)^4} e^{-y/(10)^4}\left(1 - e^{-2}\right), & 0 < y \end{cases}$$

Chapter 6

6.1-1 **a** 1/2 **b** $1/\sqrt{2}$
c $F(x) = 1 - e^{-2x},\ x > 0$

6.1-3 **a** 6, 18.5 **c** .04

6.1-5 .9502

6.2-1 **a** .3015 **c** .3970 **d** .0281
f .3231

6.2-3 **a** .011 **b** .0415 **c** .0228

6.2-5 1.239 inch

6.2-7 .2546

6.2-9 .2843

6.3-1 **a** 24 **b** .886

6.3-3 **a** .567 **b** 8, 16

6.3-5 .95

6.3-7 .151

6.4-1 **b** 1.2, 99.4 **c** .12

6.4-3 7.9 to 12.1 inches

6.5-1 2/7

6.5-3 $M(t) = 1/(1 - \mu t),\quad t < 1/\mu$

6.5-7 **b** $\hat{\lambda} = 2.2,\quad \hat{\sigma}^2 = 1/\hat{\lambda}^2 = .21$

6.5-11 **a** 31.6, 2.14 **b** 36, 2.60

Chapter 7

7.1-3 400 frames/hour

7.1-5 **a** .4457 **b** 2.4 **c** 25 minutes
d (0, 75.5 minutes)

7.1-7 **a** $.1,\ .2,\ .6\overline{6}$ **c** .00724
d 20 minutes, 10 minutes, 30 minutes

7.1-9 **a** 36 **b** (25, 47) **c** (0, 38 minutes)

7.2-1 **a** .04203 **b** .3114

7.2-3 **a** .01438 **b** .13534

7.2-5 (20.67, 33.31)

7.3-3 $f_S(t) = \frac{t^4 e^{-t}}{4!}$, 5, 2.236

7.3-5 .96622

7.4-1 **a** 6.4 **b** .125, .2

7.4-3 **a** $.3\overline{3},\ .5$ **b** $.05\overline{5},\ .08\overline{3}$

7.4-7 **a**

$$P = \begin{array}{c|cccc} & 0 & 1 & 2 & 3 \\ \hline 0 & .9 & .1 & 0 & 0 \\ 1 & .18 & .74 & .08 & 0 \\ 2 & 0 & .18 & .74 & .08 \\ 3 & 0 & 0 & .18 & .82 \end{array}$$

b (.523, .2905, .1291, .0574)

7.5-3 21.5

7.5-5 **a** 5 **b** 5/6 **c** 12 **d** 1 hour

7.5-7 **a** 10 **b** 12

7.5-9 .512

7.6-1

$$\begin{array}{c|ccccccc} & 0 & 1 & 2 & 3 & 4 & 5 & \cdots \\ \hline 0 & .900 & .100 & 0 & 0 & 0 & 0 & \cdots \\ 1 & .135 & .780 & .085 & 0 & 0 & 0 & \cdots \\ 2 & .020 & .232 & .676 & .072 & 0 & 0 & \cdots \\ 3 & .003 & .052 & .298 & .585 & .061 & \cdots & \cdots \\ 4 & 0 & .003 & .052 & .298 & .585 & .061 & \cdots \\ \vdots & \vdots & \vdots & \vdots & \vdots & \vdots & \vdots & \end{array}$$

7.7-1

	0	1	2	3	4	5	6	MEAN	STD
$r = .5$	.607	.303	.076	.013	.002	.000	.000	0.50	0.71
$r = .9$	.406	.366	.165	.049	.011	.002	.001	0.90	0.96

7.7-3

	MEAN	STD	TIME
1 copier	9.0	9.5	1 hour
2 copiers	1.1	1.3	0.1 hours

7.7-7 .84

7.8-1 **a**

0	1	2	3	4
.297	.238	.190	.152	.122

b 1.56, 1.37

7.8-3 Poisson $\mu = 8/3$

7.8-5 **a** .18 **b** 20

Chapter 8

8.1-1 **a** 213, 3.1 **b** 206.8 to 219.2 **c** .0019

8.1-3 .1190

8.1-5 $M_{\sum X_i}(\theta) = 1/(1-\theta/\lambda)^n, \quad \theta < \lambda$

8.1-7 $M_{\sum X_i}(\theta) = e^{\mu(e^\theta - 1)}, \text{ where } \mu = \sum_{i=1}^{n} \mu_i$

8.2-1 **a** 18000 **b** .1611 **c** .3085

8.2-3 **a** .0336 **b** .1357

8.2-5 **a** .0026 **b** .6568

8.2-9 .1922

8.3-1 .83 to .97 hours

8.3-3 (11.92, 13.28), (11.71, 13.49), (11.13, 14.07)

8.3-5 (.15, .29)

8.4-1 \$2880, \$354.40

8.4-3 **a** 180 hours, 16.66 hours **b** .8849

Chapter 9

9.1-1 **a** $F_{max}(t) = [(t-35)/25]^5, \quad 35 < t < 60$ **b** 55.8 minutes

9.1-5 27.6 months

9.2-1 **a** $T = \max\{T_1 + T_2, T_3 + T_4, T_5 + T_6\}$

9.3-1 $\pi_1 = 1/15, \pi_2 = 2/15, \pi_3 = 4/15, \pi_4 = 8/15$

9.3-3 $\pi_0 = \cdots = \pi_9 = 2/23, \quad \pi_{10} = 5/23$

9.3-5 $\pi_0 = .233, \quad \pi_1 = .157, \quad \pi_2 = .262,$ $\pi_3 = .174, \quad \pi_4 = .087, \quad \pi_5 = .087$

9.4-1 **b** 740 seconds, 197.2 seconds

9.4-3 $E(T) = 55 - 20p,$ $\text{STD}(T) = \sqrt{425.3 - 400p^2}, \quad 0 \le p \le 1$

Chapter 10

10.1-1 $R(t) = 1 - t/5, \quad 0 \le t \le 5$

10.1-3 $R(t) = e^{-t^2/32}, \quad 0 \le t < \infty$

10.1-7 146.7 hours

10.2-3 $h(t) = 1/(5-t), \quad 1 < t < 5$

10.2-5 .075

10.2-7 **a**

$$R(t) = \begin{cases} e^{-.002t}, & 0 \le t \le 200, \\ e^{-.2-.001t}, & t > 200 \end{cases}$$

b 835 hours

10.3-1 **a** .2084 **b** .735

10.3-3 51 to 69

10.3-5 1.34 hours

10.4-1 .97

10.4-3 50/61

10.4-5 $1 - \left(\frac{\mu_D}{\mu_U + \mu_D}\right)^k$

10.5-1 **a** .4187 **b** .97 **c** .07 **d** 5 days

10.5-3 **a** $m(t) = \ln(1 + t)$ **b** .1 **c** 8 months

INDEX

Web ID - RMudge.5517-1
Password - 8 digit Tech ID